Recent Developments on Introducing a Historical Dimension in Mathematics Education

© 2011 by the Mathematical Association of America, Inc.

Library of Congress Catalog Card Number 2011932375

Print edition ISBN 978-0-88385-188-3

Electronic edition ISBN 978-1-61444-300-1

Printed in the United States of America

Current Printing (last digit):
10 9 8 7 6 5 4 3 2 1

Recent Developments on Introducing a Historical Dimension in Mathematics Education

Edited by
Victor Katz and Constantinos Tzanakis

Published and Distributed by
The Mathematical Association of America

The MAA Notes Series, started in 1982, addresses a broad range of topics and themes of interest to all who are involved with undergraduate mathematics. The volumes in this series are readable, informative, and useful, and help the mathematical community keep up with developments of importance to mathematics.

Council on Publications and Communications
Frank Farris, *Chair*

Committee on Books
Gerald Bryce, *Chair*

Notes Editorial Board
Stephen B Maurer, *Editor*

Deborah J. Bergstrand	Thomas P. Dence
Donna L. Flint	Theresa Jeevanjee
Michael K. May	Judith A. Palagallo
Mark Parker	Susan Pustejovsky
David Rusin	David J. Sprows
Joe Alyn Stickles	Andrius Tamulis

MAA Notes

14. Mathematical Writing, by *Donald E. Knuth, Tracy Larrabee, and Paul M. Roberts.*
16. Using Writing to Teach Mathematics, *Andrew Sterrett,* Editor.
17. Priming the Calculus Pump: Innovations and Resources, Committee on Calculus Reform and the First Two Years, a subcommittee of the Committee on the Undergraduate Program in Mathematics, *Thomas W. Tucker,* Editor.
18. Models for Undergraduate Research in Mathematics, *Lester Senechal,* Editor.
19. Visualization in Teaching and Learning Mathematics, Committee on Computers in Mathematics Education, *Steve Cunningham and Walter S. Zimmermann,* Editors.
20. The Laboratory Approach to Teaching Calculus, *L. Carl Leinbach et al.,* Editors.
21. Perspectives on Contemporary Statistics, *David C. Hoaglin and David S. Moore,* Editors.
22. Heeding the Call for Change: Suggestions for Curricular Action, *Lynn A. Steen,* Editor.
24. Symbolic Computation in Undergraduate Mathematics Education, *Zaven A. Karian,* Editor.
25. The Concept of Function: Aspects of Epistemology and Pedagogy, *Guershon Harel and Ed Dubinsky,* Editors.
26. Statistics for the Twenty-First Century, *Florence and Sheldon Gordon,* Editors.
27. Resources for Calculus Collection, Volume 1: Learning by Discovery: A Lab Manual for Calculus, *Anita E. Solow,* Editor.
28. Resources for Calculus Collection, Volume 2: Calculus Problems for a New Century, *Robert Fraga,* Editor.
29. Resources for Calculus Collection, Volume 3: Applications of Calculus, *Philip Straffin,* Editor.
30. Resources for Calculus Collection, Volume 4: Problems for Student Investigation, *Michael B. Jackson and John R. Ramsay,* Editors.
31. Resources for Calculus Collection, Volume 5: Readings for Calculus, *Underwood Dudley,* Editor.
32. Essays in Humanistic Mathematics, *Alvin White,* Editor.
33. Research Issues in Undergraduate Mathematics Learning: Preliminary Analyses and Results, *James J. Kaput and Ed Dubinsky,* Editors.
34. In Eves Circles, *Joby Milo Anthony,* Editor.
35. Youre the Professor, What Next? Ideas and Resources for Preparing College Teachers, The Committee on Preparation for College Teaching, *Bettye Anne Case,* Editor.
36. Preparing for a New Calculus: Conference Proceedings, *Anita E. Solow,* Editor.
37. A Practical Guide to Cooperative Learning in Collegiate Mathematics, *Nancy L. Hagelgans, Barbara E. Reynolds, SDS, Keith Schwingendorf, Draga Vidakovic, Ed Dubinsky, Mazen Shahin, G. Joseph Wimbish, Jr.*
38. Models That Work: Case Studies in Effective Undergraduate Mathematics Programs, *Alan C. Tucker,* Editor.
39. Calculus: The Dynamics of Change, CUPM Subcommittee on Calculus Reform and the First Two Years, *A. Wayne Roberts,* Editor.
40. Vita Mathematica: Historical Research and Integration with Teaching, *Ronald Calinger,* Editor.
41. Geometry Turned On: Dynamic Software in Learning, Teaching, and Research, *James R. King and Doris Schattschneider,* Editors.

42. Resources for Teaching Linear Algebra, *David Carlson, Charles R. Johnson, David C. Lay, A. Duane Porter, Ann E. Watkins, William Watkins,* Editors.
43. Student Assessment in Calculus: A Report of the NSF Working Group on Assessment in Calculus, *Alan Schoenfeld,* Editor.
44. Readings in Cooperative Learning for Undergraduate Mathematics, *Ed Dubinsky, David Mathews, and Barbara E. Reynolds,* Editors.
45. Confronting the Core Curriculum: Considering Change in the Undergraduate Mathematics Major, *John A. Dossey,* Editor.
46. Women in Mathematics: Scaling the Heights, *Deborah Nolan,* Editor.
47. Exemplary Programs in Introductory College Mathematics: Innovative Programs Using Technology, *Susan Lenker,* Editor.
48. Writing in the Teaching and Learning of Mathematics, *John Meier and Thomas Rishel.*
49. Assessment Practices in Undergraduate Mathematics, *Bonnie Gold,* Editor.
50. Revolutions in Differential Equations: Exploring ODEs with Modern Technology, *Michael J. Kallaher,* Editor.
51. Using History to Teach Mathematics: An International Perspective, *Victor J. Katz,* Editor.
52. Teaching Statistics: Resources for Undergraduate Instructors, *Thomas L. Moore,* Editor.
53. Geometry at Work: Papers in Applied Geometry, *Catherine A. Gorini,* Editor.
54. Teaching First: A Guide for New Mathematicians, *Thomas W. Rishel.*
55. Cooperative Learning in Undergraduate Mathematics: Issues That Matter and Strategies That Work, *Elizabeth C. Rogers, Barbara E. Reynolds, Neil A. Davidson, and Anthony D. Thomas,* Editors.
56. Changing Calculus: A Report on Evaluation Efforts and National Impact from 1988 to 1998, *Susan L. Ganter.*
57. Learning to Teach and Teaching to Learn Mathematics: Resources for Professional Development, *Matthew Delong and Dale Winter.*
58. Fractals, Graphics, and Mathematics Education, Benoit Mandelbrot and Michael Frame, Editors.
59. Linear Algebra Gems: Assets for Undergraduate Mathematics, *David Carlson, Charles R. Johnson, David C. Lay, and A. Duane Porter,* Editors.
60. Innovations in Teaching Abstract Algebra, *Allen C. Hibbard and Ellen J. Maycock,* Editors.
61. Changing Core Mathematics, *Chris Arney and Donald Small,* Editors.
62. Achieving Quantitative Literacy: An Urgent Challenge for Higher Education, *Lynn Arthur Steen.*
64. Leading the Mathematical Sciences Department: A Resource for Chairs, *Tina H. Straley, Marcia P. Sward, and Jon W. Scott,* Editors.
65. Innovations in Teaching Statistics, *Joan B. Garfield,* Editor.
66. Mathematics in Service to the Community: Concepts and models for service-learning in the mathematical sciences, *Charles R. Hadlock,* Editor.
67. Innovative Approaches to Undergraduate Mathematics Courses Beyond Calculus, *Richard J. Maher,* Editor.
68. From Calculus to Computers: Using the last 200 years of mathematics history in the classroom, Amy Shell-Gellasch and Dick Jardine, Editors.
69. A Fresh Start for Collegiate Mathematics: Rethinking the Courses below Calculus, *Nancy Baxter Hastings,* Editor.
70. Current Practices in Quantitative Literacy, *Rick Gillman*, Editor.
71. War Stories from Applied Math: Undergraduate Consultancy Projects, *Robert Fraga*, Editor.
72. Hands On History: A Resource for Teaching Mathematics, *Amy Shell-Gellasch*, Editor.
73. Making the Connection: Research and Teaching in Undergraduate Mathematics Education, *Marilyn P. Carlson and Chris Rasmussen*, Editors.
74. Resources for Teaching Discrete Mathematics: Classroom Projects, History Modules, and Articles, *Brian Hopkins*, Editor.
75. The Moore Method: A Pathway to Learner-Centered Instruction, *Charles A. Coppin, W. Ted Mahavier, E. Lee May, and G. Edgar Parker.*
76. The Beauty of Fractals: Six Different Views, *Denny Gulick and Jon Scott*, Editors.
77. Mathematical Time Capsules: Historical Modules for the Mathematics Classroom, *Dick Jardine and Amy Shell-Gellasch*, Editors.
78. Recent Developments on Introducing a Historical Dimension in Mathematics Education, *Victor J. Katz and Costas Tzanakis*, Editors.
79. Teaching Mathematics with Classroom Voting: With and Without Clickers, *Kelly Cline and Holly Zullo*, Editors.

MAA Service Center
P.O. Box 91112
Washington, DC 20090-1112
1-800-331-1MAA   FAX: 1-301-206-9789

Preface

In 2000, the International Commission on Mathematical Instruction published the ICMI Study entitled *History in Mathematics Education*, edited by John Fauvel and Jan van Maanen [1]. This volume presented the state of the art in the use of the history of mathematics in mathematics education and made numerous suggestions for further research in the field. One of the major suggestions was that there should be more empirical studies of the use of history in the mathematics classroom to get more insight into its educational implications. In recent years many researchers have followed up on the recommendations of the ICMI Study and have presented their findings in numerous international meetings, including the meetings of the *International Study Group on the Relations Between History and Pedagogy of Mathematics* (the HPM Group) and the meetings of the *European Summer University on the History and Epistemology in Mathematics Education* (ESU)[1]. To further the goal of making these new results available to a wider audience, including mathematics educators, mathematics faculty in secondary schools and universities, and historians of mathematics, the contributors to these meetings were invited to submit papers based on their presentations, which were peer-reviewed to international standards by two independent referees and then revised before final acceptance. We are indebted to the referees for their willingness to support this project and provide useful reports on the 50 submitted manuscripts. This volume consists of the 24 papers that were finally accepted (coming from 13 countries worldwide) and aims to constitute an all-embracing outcome of recent activities within the HPM Group. We believe these articles will move the field forward and provide faculty with many new ideas for incorporating the history of mathematics into their teaching at various levels of education. The book is organized into four parts. The first deals with theoretical ideas for integrating the history of mathematics into mathematics education. The second part contains research studies on the use of the history of mathematics in the teaching of numerous mathematics topics at several levels of education. The third part concentrates on how history can be used with prospective and current teachers of mathematics. We also include a special fourth part containing three purely historical papers based on invited talks at the HPM meeting of 2008. Two of these articles provide an overview of the development of mathematics in the Americas, while the third is a study of the astronomical origins of trigonometry.

Theoretical Ideas for Integrating History of Mathematics into Mathematics Education

Part I contains seven articles, from five countries, dealing with theoretical aspects of using the history of mathematics classroom. In the first, David Pengelley from the United States argues that virtually all courses should be taught using original sources, and gives various reasons in support of his position. Evelyne Barbin of France then introduces the notion of dialogism in considering how one can use such sources in the classroom. In other words, she suggests that students consider an original mathematical writing as a dialogue between the author and the intended audience. Next, Gustavo Martinez Sierra and Rocio Antonio Antonio from Mexico study how the production of new mathematical knowledge emerges from the necessity of new ideas agreeing with ones that are already part of our knowledge base. Luis Puig from Spain, on the other hand, takes a look at three historical sources in algebra, considering how the authors thought about cut and paste methods in solving quadratic equations. Giorgio Bagni, an Italian who was tragically killed in a bicycle accident during the preparation of this book, considers Bombelli's *Algebra* and how its content reflects an

---

[1]These are: ESU 5, Prague, Czech Republic, 2007; the Topic Study Group 23 of ICME 11 on *The Role of History of Mathematics in Mathematics Education*, and the Affiliated Study Group-meeting of the HPM at ICME 11 in Monterrey, Mexico, 2008; the *HPM Group Satellite Meeting of ICME 11* (HPM 2008) in Mexico City, Mexico, 2008; the Working Group 15 on *The Role of History of Mathematics in Mathematics Education: Theory and Research* of the Congress of the European Society for Research in Mathematics Education (CERME 6), in Lyon, France, 2009.

emerging change in the nature of algebra. Renaud Chorlay of France considers four viewpoints in elementary function theory and shows how a historical consideration of these viewpoints can impact the education of students. Finally, again from Mexico, Gabriela Buendia Abalos and Gisela Montiel Espinosa consider trigonometry as an example of how the ambient culture affects the teaching and learning of significant mathematical ideas.

Implementing the History of Mathematics in Mathematics Education

In Part II, there are ten articles dealing with concrete uses of the history of mathematics in teaching mathematics at levels ranging from elementary school through university. In each case, the authors have conducted at least some preliminary research to test the effectiveness of their teaching methods and conclude that the use of history has a positive effect on their students. These ten papers come from seven countries on three continents, reflecting the international nature of the research. In many of these cases, teachers elsewhere should be able to modify the methods of the original authors for use in their own classrooms. We begin with an article by Man-Keung Siu from Hong Kong, China, who discusses a course for secondary students integrating physics and mathematics. Since physics and mathematics developed together to a large extent, this course is historically based. But students today are frequently not aware that mathematics has a history or even that anything new can be discovered. Next, Batya Amit, Nitsa Movshovitz-Hadar and Avi Berman in Israel show how one can introduce news from contemporary mathematics with its historical background into a secondary classroom, including such topics as Fermat's Last Theorem, Kepler's conjecture, and the mathematics of Sudoku. From Italy, we have a study by Adriano Dematté and Fulvia Furinghetti on the use of pictures of ancient documents in teaching secondary students. Then, there is a study by Luis Casas and Ricardo Luengo of Spain on the use of historical weights and measures in an elementary classroom. This activity involved the students not only in mathematics and its history, but also in the collecting of information from older members of their community. Next, Uffe Jankvist in Denmark studied how his students thought about the history of mathematics. Moving on to the university level, we have a study by Beverly Reed of the United States on how studying the history of the idea of a function positively affected her students' understanding of the concept. From Greece, we have a paper by Theodorus Paschos and Vassiliki Farmaki on how the study of the idea of motion in the later Middle Ages can help students better understand the concepts of the definite integral and the fundamental theorem of calculus. And then Tinne Kjeldsen from Denmark shows how original sources from the eighteenth century can be used in demonstrating to students the influence of physics on the development of differential equations. Returning to Greece, Michael Kourkoulos and Constantinos Tzanakis examine their statistics students' difficulties in understanding the concept of variance, i.e., why one uses the square root of the sum of the squares, and show how this difficulty is related to the historical development of the idea of variance and standard deviation. Finally, back in the United States, we have an article by Janet Barnett and others discussing the use of original sources in teaching courses in discrete mathematics.

Using the History of Mathematics in Teacher Education

Although teacher education was relevant to some of the earlier papers, in part III we present four articles dealing directly with the education of prospective teachers of mathematics. Each article shows the importance of the use of history in educating future teachers so they can meet the challenges of their own classrooms. In the first paper in this part, Bjørn Smestad of Norway discusses how to use history of mathematics in pre-service courses for elementary and lower secondary school teachers. Kathleen Clark of the United States discusses her history of mathematics course specifically designed to show prospective secondary teachers how to use history in their own teaching. Leo Rogers of the UK reflects on how recent changes in the English mathematics curriculum are now forcing university professors teaching pre-service teachers to figure out new ways to introduce relevant historical material. And finally, in a second article, Bjørn Smestad discusses a study of some Norwegian secondary teachers to understand how they actually use the history of mathematics in their classrooms.

Invited Papers on the History of Mathematics

The final part contains three important articles based on invited presentations on the history of mathematics at the HPM meeting in Mexico City in July, 2008. In the first, Karen Parshall of the United States discusses how the mathematical research community in North America evolved from its beginnings in colonial times up to the middle of the twentieth century. Ubiratan D'Ambrosio from Brazil complements this article with a study of the evolution of the mathematics community in Latin America from the time of independence again up to the middle of the last century. Finally, we conclude with an article by Glen van Brummelen of Canada on the history of trigonometry, showing that its primary motivation was the study of the heavens.

Acknowledgements

The editors wish to thank all the referees who helped make this volume possible: Amy Ackerberg-Hastings, Abraham Arcavi, Evelyne Barbin, George Booker, Benjamin V. C. Collins, Daniel Curtin, Ubi D'Ambrosio, Rosalie Dance, Abdellah El Idrissi, Florence Fasanelli, Gail FitzSimons, Michael Fried, Fulvia Furinghetti, Mary Garner, Mary Gray, Michel Helfgott, Wann-Sheng Horng, Ilhan Izmiri, Uffe Jankvist, Barbara Jur, Herbert Kasube, Michael Kourkoulos, Manfred Kronfellner, Ewa Lakoma, Snezana Lawrence, Kazem Mahdavi, John McCleary, Walter Meyer, David Pengelley, Youyu Phillips, Luis Radford, Leo Rogers, Hector Rosario, Ed Sandifer, Gert Schubring, Man-Keung Siu, Yannis Thomaidis, Homer White, Greisy Winicki, and Maria Zack. Like the authors, these referees come from numerous countries on six continents and help demonstrate the international nature of the interest in using the history of mathematics in the teaching of mathematics. The editors also wish to thank the publication staff of the Mathematical Association of America, who were responsible for final production of the book: Don Albers, Carol Baxter, Bev Ruedi, and Rebecca Elmo. Victor Katz would also like to thank his wife Phyllis for her loving help throughout the production of this book.

Bibliography

[1] J. Fauvel & J. van Maanen, 2000, *History in Mathematics Education: The ICMI Study*, Kluwer Academic Publishers, Dordrecht.

Contents

Preface   **vii**
    Bibliography ix

1 Teaching with Primary Historical Sources: Should it Go Mainstream? Can it?, David Pengelley   **1**
    1.1 Introduction 1
    1.2 A Personal Odyssey as an Illustration of Issues 1
    1.3 Motivations: Why or Why Not? 3
    1.4 Logistical Obstacles 3
    1.5 A Sample Project: Pascal on Induction and Combinatorics 4
    1.6 Finale 7
    Bibliography 7

2 Dialogism in Mathematical Writing: Historical, Philosophical and Pedagogical Issues, Evelyne Barbin   **9**
    2.1 What is Dialogism? 9
    2.2 Dialogism in Mathematics: A Paradox? 10
    2.3 Dialogism and Innovation in Mathematics 11
    2.4 Dialogism and History of Mathematics: The Question of Addressivity 12
    2.5 Dialogism and History of Mathematics: Reading Original Sources in Classroom 13
    2.6 Conclusion 14
    Bibliography 15

3 The Process of Mathematical Agreement: Examples from Mathematics History and an Experimental Sequence of Activities, Gustavo Martinez-Sierra and Rocío Antonio-Antonio   **17**
    3.1 Introduction 17
    3.2 The Process of Mathematical Agreement 17
    3.3 The Fractional Exponents 18
    3.4 The Square Root of Negative Numbers 19
    3.5 The Radian and the Trigonometric Functions 20
    3.6 An Experimental Sequence of Activities: The Square Root of Negative Numbers 22
    3.7 In Conclusion 26
    Bibliography 26

4 Researching the History of Algebraic Ideas from an Educational Point of View, Luis Puig   **29**
    4.1 Introduction 29
    4.2 Babylonian Cut and Paste Operations: A Tool for Analysis 30
    4.3 Cut and Paste Operations in Al-Khwārizmī's Algebra 33
    4.4 A Reading of Jordanus de Nemore' *De Numberis Datis* 36
    4.5 Didactical Comments 39
    4.6 By Way of Conclusion 41
    Bibliography 41

5 Equations and Imaginary Numbers: A Contribution from Renaissance Algebra, Giorgio T. Bagni — 45
- 5.1 Editors' Introductory Comments — 45
- 5.2 Introduction — 46
- 5.3 Theoretical Framework — 46
- 5.4 History of Mathematics and Imaginary Numbers — 48
- 5.5 Imaginary Numbers from History to Mathematics Education — 51
- 5.6 The Semiosic Chain — 51
- 5.7 Final Reflections — 54
- Bibliography — 55

6 The Multiplicity of Viewpoints in Elementary Function Theory: Historical and Didactical Perspectives, Renaud Chorlay — 57
- 6.1 Introduction — 57
- 6.2 The Interplay Between Four Viewpoints — 58
- 6.3 A Historical Case-Study of f' and Variations of f According to Cauchy — 59
- 6.4 Ways of World-Making: The Case of Functions — 61
- 6.5 From Historical Research to Didactic Engineering and Research in Didactics — 63
- 6.6 Conclusion — 64
- Bibliography — 65

7 From History to Research in Mathematics Education: Socio-Epistemological Elements for Trigonometric Functions, Gabriela Buendia Abalos and Gisela Montiel Espinosa — 67
- 7.1 Introduction — 67
- 7.2 Socio-Epistemological Theoretical Approach — 67
- 7.3 Didactic Difficulties with the Trigonometric Functions — 68
- 7.4 Socio-Epistemological Review: Towards Trigonometric Functionality — 70
- 7.5 A Socio-Epistemological Approach to the *Introductio in Analysin Infinitorium* — 72
- 7.6 A Socio-Epistemology Based on Uses and Significations Developed in a Historic Setting — 75
- 7.7 Problem Situations — 77
- 7.8 Final Reflection — 80
- Bibliography — 81

8 Harmonies in Nature: A Dialogue Between Mathematics and Physics, Man-Keung Siu — 83
- 8.1 Why is an Enrichment Course on Mathematics-Physics Designed? — 83
- 8.2 How is Such a Course Run? — 84
- 8.3 A Sketch of the Content of the Course — 84
- 8.4 Some Sample Problems in Tutorials — 86
- 8.5 Conclusion — 89
- Bibliography — 89

9 Exposure to Mathematics in the Making: Interweaving Math News Snapshots in the Teaching of High-School Mathematics, Batya Amit, Nitsa Movshovitz-Hadar, and Avi Berman — 91
- 9.1 Introduction: The Ever Growing Nature of Mathematics — 91
- 9.2 The Problem: A Gap Between School Mathematics and Mathematics — 92
- 9.3 A Proposed Solution and its Rationale — 92
- 9.4 The Challenge for the Teacher — 93
- 9.5 The Study — 93
- 9.6 A Snapshot: "The Search for Prime Numbers" — 94
- 9.7 Analysis of Observed Reactions to the Snapshot — 98
- 9.8 Closing Remarks — 99
- Bibliography — 100

10 History, Figures and Narratives in Mathematics Teaching, Adriano Demattè and Fulvia Furinghetti — 103
- 10.1 Introduction — 103
- 10.2 Using Pictures as a Special Case of Using Original Sources — 104
- 10.3 The Experiment — 105
- 10.4 Didactical Implications and Conclusions — 109
- Bibliography — 110

11 Pedagogy, History, and Mathematics: Measure as a Theme, Luis Casas and Ricardo Luengo — 113
- 11.1 Introduction — 113
- 11.2 The Innovation Project — 114
- 11.3 Evaluation of the Project and Conclusions — 119
- Bibliography — 120

12 Students' Beliefs About the Evolution and Development of Mathematics, Uffe Thomas Jankvist — 123
- 12.1 Introduction — 123
- 12.2 The Danish Context — 124
- 12.3 Students' Beliefs About the 'Identity' of Mathematics — 125
- 12.4 Evaluating Students' Beliefs Against the 'Goals' — 127
- 12.5 Final Remarks and Reflections — 129
- Bibliography — 130

13 Changes in Student Understanding of Function Resulting from Studying Its History, Beverly M. Reed — 133
- 13.1 Introduction — 133
- 13.2 Theoretical Basis for the Study: APOS Theory — 133
- 13.3 Procedure for the Study — 134
- 13.4 The Research Instruments — 134
- 13.5 The Worksheets and Readings — 135
- 13.6 Findings — 138
- 13.7 Discussion — 142
- 13.8 Conclusion and Summary — 143
- Bibliography — 143

14 Integrating the History of Mathematics into Activities Introducing Undergraduates to Concepts of Calculus, Theodorus Paschos and Vassiliki Farmaki — 145
- 14.1 Introduction — 145
- 14.2 The Historical Background of the Teaching Experiment — 146
- 14.3 Designing Didactic Activities Inspired by History of Mathematics — 149
- 14.4 The Multiple Linked Representations Between Rates and Totals — 159
- 14.5 Analysis of the Data Collected–Results — 160
- Bibliography — 163

15 History in a Competence Based Mathematics Education: A Means for the Learning of Differential Equations, Tinne Hoff Kjeldsen — 165
- 15.1 Introduction — 165
- 15.2 Mathematical Competence and the Role of History — 166
- 15.3 A Multiple Perspective Approach to History of Mathematics — 166
- 15.4 A Student Project on Physics' Influence on the History of Differential Equations — 167
- 15.5 Some Conclusions and Critical Remarks — 171
- Bibliography — 172

16 History of Statistics and Students' Difficulties in Comprehending Variance, Michael Kourkoulos and Constantinos Tzanakis — 175
- 16.1 Introduction 175
- 16.2 Classroom Observations 175
- 16.3 Looking for Solutions: History Enters the Scene 178
- 16.4 An Important Predecessor 180
- 16.5 The Normal Distribution and the Central Limit Theorem 182
- Bibliography 182

17 Designing Student Projects for Teaching and Learning Discrete Mathematics and Computer Science via Primary Historical Sources, Janet Heine Barnett, Jerry Lodder, David Pengelley, Inna Pivkina and Desh Ranjan — 187
- 17.1 Introduction 187
- 17.2 Pedagogical Goals and Design Principles 188
- 17.3 Incorporating Pedagogical Design Goals: Two Sample Projects 189
- 17.4 Implementation 197
- 17.5 Conclusion 198
- Bibliography 198

18 History of Mathematics for Primary School Teacher Education Or: Can You Do *Something* Even if You Can't Do Much?, Bjørn Smestad — 201
- 18.1 Introduction 201
- 18.2 Background 201
- 18.3 Ways of Working with History of Mathematics 202
- 18.4 Discussion 209
- Bibliography 209

19 Reflections and Revision: Evolving Conceptions of a *Using History* Course, Kathleen Clark — 211
- 19.1 Introduction 211
- 19.2 Course Context 212
- 19.3 The Capstone Project: The First Three Semesters 213
- 19.4 Reflections for Further Course Revision 218
- 19.5 Conclusion 219
- Bibliography 220

20 Mapping Our Heritage to the Curriculum: Historical and Pedagogical Strategies for the Professional Development of Teachers, Leo Rogers — 221
- 20.1 The English Curriculum: 2008–2009 221
- 20.2 A Pedagogical Tradition 222
- 20.3 The History of Mathematics and our Mathematical Heritage 222
- 20.4 Mapping Our Heritage 224
- 20.5 Trialling, Qualitative Results, and Development 225
- 20.6 Heritage Maps and Canonical Images 226
- Bibliography 228

21 Teachers' Conceptions of History of Mathematics, Bjørn Smestad — 231
- 21.1 Introduction 231
- 21.2 Background—Norway 231
- 21.3 Background—International 232
- 21.4 The Research Question 232
- 21.5 Method 233
- 21.6 The Participants 234

21.7 Discussion 235
21.8 Conclusion 238
Bibliography 238

22 The Evolution of a Community of Mathematical Researchers in North America: 1636–1950, Karen Hunger Parshall   241
22.1 The Seventeenth and Eighteenth Centuries: Mathematics in Colonial Settings 241
22.2 The Nineteenth Century: A Period of General Structure-Building in Higher Education and in Science   243
22.3 A Mathematical Research Community Emerges in the United States: 1876-1900 244
22.4 The Twentieth Century: The Consolidation and Growth of Research-Level Mathematics 247
22.5 The North American Mathematical Landscape by 1950 248
Bibliography 249

23 The Transmission and Acquisition of Mathematics in Latin America, from Independence to the First Half of the Twentieth Century, Ubiratan D'Ambrosio   251
23.1 Introduction 251
23.2 Independence 252
23.3 The 20th Century up to the End of World War II 253
23.4 After the End of the Second World War 255
23.5 Concluding Remarks on Contemporary Developments 257
Bibliography 258

24 In Search of Vanishing Subjects: The Astronomical Origins of Trigonometry, Glen Van Brummelen   261
24.1 Introduction 261
24.2 False Beginnings 262
24.3 Heavenly Foreshadowing 263
24.4 A Union of Opposites 264
24.5 Cultural Divergence: Trigonometry in India 268
24.6 Conclusion 270
Bibliography 270

About the Editors   273

1

Teaching with Primary Historical Sources: Should It Go Mainstream? Can It?[1] [2]

David Pengelley
New Mexico State University, United States of America

1.1 Introduction

I am truly honored to be asked to speak on integrating the history of mathematics in mathematics education. Advocating the teaching of mathematics using history is presumably not very controversial at this conference, more like "preaching to the choir", as one says in English. But I wish to be somewhat provocative, perhaps even controversial, by suggesting a dream I have had for some time, that all students should learn the principal content of their mathematics directly from studying primary sources, i.e., from the words of the original discoverers or creators of new mathematics, as is done in the humanities, where students read the great original literature, not just about the great literature. In other words, I propose that we rebuild the entire mathematics curriculum at all levels around translated primary sources studied directly by our students. If you think this is extreme, then at least I am fulfilling the role of being a provocative speaker.

My belief that we should and can aim for a mathematics curriculum that is rich throughout in primary sources has developed only very slowly from my own experiences in the past twenty years. First I would like to describe this personal evolution, because it reflects very clearly some of the important challenges involved in implementing my dream.

1.2 A Personal Odyssey as an Illustration of Issues

First I co-developed two one-semester courses for beginning and advanced undergraduate university students, based entirely on primary historical sources. Somewhat ironically, I was motivated by William Dunham's description of a great theorem enrichment course for teachers in which he rewrote the original source material in his own words, but I and my collaborator Reinhard Laubenbacher decided to skip the rewriting step and toss the original sources at our students, partly because it seemed like too much work to rewrite things; of course in retrospect this became my chief pedagogical goal, to have students read original sources themselves, the only compromise being translation into English. These courses each follow several great mathematical themes and problems through millennia via primary

---
[1] Based on a plenary address presented at the HPM Group Satellite Meeting of ICME 11 (HPM 2008) in Mexico City, Mexico, 14–18 July 2008.
[2] This material is based upon work supported by the U.S. National Science Foundation under Grant No. 0717752.

sources. The courses have been continually successful now for two decades, and have led to two books [7, 5] each with multiple chapters built entirely around primary sources, from which different chapters can be taught in different incarnations of the course [9]. They fully embody my vision of courses and books in which primary sources are the principal objects of study and learning.

However, while they are solid mathematics courses, they are not focused on syllabi in the standard curriculum, i.e., they do not fall into the category of a course in calculus, or discrete mathematics, or real analysis, or abstract algebra, or the many other compartmentalized topics in the typical institutional curriculum. They were instead designed around historical development of great ideas viewed through primary sources, not just a purely modern vision of mathematics; and they are flexible, with different subjects covered in different semesters. In other words, they were designed and implemented in total freedom, rather than under the constraint of an existing course syllabus. So while these courses are taken by many university mathematics students as an elective, and by students studying other disciplines, they are not in the "mainstream" of the curriculum. Moreover, students and colleagues alike tend to consider them as "history of mathematics" courses, simply because no other mathematics courses have any meaningful history in them. Together these features leave the two courses somewhat outside the main path of a standard undergraduate mathematics student's degree course work, and hinder the adoption of the courses elsewhere.

At about the same time, I also became heavily involved in collaborative developing, teaching, and publishing of student projects for calculus courses [6]. While these were not historical in nature, this began my very slow process of moving away from lecturing in regular teaching, towards a more student-centered, problem-driven classroom, which I personally find prerequisite to engaging primary historical sources as the principal objects of study. Only by combining student project activity with an active classroom and historical materials have I ultimately managed to even begin building a standard curriculum around primary sources. I have recently written on my current thoughts [11] about creating a classroom dynamic in which students are engaged in high-level active work, rather than listening to lecture.

Then about thirteen years ago, I co-created a graduate level mathematics course on the role of history in teaching mathematics, in which each graduate student develops a written teaching module based on historical material. While this is a successful mathematics education graduate course, it is a course on mathematics education, not a mathematics course based on historical material.

Around this time I also made my first attempt to inject a primary historical source in a substantial way into a regular course, namely Arthur Cayley's first paper on group theory in an abstract algebra course when students first encounter groups [12]. I began to realize that my students could benefit tremendously from having their very first encounters with the notion of an abstract group be through the wonderful mathematics emerging in the nineteenth century that motivated Cayley to define and develop the abstract idea and first steps towards a theory of groups. This was my first indication that one can very profitably simply start using primary sources as key documents of study in a regular course, without dramatically changing the "content" of the course.

This idea has expanded greatly during recent years into an increasing collaboration with an expanding circle of colleagues. Some of us had experience both teaching with primary sources and with designing substantial student projects, which engage students in large multi-step assignments and written reports on their investigation, and which may last from one to several weeks at a time. We decided to combine these pedagogical approaches, and to focus on regular course content on the discrete side of mathematics, very broadly conceived. So with support from the U.S. National Science Foundation, I am now part of a team of seven faculty with additional collaborating writers and testers at numerous institutions, who are developing, testing, and evaluating student projects based on primary historical sources for teaching regular course syllabi in discrete mathematics, abstract algebra, graph theory, combinatorics, logic, and computer science (e.g., courses on algorithms or automata theory). We hope also that a useful statistical evaluation of the effects of our historical projects will emerge from the nature and scope of this endeavor. Details of our 20–30 student projects based on primary sources, some completed, tested, and published, some yet to be written, are available in a resource book [2] and at our web sites [4, 1]. Below I will use one of these projects to illustrate how I believe primary sources can be central to the curriculum.

Most recently, in teaching an upper undergraduate level geometry course, I realized that I could have students learn most of the course content on the hyperbolic non-Euclidean plane from the original sources by Euclid, Legendre, Lobachevsky, and Poincaré presented in the geometry chapter of my first book [7] of annotated primary sources, so these pre-prepared primary sources fit well and easily into the course.

I finally feel I am on a route towards a standard course curriculum in which primary historical sources play a core role, but you can see that it has been a long, slow road, that I am benefiting from working in collaboration with others, and that so far I still only have a part of some courses based this way. Nonetheless, I now actually see that this could grow into courses built entirely around primary sources. Of course I realize that I am only one of many around the world who are working to incorporate primary sources in key ways into the regular curriculum, and I would like to acknowledge and applaud everyone else's efforts as well; this gives me the inspiration of community to continue, and I hope that together we may have an impact. My intent in this Section has simply been to show by example what some of the challenges are in basing courses on primary sources, but that it may be possible.

1.3 Motivations: Why or Why Not?

Why should we use primary sources at the foundations of our teaching? Or why not?

The reasons for doing this have already been enunciated by many others over the years, but I will merely mention here motivation and deep connections along time, understanding essence, origin, and discovery, mathematics as a humanistic endeavor; practice moving from verbal to modern mathematical descriptions; reflection on present-day standards and paradigms; participating in the process of doing mathematics through experiment, conjecture, proof, generalization, publication and discussion; more profound technical comprehension from initial simplicity; also dépaysement (disorientation, cognitive dissonance, multiple points of view); and a question-based curriculum that knows where it came from and where it might be going. Questions before answers, not answers to questions that have not been asked. In [3], we illustrate specifically how pedagogical design principles like these can be built into student projects based on primary sources.

On the other hand, one can think of reasons why teaching with primary sources at the core might not be good to aim for. In his intentionally "devil's advocate" article [13], Man-Keung Siu lists possible unfavorable factors a teacher might express, some of which could apply to primary sources. There are first some pedagogical ones: "How can you set questions on it on a test?; It can't improve the student's grade; Students don't like it; Students regard it as just as boring as the subject mathematics itself!; Students do not have enough general knowledge of culture to appreciate it!; Progress in mathematics is to make difficult problems routine, so why bother to look back?; What really happened can be rather tortuous. Telling it as it was can confuse rather than enlighten!; Does it really help to read original texts, which is a very difficult task?; Is it liable to breed cultural chauvinism and parochial nationalism?; Is there any empirical evidence that students learn better when history of mathematics is made use of in the classroom?"

I now have enough experience actually teaching with primary sources to say that I personally have found all these concerns to be either untrue or irrelevant with my students and my chosen primary sources. I could elaborate and explain, but here I will only affirm my clear experience that with carefully selected and prepared primary source material, and the right pedagogical method in the classroom, these objections or concerns can and should be rejected. Thank you for raising them, Man-Keung!

Man-Keung also listed concerns that are logistical in nature, and I will address these next.

1.4 Logistical Obstacles

Man-Keung Siu's logistical concerns about using history certainly can apply to teaching with primary sources:

- "There is a lack of resource material on it!"
- "There is a lack of teacher training in it!"
- "I have no time for it in class!"

The good news is that the first two concerns are the things we can all work on, and doing so will influence the third!

1.4.1 Is there a lack of resource material?

Yes, but the availability of published primary sources and translations in all aspects of mathematics has been growing at great speed in the past few decades, thanks to the work of many wonderful people, and this work we should all continue. One resource bibliography for finding available primary source material is [10]. It would be wonderful to have a continually updated central listing of these sources. Providing such a central resource online is something incredibly useful that HPM could sponsor, and I believe it is necessary to widespread adoption of teaching with primary sources. I will address the important issue of convenient packaging of primary sources for teaching below.

1.4.2 Instructor training, motivation?

Is there a lack of teacher training in using primary sources? Yes, of course, and this challenge can and should be relieved by more formal training opportunities, which is another task for us. But the question, I think, really hints at a deeper issue. How do we interest other instructors in teaching with history, and in particular in using primary sources? It will only be through instructors' desire to teach with history that it will happen, not by coercion, since mathematics instructors like to make their own pedagogical decisions. Since I believe that enticement is the only way, we should entice with wonderful source material packaged to make instructors salivate at the idea of learning and teaching with them. Some teachers want or need prepared guiding materials in the form of textbooks or projects, while others like to create their own. So I believe that the solution lies in providing a variety of packaging for primary source materials, and flexibility in how they can be used, along with our own leading by example in our teaching. Let us discuss packaging.

1.4.3 Packaging into textbooks and projects versus time and pedagogical style

In many parts of the world, textbooks are the driving force behind curriculum and pedagogy, whether we like it or not. So I believe that success in attracting others to teaching mainly with primary sources will require us to create textbooks that have this as their theme. But student projects are also playing a more substantial role in teaching these days, so projects based on primary sources can complement or even supplant portions of a standard textbook, and thus play an intermediate role as stepping stones in the direction I am advocating. This is the approach I am currently working on, as mentioned earlier [4, 1, 2]. In fact, a course could be built entirely on a sequence of projects based on primary sources, and I am working in that direction.

These issues cannot be divorced from the remaining question of whether there is time in class for teaching with historical sources. My personal experience is a resounding yes, there is time, and I constantly become stronger in that conviction, by changing my teaching in two respects.

First, one can move entirely away from lecturing, whether using historical sources or not, by having students first do advance reading of all new material at home, and working and writing profitably about it, entirely before initial class contact with the material. This makes lecture totally unnecessary and unfruitful, and means that class time is spent on student work and interaction with the instructor and each other, and some whole class discussion, that already starts at a higher level; the time thus saved and redirected from lecture is enormous. I have written about the details of how I implement this [11].

Second, one can implement learning from primary sources through projects in such a way that it literally takes over core topics from the textbook, i.e., one can find and develop primary sources to teach core material of the course. Then the textbook becomes at most ancillary, perhaps a source of modern notation, extra exercises, and an alternative, more modern point of view. The time otherwise spent with the textbook will instead be spent learning the same material from primary sources, and the textbook becomes a supplement, not the other way around. I would like to elaborate on one example of this.

1.5 A Sample Project: Pascal on Induction and Combinatorics

To see a detailed example of how core syllabus material can be taught directly from a historical project, and its effect on students, consider an introductory discrete mathematics course intended to have students start learning to make proofs in mathematics, in which some key content is to learn mathematical induction as a proof technique, and to become comfortable with index notation, binomial coefficients, combination numbers, factorials, and some elementary number theory. I have combined all these core topics in a three week class project [2, 4] centered on Blaise Pascal's *Treatise on the Arithmetical Triangle* [8]. This large student project is pedagogically analyzed in some detail in (Barnett et al, this volume). Here I present just a few key excerpts, intending only to demonstrate the power of the primary source for covering core material, and the kinds of challenges I give students to achieve this.

1.5. A Sample Project: Pascal on Induction and Combinatorics

Pascal's treatise expounds the principle of mathematical induction, and his triangle leads into combinatorics. In fact this is the first place in the mathematical literature where the principle of mathematical induction is enunciated so completely and generally, as a means of establishing the indefinite persistence of an observed pattern.[See Pascal's version of his triangle on p. 190 of this volume.]

After a good bit of historical background and context, students begin studying Pascal's highly verbal definitions for the triangle. Since Pascal's work involves no index notation, students learn naturally about double-indexing from translating Pascal's description into modern terminology. The project continues with exercises for students based on the first several of Pascal's 23 *consequences* (theorems) and his proofs, connecting them to modern notation, indexing, summation notation, terminology, and the adequacy of Pascal's proofs by iteration or generalizable example. Pascal's consequences actually ease slowly and totally naturally into the concept of proof by mathematical induction in order to prove symmetry in the triangle, allowing the concept of induction to evolve in students minds, rather than being presented abstractly out of nowhere. Pascal proves his claims, even his mathematical induction, by generalizable example, largely because he has no indexing notation to deal conveniently with arbitrary elements. Having students make all this precise in full generality with modern notation enables them to begin to think in terms of induction before it is formally introduced, and to powerfully appreciate the efficacy of indexing notation.

The crowning consequence in Pascal's treatise is the twelfth, in which Pascal derives a formula for the ratio of consecutive numbers in a base. From this he will obtain an elegant and efficient "closed" formula for all the numbers in the triangle, a powerful tool for much future mathematical work. And it is right here that Pascal enunciates the general proof principle we call induction. Again we ask students to translate Pascal's proof by generalizable example into a modern and completely general proof. This is far from trivial, and even involves an understanding of a property of proportions that is largely lost today. We highlight the following excerpt from the middle of the project, consisting of primary source material (in sans serif font) and exercises for students, to illustrate the level of challenge and richness of content of the source for teaching core course material.

Twelfth Consequence

In every arithmetical triangle, of two contiguous cells in the same base the upper is to the lower as the number of cells from the upper to the top of the base is to the number of cells from the lower to the bottom of the base, inclusive.

Let any two contiguous cells of the same base, E, C, be taken. I say that

E : C :: 2 : 3
the lower the upper because there are two cells from E to the bottom, namely E, H, because there are three cells from C to the top, namely C, R, μ.

Although this proposition has an infinity of cases, I shall demonstrate it very briefly by supposing two lemmas:

The first, which is self-evident, that this proportion is found in the second base, for it is perfectly obvious that $\varphi : \sigma :: 1 : 1$;

The second, that if this proportion is found in any base, it will necessarily be found in the following base.

Whence it is apparent that it is necessarily in all the bases. For it is in the second base by the first lemma; therefore by the second lemma it is in the third base, therefore in the fourth, and to infinity.

It is only necessary therefore to demonstrate the second lemma as follows: If this proportion is found in any base, as, for example, in the fourth, $D\lambda$, that is, if $D : B :: 1 : 3$, and $B : \theta :: 2 : 2$, and $\theta : \lambda :: 3 : 1$, etc., I say the same proportion will be found in the following base, $H\mu$, and that, for example, $E : C :: 2 : 3$.

For $D : B :: 1 : 3$, by hypothesis.

Therefore $\underbrace{D + B}_{E} : B :: \underbrace{1 + 3}_{4} : 3$

$E : B :: 4 : 3$

Similarly $B : \theta :: 2 : 2$, by hypothesis.

Therefore $\underbrace{B + \theta}_{C} : B :: \underbrace{2 + 2}_{4} : 2$

$C : B :: 4 : 2$

But $B : E :: 3 : 4$

Therefore, by compounding the ratios, $C : E :: 3 : 2$. Q.E.D.

The proof is the same for all other bases, since it requires only that the proportion be found in the preceding base, and that each cell be equal to the cell before it together with the cell above it, which is everywhere the case.

Project Question 6. Pascal's Twelfth Consequence: the key to our modern factorial formula

(a) Rewrite Pascal's Twelfth Consequence as a generalized modern formula, entirely in our $T_{i,j}$ terminology. Also verify its correctness in a couple of examples taken from his table in the initial definitions section.

(b) Adapt Pascal's proof by example of his Twelfth Consequence into modern generalized form to prove the formula you obtained above. Use the principle of mathematical induction to create your proof.

From his Twelfth Consequence Pascal can develop a "formula" (essentially the modern factorial formula) for the numbers in the triangle, which can then be used in future work on combinatorics, probability, and algebra. In the following project excerpt, we have students follow Pascal's generalizable example to do so in modern form.

Problem

Given the perpendicular and parallel exponents of a cell, to find its number without making use of the arithmetical triangle.

Let it be proposed, for example, to find the number of cell ξ of the fifth perpendicular and of the third parallel row.

All the numbers which precede the perpendicular exponent, 5, having been taken, namely 1, 2, 3, 4, let there be taken the same number of natural numbers, beginning with the parallel exponent, 3, namely 3, 4, 5, 6.

Let the first numbers be multiplied together and let the product be 24. Let the second numbers be multiplied together and let the product be 360, which, divided by the first product, 24, gives as quotient 15, which is the number sought.

For ξ is to the first cell of its base, V, in the ratio compounded of all the ratios of the cells between, that is to say, $\xi : V$

in the ratio compounded of $\quad \xi : \rho, \quad \rho : K, \quad K : Q, \quad Q : V$
or by the twelfth consequence $\quad 3 : 4 \quad\quad 4 : 3 \quad\quad 5 : 2 \quad\quad 6 : 1$

Therefore $\xi : V :: 3 \cdot 4 \cdot 5 \cdot 6 : 4 \cdot 3 \cdot 2 \cdot 1$.

But V is unity; therefore ξ is the quotient of the division of the product of $3 \cdot 4 \cdot 5 \cdot 6$ by the product of $4 \cdot 3 \cdot 2 \cdot 1$.

N.B. If the generator were not unity, we should have had to multiply the quotient by the generator.

Project Question 7. Pascal's formula for the numbers in the Arithmetical Triangle

(a) Write down the general formula Pascal claims in solving his "Problem." Your formula should read $T_{i,j}=$ "some formula in terms of i and j." Also write your formula entirely in terms of factorials.

(a) Look at the reason Pascal indicates for his formula for a cell, and use it to make a general proof for your formula above for an arbitrary $T_{i,j}$. You may try to make your proof just like Pascal is indicating, or you may prove it by mathematical induction.

The project continues on perfectly naturally to integrate combinatorics, the binomial theorem, Fermat's Theorem (proof by induction on the base using Pascal's formula for the binomial coefficients and uniqueness of prime factorization), and to end with the RSA cryptosystem. This goes far beyond the historical source, but shows how the source serves as a natural foundation for the flow of important core topics.

I make one final comment on the efficacy of a core project like this. On part of a final exam I gave my students a choice between a proof by induction of a standard homework-like summation formula from their textbook or digesting, explaining, and adapting a modern proof by induction from a Consequence in Pascal's treatise that they had never seen before. Half the students chose to do new interpretation and modern proof work from the Pascal treatise!

1.6 Finale

I must end with an exhortation of one more reason to teach core material from primary sources: It is inspiring, fun, lively, rewarding and enriching for instructors as well as students. It will keep you happy, excited, and alive.

Bibliography

[1] Barnett, J., G. Bezhanishvili, H. Leung, J. Lodder, D. Pengelley, I. Pivkina, and D. Ranjan, 2008, *Learning Discrete Mathematics and Computer Science via Primary Historical Sources*, www.cs.nmsu.edu/historical-projects/.

[2] Barnett, J., G. Bezhanishvili, H. Leung, J. Lodder, D. Pengelley, D. Ranjan, 2009, "Historical Projects in Discrete Mathematics and Computer Science," in *Resources for Teaching Discrete Mathematics*, B. Hopkins (ed.), Washington, D.C.: Mathematical Association of America, MAA Notes volume 74.

[3] Barnett, J., Lodder, J., Pengelley, D., Pivkina, I., Ranjan, D., this volume, "Designing student projects for teaching and learning discrete mathematics and computer science via primary historical sources," in *Recent developments on introducing a historical dimension in mathematics education*, C. Tzanakis & V. Katz (eds.),Washington, D.C.: Mathematical Association of America.

[4] Bezhanishvili, G., H. Leung, J. Lodder, D. Pengelley, and D. Ranjan, 2003, *Teaching Discrete Mathematics via Primary Historical Sources*, www.math.nmsu.edu/hist_projects/.

[5] Knoebel, A., R. Laubenbacher, J. Lodder, and D. Pengelley, 2007, *Mathematical Masterpieces: Further Chronicles by the Explorers*, New York: Springer-Verlag. (Excerpts and reviews at www.math.nmsu.edu/~history/.)

[6] Lakey, J. and D. Pengelley, 1993, *Evolution of Calculus Courses at New Mexico State University*, www.math.nmsu.edu/evolution_calculus.html.

[7] Laubenbacher, R. and D. Pengelley, 1999, *Mathematical Expeditions: Chronicles by the Explorers*, New York: Springer-Verlag, revised second printing, 2000. (Excerpts and reviews at www.math.nmsu.edu/~history/.)

[8] Pascal, B., 1991, "Treatise on the Arithmetical Triangle," in *Great Books of the Western World*, Mortimer Adler (ed.), Chicago: Encyclopædia Britannica, Inc.

[9] Pengelley, D., 1999, *Teaching with original historical sources in mathematics*, a resource website, www.math.nmsu.edu/~history/.

[10] ——, 2003, *Some Selected Resources for Using History in Teaching Mathematics*, www.math.nmsu.edu/~davidp/.

[11] ——, 2008, *Comments on Classroom Dynamics*, at www.math.nmsu.edu/~davidp/.

[12] ——, 2005, "Arthur Cayley and the first paper on group theory," in *From Calculus to Computers: Using the Last 200 Years of Mathematical History in the Classroom,* R. Jardine and A. Shell (eds.), Washington: Mathematical Association of America, pp. 3–8, and at www.math.nmsu.edu/~davidp/.

[13] Siu, Man-Keung, 1995, No, I don't use history of mathematics in my class. Why?, hkumath.hku.hk/~mks/.

About the Author

David Pengelley's mathematical research is in algebraic topology and the history of mathematics, and he develops the pedagogies of teaching with primary historical sources, with student projects, and with student advance reading, writing, and active classroom work replacing lectures. He has developed a mathematics education graduate course on the role of history in teaching mathematics. He recently authored the articles "Dances between continuous and discrete: Euler's summation formula," "Did Euclid need the Euclidean algorithm to prove unique factorization?," "Voice ce que j'ai trouvé: Sophie Germain's grand plan to prove Fermat's Last Theorem," and the book *Mathematical Masterpieces: Further Chronicles by the Explorers*. David was the 2009 recipient of the Mathematical Association of America's national Haimo Award for Distinguished College or University Teaching of Mathematics.

2

Dialogism in Mathematical Writing: Historical, Philosophical and Pedagogical Issues

Evelyne Barbin
Université de Nantes, France

2.1 What is Dialogism?

Mikhail Bakhtin (1895–1975) was a well-known Russian literary critic and semiotician. The notion of dialogism was mainly developed in his paper entitled "The Problem of Speech Genres" and written in the years 1952–1953, which was translated into French in 1984 and English in 1986. Here, Bakhtin explained that every utterance may be considered as a rejoinder in dialogue. Firstly, every utterance must be regarded as a response to preceding utterances in a given sphere of communication. Secondly, every utterance is oriented toward the response of the others. This notion of dialogism is linked to three major notions named by Bakhtin; *responsive attitude of the listener*, *addressivity* and *speech genres*.

For Bakhtin, when a listener understands the meaning of speech, he takes an active *responsive attitude* towards it: "he either agrees or disagrees with it (completely or partially), augments it, applies it, prepares for its execution, and so on. [...] Any understanding of live speech, a live utterance, is inherently responsive, although the degree of this activity varies extremely" [4, p. 68]. Bakhtin notes that this also pertains to written and read speech. Thus, as he explains, the speaker himself is oriented toward an active responsive attitude: he expects some response, agreement, sympathy, objection, and so on. That means that the utterance (i.e., what is said) is related not only to preceding utterances, but also to subsequent links in a chain of statements. From the beginning, the utterance is constructed while taking into account possible responsive reactions of an audience.

So, for Bakhtin, an essential marker of the utterance is its *addressivity*: the utterance has both an author and an addressee. "This addressee can be an immediate participant-interlocutor in an everyday dialogue, a differentiated collective of specialists in some particular area of cultural communication, a more or less differentiated public, ethnic group, contemporaries, like-minded people, opponents and enemies, a subordinate, a superior, someone who is lower, higher, familiar, foreign and so forth. And it can also be an indefinite, unconcretized *other*" [4, p. 95].

These considerations about the addressee determine a choice of a *genre of speech* by the author: the choice of compositional devices, the choice of language vehicles, the *style* of his or her utterance. Bakhtin gives as examples popular scientific literature addressed to a particular group of readers with a particular background of responsive understanding, special educational literature addressed to another kind of reader, and special research work addressed to an entirely different group.

The choice of a particular speech genre "is determined by the specific nature of the given sphere of speech communication, semantic (thematic) considerations, the concrete situation of the speech communication, the personal composition of its participants, and so on" [4, p. 78]. The genre of speech depends on the sphere of speech communication according as the sphere is reduced to one colleague or to the readers of a scientific review, or enlarged to contemporaries for a manual, or more when a text is intended for posterity.

So, with the notion of dialogism, we have the idea that an utterance is not only a product of an author about a certain subject. Bakhtin writes: "Each utterance is filled with echoes and reverberations of other utterances to which it is related by the communality of the sphere of speech communication. Every utterance must be regarded primarily as a response to preceding utterances of the given sphere (we understand response here in the broadest sense). Each utterance refutes, affirms, supplements, and relies on the others, or supposes them to be known, and somehow takes them into account [...]. The utterance is filled with *dialogic overtones*, and they must be taken into account in order to understand fully the style of the utterance" [4, p. 91–92]. He explains that our thought itself, our philosophical, scientific and artistic thoughts, are born and shaped in the process of interaction with others.

2.2 Dialogism in Mathematics: A Paradox?

It seems a paradox to speak of dialogism about a subject as mathematics. Indeed, a mathematical text is generally thought as universal: today, a mathematician does not write "I" and he or she is not seen as an author, they have no addressee because they write for 'eternity.' No author, no addressee: a mathematical text seems to belong to everybody. But we know, as teachers and as historians, that the risk is that it belongs to nobody, or only to a restricted sphere of persons (phenomena of exclusion and ethnocentrism for instance).

So, let us examine such a paradox with an example. "That all right angles are equal to one another": it is postulate 4 of Euclid's *Elements* [11, p. 200]. As a postulate in Aristotle's sense, the reader has to agree to it, as true and obvious by itself. The reader should not have any "active responsive attitude." So, this sentence is not accompanied by other sentences: *The Elements* did not say why we should take this postulate, how it is chosen, and so on.

But, now, let us take the *Commentary of the first book of Euclid's Elements* of Proclus written in the fifth century. Here, we learn about many readers who "refuted, affirmed, supplemented" Euclid's postulate. Proclus begins with comments of Geminus (around first century), who discusses if Euclid's sentence can be or not taken as a postulate. Following this, Proclus gives one "proof" of the postulate. Then, he mentions other proofs given by Apollonius (third century BCE), Porphyre (third century) and Pappus (fourth century). All these texts are lost, especially the commentary of Pappus about the reciprocal of the postulate, "each angle equal to a right angle is right," which is also commented on by Proclus. All these readers had an "active responsive attitude."

From our point of view, the commentary on Euclid's *Elements* by Al-Narizii is very interesting because it is given in a style of a dialogue. For postulate 4, there is a dialogue between Euclid and Simplicius: "Euclid says: All right angles are equal to one another. Simplicius says: "The truth will appear clearly to who uses a logical reasoning in this matter. If a right angle comes from a perpendicular line without any inclination, as this perpendicular line does not admit any increase or decreasing, but stays always in the same state, then the right angles are always equals." Further, there is a new dialogue about postulate 5 where a person is Al-Narizii himself. "Euclid says: That, if a straight line falling on two straight lines make the interior angles on the same side less than two right angles, the two straight lines, if produced indefinitely, meet on that side on which are the angles less than the two right angles. Simplicius says: This postulate is not quite clear. We need a proof by lines, as Anthimathus and Diodorus did with many propositions. Al-Narizii says: we gave an explanation for that and the additions of Geminus after the proof of proposition 26 of the first book" [1, p. 23–29].

Now, if we turn to commentaries of postulate 4 by an historian like Heath, we will find commentaries of Geminus according to Proclus, but also there are other more recent comments like those of Saccheri, Veronese, Ingrami, Enriques, Amaldi. Heath finishes with Hilbert: "Hilbert takes quite a different line. He considers that Euclid was wrong in placing postulate 4 among the axioms. He himself, [...] proves several theorems about the congruence of triangles and angles, and then deduces our postulate" [11, p. 201]. So, Heath considers that the work of Hilbert is like a response to Euclid. Then, he comments on Euclid as if he was able to know the thought of Euclid : "It was essential from Euclid's point of view that it [postulate 4] should come before postulate 5 since the condition in the latter [...]." Indeed, now the utterance "all right angles are equal to one another" is filled for us with echoes and reverberations of other utterances. Using the word of Bakhtin, we can say it is a "polyphony."

2.3 Dialogism and Innovation in Mathematics

The Bakhtinian approach offers to the historian of mathematics a way to understand the question of innovation in science, which is completely different from that of Kuhn [7]. Kuhn treats the question of innovation in the homogeneous context of tradition [15], while the Bakhtinian approach permits us to incorporate an innovation in the heterogeneous context of the exchanges between a mathematician and his contemporaries. For instance, Bakhtin explains that in some cases, an author can choose a special speech genre, which is the intimate style. He writes: "when the task was to destroy traditional official styles and world views that faded and become conventional, familiar styles became very significant in literature" [4, p. 97].

Let us examine the case of Archimedes' works. Archimedes proves his results on areas and volumes by *reductio ad absurdum*. For instance, he proves that a parabolic segment is equal to a third of a certain triangle by showing that in each supposition, if the segment is either greater or smaller than the third of the triangle, this situation leads to a contradiction.

This kind of proof has the advantage of avoiding any consideration of the infinite or of the composition of the continuum. But, as mathematicians of the seventeenth century noticed, this kind of proof does not explain how the value of one third is obtained. So, these mathematicians thought that Archimedes had a "method of invention" but that he had hidden it to appear more admirable [5]. We have known for a hundred years that it is true that Archimedes had a "method of invention," but it is false that he had hidden it. Indeed, the method of Archimedes to obtain the value of one third and other results appears in a palimpsest discovered by Heiberg. In a recent book, Reviel Netz gives a commentary on the translated text of Archimedes, where he shows the original function of the text to be an act of communication [17]. From our point of view on speech genre, it is interesting to point out that this method appears in a letter of Archimedes to his friend Eratosthenes.

In this letter, Archimedes writes: "Seeing moreover in you, as I say, an earnest student, a man of considerable eminence in philosophy, and an admirer [of mathematical inquiry], I thought fit to write out for you and explain in detail in the same book the peculiarity of a certain method, by which it will be possible for you to get a start to enable you to investigate some of the problems in mathematics by means of mechanics." So, he addresses Eratosthenes in an intimate style, which expresses his respect. Then, Archimedes explains why he writes out this method for him. "This procedure is, I am persuaded, no less useful even for the proof of the theorems themselves; for certain things first became clear to me by a mechanical method, although they had to be demonstrated by geometry afterwards because their investigation by the said method did not furnish an actual demonstration" [2]. So, this method is useful, but it is a mechanical procedure and not a geometrical proof. We can understand that a letter to a friend is a better speech genre to use in writing about this method than a treatise.

The mechanical procedure of Archimedes can be compared with the method of indivisibles introduced in the seventeenth century by Cavalieri and Roberval. This method is a method for inventing propositions on areas and volumes. It consists of seeing an area or a volume as a sum of its indivisibles, lines or planes. All the mathematicians used it, because it was necessary to invent new results. But it was not an orthodox geometrical procedure, and the question arose: can some result obtained by such a method be considered as proved? [5]. Consequently, it is noteworthy that the story of the letter above is repeated. Now the author of the letter is the French mathematician Blaise Pascal.

In 1658, Pascal found a method to obtain centers of gravity of areas or volumes defined by a cycloid. He organized a competition between mathematicians of Europe where the challenge consisted of solving four problems, but nobody found any solutions. In December 1658, Pascal gave the results as an application of his method, but he chose not to write a treatise, but a letter. This method used indivisibles and the sum of indivisibles. So, the style of a letter permitted him to give a legitimacy to his method as the true way used by him to find results. He wrote in his letter to Carcavi: "I would like that you had been rewarded by this discourse I give to you: where you will see not only the resolution of these problems, but also the methods I used, and the way by which I arrived at them. As you told me, it is what you mainly wish, and what you often regret about the Ancients, that they do not do the same, giving us only their results without teaching us the ways by which they obtained them, as if they had been jealous of this knowledge" [18, p. 131].

The style of a letter permits Pascal to have a direct addressee when he introduces the indivisibles: "I would not make any difficulty to use the language of indivisibles, [...] which seems not geometrical only to those who do not understand the doctrine of indivisibles, and imagine that it is to sin against geometry to use it" [18, p. 135]. This kind of utterance is not possible in a treatise. Further, Pascal explains why it is not a sin to use indivisibles, appealing to the intelligence of his reader. We can see here that Pascal chooses the familiar style of a letter "to destroy the traditional official style," as Bakhtin writes, of the Archimedean style of proof of *reductio ad absurdum*. Pascal did not know the other letter, the one of Archimedes to Eratosthenes.

We see, with these two examples, that the dialogical approach is an interesting way to address the introduction of an innovation in history of mathematics. Indeed, an invention concerns a new way, for instance, to obtain a result or to understand a notion. So, the choice of a familiar style allows the possibility of explaining it in terms of a personal procedure, but with usual terms admitted by the other. So, as now the invention is written, it can be read by the others and it can become an innovation for all. *Dialogism concerns the transition from invention to innovation.*

2.4 Dialogism and History of Mathematics: The Question of Addressivity

There are many works and papers on the Bakhtinian approach in education (but not too much in France). They particularly concern the English-teaching field. Caryl Emerson writes about these numerous works: "Here as elsewhere in the humanities, Bakhtin was applied like a talisman, quoted, misquoted, paraphrased, carnivalized into a legend, and soon he became a familiar presence." [10, p. 21]. Some works also concern interactions in the classroom. For instance, Elsie Rockwell develops a Bakhtinian approach in her ethnographic research on a lesson observed in a rural classroom in central Mexico. She wrote about dialogism: "teaching is inherently dialogical, in the Bakhtinian sense; response is constantly required, awaited, elicited, expected. Even when there is no immediate verbal response, the teacher's utterance begs reaction or remains pending another moment when it may again be taken up, admitted or subtly refuted by students" [20, p. 264].

Indeed, the mathematical writings of pupils are special in terms of dialogism. They are answers to only one person, their teacher, and there is only one addressee, the teacher. We know an effect of this: a pupil writes a text for his teacher, a text written as his teacher wants and accepts as good. So, we often obtain texts that look like mathematical texts but which do not have any mathematical sense. So, it would be interesting to vary the correspondents of the pupils.

Correspondence with others is a part of the activities of a mathematician. It is the moment where a mathematician has to make his ideas explicit; he has to understand better what he does because he has to explain it to someone else. It is also the moment where he can write how he found his result, why he appreciates the value of this result and so on. He can also compare his result with another one, or he can explain his agreement or disagreement with regard to other results, notions, methods and proofs. Correspondence between individuals are intermediate writings between personal papers and papers or books intended for a larger sphere. The speech genre of correspondence permits us to write about mathematics and not only mathematics.

The studies of historical correspondence are often interesting, and it is the case in particular for the letters between Christian Huygens and Gottfried Leibniz. Leibniz learned mathematics when he was in Paris during the years 1672–1676 [13], and it is Huygens who initiated him into the mathematics of the XVIIth century. In 1684, Leibniz published his famous paper, *Nova methodus*, where he presented his new calculus to find maxima, minima and tangents [16]. This calculus is far from the practices of Huygens, who was a great virtuoso of geometrical methods.

Six years after the publication of the *Nova methodus*, Leibniz wrote in July 1690 to his ancient master to ask him what he thinks about his calculus: "I do not know, Sir, if you saw in Leipzig Acta a manner I propose to calculate, to submit to analysis what Descartes himself accepted. [...] Now I use sums and differences like dy, ddy, $dddy$, that mean differences of magnitude of y, or the differences of differences, or the differences of the differences of the differences" [14, p. 450–452]. In August 1690, Huygens answers : "I tried to understand your differentialis calculus, and I did so much that now I understand the examples you gave, but only since two days, one on the cycloid, the other on the research of the Theorem of M. de Fermat. And I even recognized foundations of this calculus and all your method as very good and very useful. Nevertheless, I think to have something equivalent" [14, p. 497–498]. It means that Huygens thinks that his usual methods are comparable to the calculus of Leibniz or near to them. Of course,

Leibniz believes that this is not the case, so his answer to Huygens is amusing, when he writes in October 1690 : "I conceive you have an equivalent method to my calculation of differences. Because what I call dx or dy you can write it with some other letter" [14, p. 516].

Christian Huygens does not agree with all the novelties introduced by his former pupil. It is interesting to see how the mathematical knowledge of Huygens can produce an obstacle to admit new notions. For instance, he wrote in November 1690: "For me, I confess that these kinds of super-transcendental curves, where the unknown enters in the exponent, seem to me so obscure, that I would not agree to introduce them in Geometry" [14, p. 537]. Three years later, he still has the same difficulties. He wrote on the 3rd of September, 1693: "I am not sufficiently versed in Exponential calculus, which appeared to me difficult and tiring" [14, p. 492]. But two weeks later, on the 17th of September, it seems that he had surmounted his difficulties, since in a letter to Leibniz he said: "I made some progress in the geometrical subtleties and in your excellent differential calculus, of which I appreciate more and more the usefulness" [21, p. 510]. We can imagine the gladness of Leibniz when he received this letter and when he answers : "All I wanted with this new calculus, that you begin to find convenient, Sir, was to open a lane where persons more penetrating than me would discover something of importance. [...] I am glad to see by your solution of Bernoulli's problem, that you remarked what is the more beautiful in our differential calculus, as soon as you wanted to take the effort to enter in it" [14, p. 539].

In the ECCE[1] Mathematics Project, a group of INRP and IREM Nantes, we try to establish real correspondence between students (not by emails but by handwritten letters). The question of addressivity led us to establish mathematical correspondences between two students of different levels: one secondary school student and one university student. The correspondence begins with a problem to solve such that it is possible for secondary school students to solve it with tools taught in their classroom, but such that a university student knows more powerful tools to solve it. We ask secondary school students to write their attempts precisely, ideas of research and their (partial or intermediate) results, and we ask university students to help the school student to improve their research. The purpose is not to give the answer, but to encourage the school student to be clearer in expressing their problems. The results are very different from the usual mathematical texts written by students.

2.5 Dialogism and History of Mathematics: Reading Original Sources in Classroom

Here the purpose is to use the notion of dialogism about history and pedagogy of mathematics, and especially on reading original sources in classroom. I accorded three functions to this reading: replacement, reorientation and culture [6]. As the historian Paul Veyne wrote, history has the virtue to astonish us with what goes without saying [21], and it is the reason for which reading original sources can have a function of a reorientation ("dépaysement" in French). Reorientation is interesting because it offers a possibility for teacher and pupils to think about mathematics, not only as a part of the curriculum or as a scholarly task, but as a human activity.

As we see above, a simple utterance like "all right angles are equal to one another" can be filled with echoes and reverberations of other utterances. So, history of mathematics can produce a reorientation: now, we are astonished with what is not said. But to obtain this effect, we had to read Euclid's utterance with a dialogical approach.

To read an original source with a dialogical approach means to think of this source not as written by the author for us, who live in the 21st century, but for his contemporaries.

For instance, if we read a source in the classroom only to translate it into modern mathematical language, we immediately kill all the effects of reorientation. Instead, to make the familiar unfamiliar, you made the unfamiliar familiar.

An original source has to be read as a rejoinder in a dialogue. What dialogue? Firstly, it is a dialogue between the author and his or her contemporaries. To take dialogism into account is a good means whereby pupils can understand that mathematics is not a "long quiet way," but that mathematics is a struggle for spirit. We have to read the author as somebody explaining something to somebody else. So, it is also a means to establish a second dialogue, a dialogue between the teacher and his or her students. In this case, an original source could be filled also with utterances between

---

[1] ECCE means *Écrire–Chercher–Concevoir–Échanger des mathématiques: To write–to research–to conceive–to exchange mathematics.*

the teacher and the students. Perhaps, it explains the reason why students remember working on an original source in classroom for a long time, as many teachers notice.

I would like to give an example from a famous text I use in a course for students, *La géométrie* of Descartes. I begin to explain to students that contemporaries of Descartes did not understand this book and that Descartes wrote to his friend Mersenne to express his sadness about that. Here, I like to bring to the students' notice that some persons understood the purpose of Descartes, and that one of them was a woman, Princess Elisabeth [8, p. 37–50].

In reading the first pages of Descartes' book, it is necessary to understand that each sentence of Descartes is a rejoinder, opposed to other utterances, especially those of the Ancients. I take some sentences of Descartes:

> "All problems in geometry can easily be reduced to such terms that a knowledge of the lengths of certain straight lines is sufficient for their construction" [9, p. 333].

Ancients: figures like triangles or squares are seen as areas and not by their simple edges.

Descartes: in geometrical figures we have only to consider simple things, which are straight lines.

> "These simple lines can be added or multiplied like arithmetical numbers are added or multiplied"

Ancients: Geometry and Arithmetic are separated fields.

Descartes: no, we can use arithmetical operations in Geometry.

Here a special comment has to be made about the introduction of a unit in Geometry by Descartes. The idea of a unit in geometry is a very good example of a familiar thing, which Descartes' text made unfamiliar.

> "It is to be noticed that by a^2 or b^3 or similar [notation], I conceive simple lines, instead of using the usual words I call them squares or cubes, etc."

Descartes to his reader (addressivity): with "squares" and "cubes," it is just a way to say things.

Classroom: we also use these words but we do not pay attention to them.

We continue until the statement of the Cartesian method to solve problems. Here it is interesting to oppose Descartes, who solves geometrical problems by resolution of algebraic equations, with algebraists, who legitimate the resolution of algebraic equations by geometrical proofs.

When it is clear that the purpose of Descartes is not orthodox, it is interesting to examine with students whether or not he anticipates the difficulties of his reader. Each mathematical text is always a pedagogical text, and it is particularly important to read *La géométrie* with this idea in mind. In this way, a teacher and pupils can discuss a mathematical text; but it is not a common situation. This situation can continue as a reflection with students about mathematical writings.

A dialogical approach to reading original sources permits us to read with students a first dialogue where a Mathematician M of the past explains something to somebody else: his reader R (Sphere 1). It permits also to establish a second dialogue, a dialogue between the Teacher T and his or her Student S (Sphere 2). So, the classroom becomes a new sphere of communication speech: an original source could be filled also with utterances between the teacher and the students (Spheres 1 and 2).

> Sphere 1: History of mathematics $\quad M \longleftrightarrow R$
> ....
> Sphere 2: Classroom $\quad T \longleftrightarrow S$

2.6 Conclusion

There are two ideas often expressed about the use of history in mathematical teaching. The first one, is that mathematics must be seen as a "human activity" from the philosophical, epistemological and mathematical points of view. The second one is that "human activity" means that mathematics is a result of activities of men and women in the history of humanity. The dialogical approach is one manner to understand why these two ideas have to be linked and to imagine how they can become more effective. In teacher training, we see the necessity for teachers to read mathematics within its historical contexts (Sphere 1). In the classroom, we see history not as a new subject to be taught, but as integrated into the mathematical classroom (Sphere 2).

2.6. Conclusion

Acknowledgment I thank Leo Rogers for his reading of my English translation.

Bibliography

[1] Al-Narizii, 1897, *Euclidis Elementa ex interpretatione Al-Hadschschadschii cum Commentariis Al-Narizii*, Besthorn R. O. and Heiberg J. L. (ed), Libraria Gyldendaliana.

[2] Archimedes, 1909, *Geometrical Solutions derived from Mechanics*. D. E. Smith (ed), The Open Court Publishing Company.

[3] Bakhtin, M. M., 1981, *The Dialogic Imagination: Four Essays*, translated by C. Emerson and M. Holquist. Austin: University of Texas Press.

[4] ———, 1986, *Speech genres and Other Late Essays*, translated by Vern W. McGee. C. Emerson, and M. Holquist (eds), Austin: University of Texas Press.

[5] Barbin, E., 1992, "Démontrer: convaincre ou éclairer? Signification de la démonstration mathématique au XVIIe siècle," *Cahiers d'histoire et de philosophie des sciences*," 40, 29–49

[6] ———, 1997, "Histoire et enseignement des mathématiques : pourquoi? comment?," *Bulletin de l'AMQ* (Association Mathématique du Québec), vol. XXXVII, 1, 20–25.

[7] ———, 2010, "Une approche bakhtinienne des textes d'histoire des sciences," *Méthode et histoire. Quelle histoire font les historiens des sciences et des techniques?* A. L. Rey (ed), Publications de la SFHST, pp. 202-216.

[8] Descartes, R., 1996, *Œuvres*, Adam, C., Tannery, IV, P., Paris: Vrin.

[9] ———, 1987, *Discours de la méthode*, Paris: Fayard.

[10] Emerson, C., 2000, "The New Hundred Years of Mikhail Bakhtin (The View from the Classroom)," *Rhetoric review*, vol. 19, 1–2, 12–27.

[11] Euclid, 1956, *The thirteen books of Euclid's Elements*, translated by Heath, vol.1, New York: Dover.

[12] Fauvel, J. and J. van Maanen, (eds.), 2000, *History in Mathematics Education: The ICMI Study*, Dordrecht-Boston-London: Kluwer.

[13] Hoffman, J. E., *Leibniz in Paris 1672–1676*, Cambridge: Cambridge University Press, 1974.

[14] Huygens, C., *Œuvres complètes de Christian Huygens*, IX et X, La Haye:Nijhoff, 1888–1950

[15] Kuhn, T. S., 1959, "The Essential Tension; Tradition and Innovation in Scientific research," *The Third University of Utah Research Conference on the Identification of Creative Scientific Talent*, C. W. Taylor (ed.), Salt Lake City: University of Utah Press, 162–177.

[16] Leibniz, G.W., "Nova methodus," *Leibnizens Mathematische Schriften*, V, ed. Gerhardt, Hidesheim: Olms, 1971, 220–226.

[17] Netz, R. and W. Noel, 2007, *The Archimedes Codex: Revealing the Secrets of the World's Greatest Palimpsest*. London: Weidenfeld and Nicholson.

[18] Pascal, B., 1963, *Œuvres complètes*, Paris: Le Seuil.

[19] Proclus, 1970, *A commentary on the first book of Euclid's Elements*, translated by Glenn R. Morrow, Princeton: Princeton University Press.

[20] Rockwell, E., 2000, "Teaching Genres: A Bakhtinian Approach," *Anthropology & Education Quarterly*, vol. 31, 3, 260–282.

[21] Veyne, P., 1971, *Comment on écrit l'histoire. Essai d'épistémologie*. Paris: Éditions du Seuil.

About the Author

Evelyne Barbin is full professor of epistemology and history of sciences at the University of Nantes (France), a member of the Centre François Viète, coordinator of the master on history of sciences and member of the Institute on Research on Mathematical Education (IREM) of Nantes. Her research concerns the history of mathematics and the introduction of an historical perspective in teaching. As convenor of the National Committee of the IREM "Epistemology and History of Mathematics," she has edited many books, organized 17 national colloquia, 8 interdisciplinary Summer Universities, and the first European Summer University on Epistemology and History of Mathematics in Mathematics Educationin 1993. Since 1980, she is a member of the HPM Group and Chair of HPM from 2008. Her recent publications include:

"On the argument of simplicity in Elements and schoolbooks of Geometry," *Educational Studies in Mathematics*, n°66, 2, 2007, pp. 225–242.

Les discours de l'évidence mathématique, in *Histoire et enseignement des mathématiques : rigueurs, erreurs, raisonnements*, dir. Barbin, E., Bénard, D., INRP, 2007, pp. 13–28.

The notion of Magnitude in Teaching: The new Elements of Arnauld and his inheritance, *International Journal for the History of Mathematics Education*, vol.4, n°2, 2009, pp. 1–18.

Evolving Geometric Proofs in the 17th Century: From Icons to Symbols, Hanna G., Jahnke N., Pulte H., ed. *Explanation and Proof in Mathematics: Philosophical and Educational Perspectives*, Springer-Verlag, 2010, pp. 237–252.

Nombres et grandeurs : des Pythagoriciens aux algébristes de la Renaissance, in Barbin, E. (éd.), *Des défis mathématiques d'Euclide à Condorcet*, Paris, Vuibert, 2010, pp. 65–82.

3

The Process of Mathematical Agreement: Examples from Mathematics History and An Experimental Sequence of Activities

Gustavo Martinez-Sierra and Rocío Antonio-Antonio
CICATA-IPN, Mexico

3.1 Introduction

Usually systematization processes are interpreted as processes beyond the processes of mathematical discovery. For example, Mariotti [9] establishes two moments for the production of mathematical knowledge: "...the formulation of a conjecture, as the core of the production of knowledge, and the systematization of such knowledge within a theoretical corpus." In this same vein is to contrast the argumentation process of a conjecture with the process of a theorem proof [1].

We proceed from the consideration that, for purposes of learning, there are propositions whose validity can be established from the outset as true to the need to bring coherence to a system of knowledge. The truth of the statement can be interpreted as "agreed truth;" in the sense that it is set from the necessity to make a theoretical corpus.

The following intends to show a knowledge production process, that we have called the *process of mathematical agreement*, which has the characteristic of combining different moments in the production of mathematical knowledge. In this sense the process of mathematical agreement, can be interpreted as a process of systematization of knowledge.

We will present three examples from the history of mathematics that show the production of the meaning of: 1) fractional exponents, 2) the square root of negative numbers as precursor to the meaning of complex numbers and 3) the radian and the trigonometric functions. In order to validate the process of mathematical agreement we then present the results of an experimental sequence that has the objective of student acceptance of the square root of negative numbers and of the operations on them.

3.2 The Process of Mathematical Agreement

A process of *mathematical agreement* [10, 11] may be understood as a consensus-seeking process within a community that works to give unity and coherence to a corpus of knowledge. The production of consensus is possible because a *process of systematization of knowledge* exists in this community. This means that there is a mathematical activity to relate diverse pieces of knowledge in order to make a theoretical corpus.

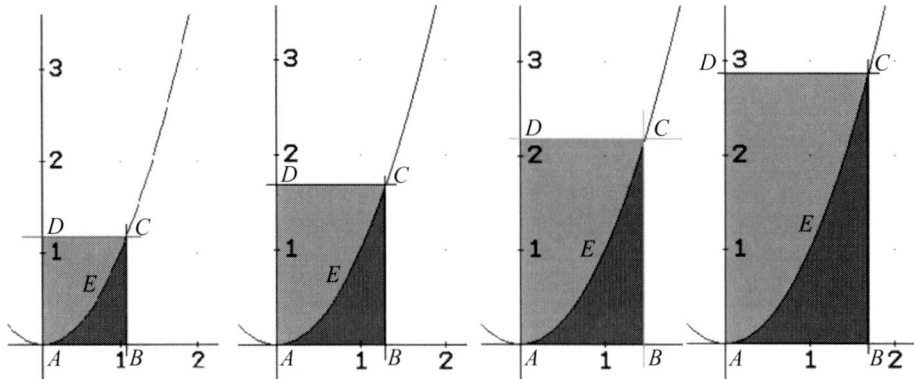

Figure 3.1. Characteristic ratio of the curve $y = x^2$.

The process of mathematical agreement brings out emergent properties unforeseen by earlier knowledge. The emergent properties could be a *statement whose truth we must agree on* (*agreed truth*), a definition, an axiom, an interpretation or a restriction, among others, in order to make a theoretical corpus. The form of the property depends on the specific, theoretical goals.

The search for systematization present in the process of mathematical agreement, which is a search for relationships, could take two paths: 1) *rupture* caused by leaving aside one meaning in favor of another which is eventually built for the task of making a theoretical corpus; that is, changing the focus of the meaning, and 2) *continuity* by conserving the meaning in the task of making a theoretical corpus. A *mathematical agreement*, then, as a product, can be interpreted as an emergent property establishing a relationship of continuity or rupture of meanings.

For example, in order to construct the statement $2^0 = 1$, we can do it through reasoning such as the following: if we *want* to have that $2^0 * 2^2 = 2^{0+2} = 2^2$ *we must agree* that $2^0 = 1$ and *we must agree* that the meaning of the symbol 2^x is no longer a repeated multiplication.

3.3 The Fractional Exponents

Towards the end of the sixteenth century it was known that the curves $y = kx^n (n = 1, 2, 3, 4 \ldots)$, called by index n, had a property called the "characteristic ratio." This knowledge was a general part of the fundamental problem of the time of the mechanical and algebraic calculation of areas defined by distinct curves [2]. Taking as an example the curve $y = x^2$ it was said that the characteristic ratio was equal to $1/3$; since, if we take an arbitrary point on the curve, C (see Figure 3.1) the area of $AECBA$ is in the proportion of $1 : 3$ to the area of the rectangle $ABCD$, in the same way as the proportion between the area of $AECBA$ and the area of $AECDA$ is $1 : 2$. In general, it was known that the characteristic ratio of the index curve n is $1/(n+1)$ for all positive whole numbers n[1].

Considerations about the complement of the areas under the graphs $y = x^n$, (n a positive integer) suggested to Wallis the concept of fractional exponent [15]. Thus, because the area under the curve $y = \sqrt{x}$ is the complement of the area under $y = x^2$, it must have a characteristic ratio of $2/3 = 1/(1 + 1/2)$ so that the index or exponent belonging to $y = \sqrt{x}$ must be $1/2$ (see Figure 3.2). The same can be shown for $y = \sqrt[3]{x}$, whose *characteristic ratio* must be $3/4 = 1/(1 + 1/3)$ so its index will be $1/3$.

In terms of the mathematical agreement process we can say that if the curve $y = x^2$ has a characteristic ratio of $1/3$, the curve $y = \sqrt{x}$ should also have a characteristic ratio and it must be equal to $2/3$ (since it can be observed that the areas under both curves complement each other to make a rectangle). If we want the curve $y = \sqrt{x}$ to have an index, *we must agree* that $1/2$ is the index.

This same interpretation gave Wallis the meaning to the zero exponent [5]: "Because $y = x^0$ must have ratio 1, it must be a horizontal line. Because 1 raised to any power is 1, this horizontal line must be at the height of 1."

---

[1]In modern terms the notion of characteristic ratio arises from the fact that $a > 0)(\int_0^a x^n dx) : a^{n+1} = 1 : (n+1)$.

3.4 The Square Root of Negative Numbers

Through history, we identify four great stages in the epistemological conceptions of complex numbers [7, 12, 13]: 1) *Algebraic.* The first appearances of square roots of negative quantities, 2) *Analytic.* The acceptance and generalization of the use of imaginary expressions thanks to the development of infinitesimal analysis, 3) *Geometric.* The introduction of an axis of imaginary numbers in which $\sqrt{-1}$ is a unit perpendicular to 1, and 4) *Formal.* The formalization of complex numbers.

In 1545 Cardano published, in his *Ars Magna*, Tartaglia's method of solution of cubic equations. This solution is known as the Cardano method, which for the case $y^3 + py + q = 0$ takes the form:

$$y = \sqrt[3]{-\frac{q}{2} + \sqrt{\frac{p^3}{27} + \frac{q^2}{4}}} + \sqrt[3]{-\frac{q}{2} + \sqrt{\frac{p^3}{27} + \frac{q^2}{4}}}$$

In modern terms the formula implies the use of complex numbers when $\frac{q^2}{4} + \frac{p^3}{27} < 0$.

Nevertheless, this cannot be considered an unsolved case, *because the cubic equation always has at least one real root*. So Cardano's formula poses the problem of agreeing on a real value, found by inspection, let us say, of an expression of the form:

$$y = \sqrt[3]{a + b\sqrt{-N}} + \sqrt[3]{a - b\sqrt{-N}} \text{ (}N\text{ being a natural number).}$$

Cardano did not confront this problem (the simplification of $\sqrt[3]{a \pm b\sqrt{-N}}$, called the irreducible case) in his *Ars Magna*, considering these numbers to be "as subtle as they are useless;" he was incapable of doing anything with the so-called "irreducible case" of the cubic equation, in which there are three real solutions that appear as the sum or difference of what we now call complex numbers.

This difficulty was resolved in the sixteenth century by Rafael Bombelli, whose Algebra appeared in 1572 [15]. Bombelli deduced [4] the formal algebra of complex numbers (eventually formulating the four operations with complex numbers in their current form) with the specific aim of reducing expressions $\sqrt[3]{a + b\sqrt{-N}}$ to the form $c + d\sqrt{-1}$; thus his method allowed him to demonstrate the "reality" of some expressions that result from Cardano's formula. For example, the solution to $y^3 = 15y + 4$, given by Cardano's formula, is

$$y = \sqrt[3]{2 + 11\sqrt{-1}} + \sqrt[3]{2 - 11\sqrt{-1}} \qquad (3.1)$$

On the other hand, inspection gives the solution $y = 4$. Bombelli suspected that the two parts of y in Cardano's formula were of the form $2 + n\sqrt{-1}, 2 - n\sqrt{-1}$ and he found these expressions formally by cubes using $(\sqrt{-1})^2 = -1$:

$$(2 + \sqrt{-1})^3 = (2)^3 + 3(2)^2(\sqrt{-1}) + 3(2)(\sqrt{-1})^2 + (\sqrt{-1})^3 = 2 + 11\sqrt{-1}$$

This result certainly implies:

$$\sqrt[3]{2 + 11\sqrt{-1}} = 2 + \sqrt{-1} \qquad (3.2)$$

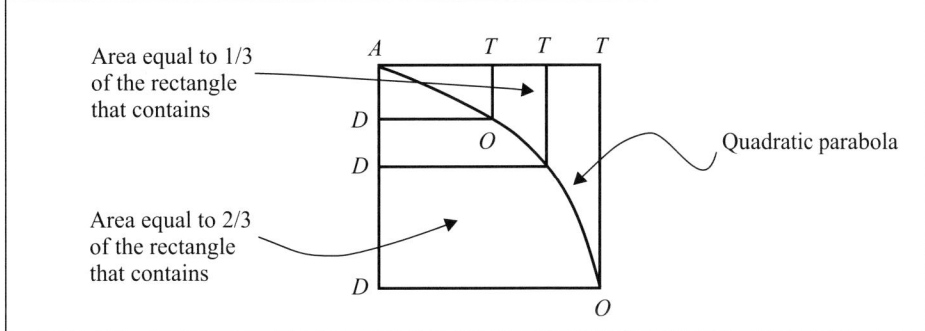

Figure 3.2. Relationship between a curve of index 2 and 1/2 using the graph of Wallis [15].

$$\sqrt[3]{2 - 11\sqrt{-1}} = 2 - \sqrt{-1} \qquad (3.3)$$

Substituting (3.2) and (3.3) in (3.1) we get

$$y = 2 + \sqrt{-1} + 2 - \sqrt{-1} = 4$$

as desired.

In summary, the first formulation in relation to numbers with form $A + B\sqrt{-N}$ (N being a positive number) was accepted in a limited algebraic domain because they appeared useful in the solution of third degree equations $y^3 + py + q = 0$. Our interpretation is that the existence of the square root of negative numbers was accepted through its operations in order to make a theoretical corpus that includes the algebraic formula

$$y = \sqrt[3]{\left(-\frac{q}{2} + \sqrt{\left(\frac{p^3}{27} + \frac{q^2}{4}\right)}\right)} + \sqrt[3]{\left(-\frac{q}{2} - \sqrt{\left(\frac{p^3}{27} + \frac{q^2}{4}\right)}\right)}$$

as a solution to that equation and the fact that a cubic equation always has at least one real root. In terms of the *mathematical agreement* process we can say: if *we want* Cardano's formula to give the real solution of $y^3 + py + q = 0$, *we must agree* on the existence of and on operations with numbers of the form $A + B\sqrt{-N}$ with $(\sqrt{-N})^2 = -N$ and $(\sqrt{-N})^3 = -N\sqrt{-N}$.

3.5 The Radian and the Trigonometric Functions

Analyzing different textbooks used in Mexican pre-university level education we can identify the presence of a common pattern in the construction of the trigonometric functions which consists of following the transition degrees → radians → real in the domain of the trigonometric functions. There are two important points to bring out from the textbook analysis. The first is the observation that the reason for the sudden appearance of radians as a measurement of angle is never made explicit. The second is the observation that the transition from radians to real numbers as an argument of the trigonometric functions is not considered as an object of study from the conceptual point of view. This can be perceived in the phrases presented in the textbooks, for example: *"it is commonplace to omit the word radians," "when the value of an angle is used in radians, the units are not normally given," "for convenience and simplicity we will omit the word radians."*

Our interpretation is that a radian is a concept whose function is to supply *continuity* between the use of degrees in trigonometric ratios and the use of real numbers in trigonometric functions. This interpretation motivates us to analyze the history of trigonometric functions and analyze the role that plays the concept of radian. The most noticeable part of this historical analysis is that the concept of radian did not exist in the first definitions of trigonometric-function-like objects belonging to the theoretical corpus of calculus. According to Katz [8]:

> "...no textbook until 1748 dealt with the calculus of these functions. That is, in none of the dozen or so calculus texts written in England and the continent during the first half of the 18th century was there a treatment of the derivative and integral of the sine or cosine or any discussion of the periodicity or addition properties of these functions. This contrasts sharply with what occurred in the case of the exponential and logarithmic functions. We attempt here to explain why the trigonometric functions did not enter calculus until about 1739. In that year, however, Leonhard Euler invented this calculus. He was led to this invention by the need for the trigonometric functions as solutions of linear differential equations. In addition, his discovery of a general method for solving linear differential equations with constant coefficients was influenced by his knowledge that these functions must provide part of that solution." [8, p. 311]

In this way, Euler, in his book *Introductio in analysin infinitorum* (*Introduction to the analysis of the infinite*) [6] provides a treatment of what can be called the precalculus of the trigonometric functions. He defines them numerically and discusses several of their properties, including the formulas for addition and their development in series of powers, which gave them the status of function. We remember that for Euler *"A function of a variable quantity is an analytic expression formed arbitrarily with this variable and with numbers or constant quantities"* [6, p. 3] and that when he

3.5. The Radian and the Trigonometric Functions

establishes: "...*analytic expression formed arbitrarily...*" he is accepting the use of the usual algebraic operations such as addition, multiplication, differences, quotients and transcendent operations like exponential, logarithmic and trigonometric. He also admits the extension of these to the infinite and the solution of algebraic equations, where the constants can even be complex numbers.

In the first volume, Chapter VIII, "On Transcendental Quantities which Arise from the Circle," Article 127, Euler defines the trigonometric functions as transcendent quantities that are born from the circle and points out that π is the semi-circumference of a circle (of radius 1) and in consequence is the length of the arc of 180° and then he establishes $\sin 0\pi = 0, \cos 0\pi = 1, \sin 2\pi = 0$ and $\cos 2\pi = 1$. In Euler's words:

> 127. We always assume that the radius of the circle is 1 and let z be an arc of this circle. We are especially interested in the sine and cosine of this arc z. Henceforth we will signify the sine of the arc z by $\sin z$. Likewise, for the cosine of the arc z we will write $\cos z$. Since π is an arc of 180 degrees, $\sin 0\pi = 0$ and $\cos 0\pi = 1$. Also $\sin \pi/2 = 1, \cos \pi/2 = 0, \sin \pi = 1$ and $\cos \pi = -1, \sin 3\pi/2 = -1, \cos 3\pi/2 = 0, \sin 2\pi = 0$, and $\cos 2\pi = 1$.

Euler does not mention, why, for example, $\cos \pi = -1$. The information we have up to now only allows us to speculate that perhaps Euler used the formula for the cosine of the sum of two arcs, $\cos(A+B) = \cos A \cos B - \cos A \cos B$, to build the mathematical agreement $\cos \pi = -1$ and other statements. One possible reasoning is the following: Supposing we want to assign a meaning to the symbol $\cos \pi$. What meaning will it be? If we take the formula $\cos(A+B) = \cos A \cos B - \cos A \cos B$ as the knowledge base that we want to preserve, it must follow that

$$\cos \pi = \cos\left(\frac{\pi}{2} + \frac{\pi}{2}\right) = \cos\frac{\pi}{2} \times \cos\frac{\pi}{2} - \sin\frac{\pi}{2} \times \sin\frac{\pi}{2} = 0 - 1 = -1$$

so that we must agree that $\cos \pi = -1$.

The previous conjecture is supported when we note that in a following article Euler makes a series of calculations based on the sine and cosine formulas of the sum and difference of two arcs. In the words of Euler:

> 128. We note further that if y and z are two arcs, then $\sin(y+z) = \sin y \cos z + \cos y \sin z$ and $\cos(y+z) = \cos y \cos z - \sin y \sin z$. Likewise, $\sin(y-z) = \sin y \cos z - \cos y \sin z$ and $\cos(y-z) = \cos y \cos z + \sin y \sin z$. [...]

In the same sense we have been able to interpret that the incorporation of the trigonometric functions into Euler's theoretical corpus was possible through the following relations [3]:

(A) $\sin^2 z + \cos^2 z = 1$

(B) $\sin(y+z) = \sin y \cos z + \cos y \sin z$

(C) $\cos(y+z) = \cos y \cos z - \sin y \sin z$

Using the above relations Euler breaks down (A) into

(D) $(\cos z + i \sin z)(\cos z - i \sin z) = 1$

And using (B), (C) and (D) he finds the relation

(E) $(\cos z + i \sin z)(\cos y + i \sin y) = \cos(y+z) + i \sin(y+z)$

Using (E) he finds that

$$\cos nz = \frac{(\cos z + \sin z)^n + (\cos z - i \sin z)^n}{2}$$

Finally developing the above powers and using the notions of "infinitely large whole number" and "infinitely small quantity" he finds that

$$\cos y = 1 - \frac{y^2}{1 \cdot 2} + \frac{y^4}{1 \cdot 2 \cdot 3 \cdot 4} - \frac{y^6}{1 \cdot 2 \cdot 3 \cdot 4 \cdot 5 \cdot 6} + \cdots.$$

Thus the cosine emerges as a function in Euler's sense.

3.6 An Experimental Sequence of Activities: The Square Root of Negative Numbers

In order to experiment with the process of mathematical agreement, we designed an experimental sequence of activities for pre-university students (pupils between 15 and 18 years old). For the construction of the sequence of activities, we have *adapted* the process of mathematical agreement presented in Section 3.4 to polynomials of the form $x^n - 1 = 0$. In particular, the purpose of the sequence is to lead the student to accept square roots of negative numbers through the calculation of roots in polynomials of the form $x^n - 1 = 0$. This acceptance will eventually lead to the students' expanding their number fields from that of the real numbers to that of the complex numbers.

In the sequence of activities the students find the roots of polynomials through the explicit instruction to factor and to use the usual quadratic formula and, finally, to use the tools of the development of a binomial to corroborate that their calculation is correct. For example, for the equation $0 = x^3 - 1 = (x - 1)(x^2 + x + 1)$, the first factor gives $x - 1 = 0$. The second, via an application of the quadratic formula, yields $x = (-1 \pm \sqrt{-3})/2$. Finally, simply expand and confirm that $\left((-1 \pm \sqrt{-3})/2\right)^3 = 1$. Our hypothesis is that the idea that such polynomials have n roots can be supported by accepting operations on the square roots of negative numbers. In terms of the mathematical agreement process we can say: if *we want* $x^n - 1 = 0$ to have n roots *we must agree* on the existence of numbers of the form $\sqrt{-N}$ with $(\sqrt{-N})^2 = -N$ and $(\sqrt{-N})^3 = -N\sqrt{-N}$.

The design of the sequence of activities is grouped in four phases: I. *Review* the calculation of roots of an equation, II. *Identify* students' previous knowledge about the square root of a negative number, III. *Operate* with square roots of negative numbers in the calculation of roots of polynomials of the form $x^n - 1 = 0$.

The exploration of the sequence was done in the pre-university level campus of the city of Chilpancingo (capital of the Mexican state of Guerrero), where we worked with 10 students (6 girls and 4 boys) of first grade for three and one-half hours. The students formed three work teams: teams 1 and 3 with three students, and team 2 with 4 students. The results of teams 1 and 2 only are reported here[2]. The time taken was determined by the students' performance while participating in the sequence, which enabled them to reach Activity 10 of the third phase.

Students' Activities

Phase I. Recall the calculation of roots of an equation (only real roots).

In the first phase (Activities 1, 2, 3, 4, 5 and 6) we want the students to remember two complementary tasks 1) how to calculate the roots of an equation (working only with quadratic, cubic and fourth degree equations) by using the standard formula for second degree equations and factoring and 2) that the degree of the equation determines the number of roots. The following table shows the answers given by some of the students.

Activity 4.

Find the three roots of the equation $x^3 - 3x^2 - 4x = 0$ and verify that in fact they are roots of the equation.

Team 2. Given answers Team 2 said that the number of roots in an equation can be determined by the power and by the names quadratic or cubic. In the following dialog we present a part of it (the letter "P" represents the teacher's participation).

P: Everyone talks about the degree to determine the amount of roots? But, what degree should it be?

D: From second degree and higher, two, three (G interrupts)

G: No, from one and higher

P: (writes a 4th-degree equation $x^4 - x^3 - 7x^2 + x + 6 = 0$ and asks) on what power should we focus on?

D: On the first term?

---
[2] The given answers of some students were presented for only two of the three teams as these exemplify what this manuscript wants to show.

3.6. An Experimental Sequence of Activities: The Square Root of Negative Numbers

P: On the power of the first term? But that power, how does it compare with the rest?

E: Is greater (tried to add it to their answers)

P: Then if we have a cubic equation, How many answers are we going to find?

E: Three

P: A fourth?

E: Four

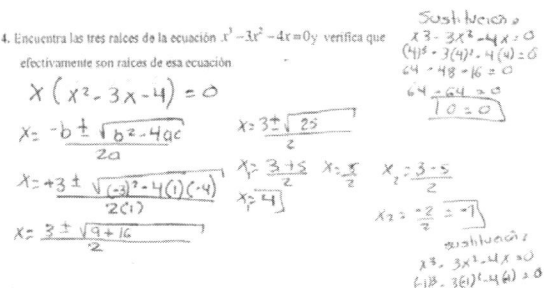

Figure 3.3. Activity 4.

Phase II. Identify students' previous knowledge about the square root of a negative number.

The second phase (Activities 7 and 8) aims to identify students' previous knowledge, (such as indicating the square root of a negative number and showing their familiarity with it), by calculating the roots of the quadratic equation $x^2 + 1 = 0$ by explicitly responding to the request to use the general quadratic formula. The general result found is that the previous knowledge identified in the two teams is that "*the square roots of negative numbers do not exist.*" In the following table we show the description and interpretation of the students' production.

Activity 8.

Use the second degree general formula to find the two roots of the equation $x^2 + 1 = 0$ *and verify that they are in fact roots of that equation.*

Team 1. Given Answers These students had to obtain the value of $\sqrt{-4}$. They key it into the calculator to obtain its value and it shows as "*error*" which prompts them to conclude that "*it doesn't exist.*"

At first it was hard for them to accept this number as a solution but after verifying the answer by using the normal operational procedures (for example, raising the power of a fraction) and recalling that the number of roots to be found depend on the degree of the equation, they concluded that it is "*one root with a minus sign and another with a plus sign.*"

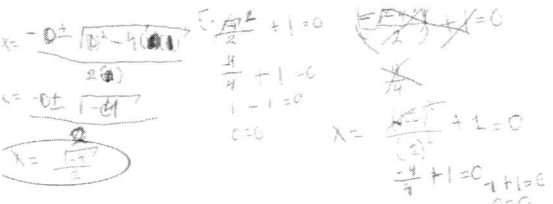

Figure 3.4. Activity 8. Team 1.

Team 2. Given Answers   This group was given two cases to obtain the roots of this equation. The first case uses the general formula just as we asked and the second case is working x out directly in the equation.

Using the second degree formula they argued that "*the root does not exist.*" They are helped in this aspect to find the solution through the usual second degree formula and conclude in each case that the results found "*are*" roots of the equation because "*they can be equaled.*"

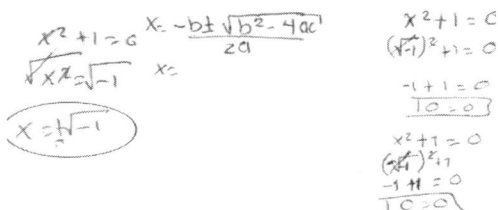

Figure 3.5. Activity 8. Team 2.

Phase III. Accept and operate with square roots of negative numbers in the calculation of roots of polynomials of the form $x^n = 1$.

In this third phase (Activities 9, 10, 11 and 12) the students were expected to operate and accept the square roots of negative numbers as solutions to these equations. To motivate them, they were asked to verify if the roots found satisfy the equation. Here they will use the tools of the development of a binomial, factorization and the usual quadratic formula. The aim of these activities is that on doing the factorizations of the polynomials in Activities 10, 11 and 12, the students will realize that they already have some of the roots, which they found in the previous activities. In the following table we show the description and interpretation of the students' production in the two teams reported here.

Activity 9

Use cubic difference factorization and the general second degree formula to find the three roots of the equation $x^3 - 1 = 0$. Also check that they are indeed roots of this equation.

Team 1. Given Answers   The reaction of these students was worse than in Activity 8 by getting this value and more so when they were asked to verify if it was the root of the cubic equation; it was not easy for them to do because of the calculations required, especially $\left(\frac{-1\pm\sqrt{-3}}{2}\right)^3$ and $(\sqrt{-3})^3 = (\sqrt{-3})^2(\sqrt{-3}) = -3\sqrt{-3}$. In the first calculation they directly eliminated the square root of the cubic exponent and also removed the minus sign of one, obtaining $\left(\frac{-1\pm\sqrt{-3}}{2}\right)^3 = \frac{1\pm(-3)}{4}$.

Team 2. Given Answers   In this team only one member did the operations of **Activity 9** and explained to the rest the operations carried out, but s/he was unable to get the others to do the verification calculations.

In the same way as in Activity 8, they accepted operating with a number arguing "there is not real value for" $(-1 \pm \sqrt{-3})/2$ using the concept they have that "the inverse of an exponent is a radical" and by calculating repeated multiplication and concluding that they can find the root of the polynomial if it "can be equaled" (its result is zero). This is referred to as the process of mathematical agreement.

Activity 10

Use cubic difference factorization and the general second degree formula to find the four roots of the equation $x^4 - 1 = 0$. Also check that they are indeed roots of this equation.

In Activity 10, we can see that in the two verifications the square root is eliminated directly with the exponent four and they raise the denominator to the square.

3.6. An Experimental Sequence of Activities: The Square Root of Negative Numbers

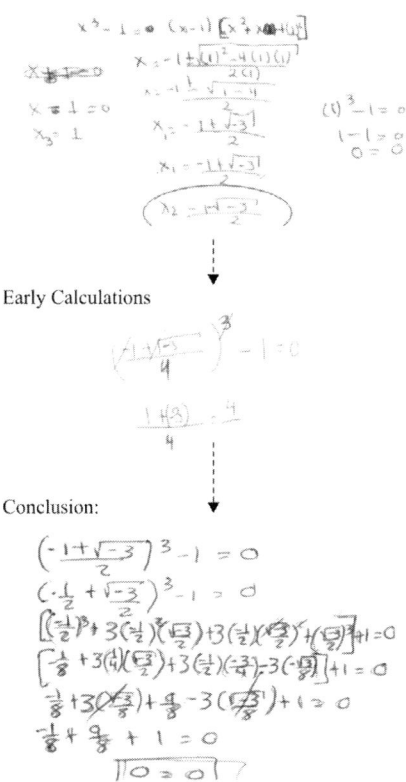

Figure 3.6. Activity 9

Team 1. Given Answers See Figure 3.6.

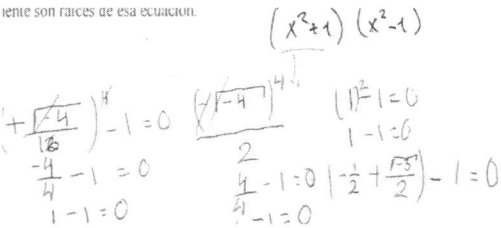

Figure 3.7. Activity 10. Team 1.

Team 2. Given Answers In this activity the calculations of a student from team number 1 is shown. He argues that: "*is root to the fourth* $(\sqrt{-1})^4 = \sqrt{-1}\sqrt{-1}\sqrt{-1}\sqrt{-1}$ *and equal roots are added* $\sqrt{-1}\sqrt{-1} = (\sqrt{-1})^2$ *and you have double root, and there are two, this one and this one would be the same* $(\sqrt{-1})^4 = (\sqrt{-1})^2(\sqrt{-1})^2$, *this one is eliminated* $(-1)(-1) = 1$, *minus times minus gives plus, it would be one, minus one, it would be zero*." Lastly, when they are asked if they are roots of this equation, they respond, "*yes, because they can be equaled*."

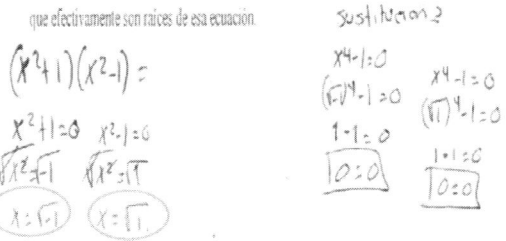

Figure 3.8. Activity 10. Team 2

About the students' given answers

Despite the arguments used by the students about there not being a real value for: $\frac{\sqrt{-4}}{2}, \sqrt{-1}, (-1 \pm \sqrt{-3})/2$ they finally accepted operations with these numbers using the concept they have of "the inverse of an exponent is a radical" and by multiplying repeatedly and concluding that this was a root of the polynomial because it "can be equaled" (its result is zero). This is referred to in this work as the meaning and operation of the complex number as a process of mathematical agreement in algebraic context: "the complex number can be considered as mathematical agreement between the degree of an equation and its solutions." We mention the above considering that they did not check all the values obtained in the polynomials proposed. We consider that the results of our sequence of activities gives evidence that it is possible to construct the meaning of the complex numbers and their operations through the process of mathematical agreement, under the calculation of roots of the form $x^n - 1 = 0$.

3.7 In Conclusion

We have presented three examples from the history of mathematics that show that it is possible to produce the meaning of: 1) fractional exponents, 2) the square root of negative numbers as precursor to the meaning of complex numbers and 3) the radian and the trigonometric functions. Thus, for example, in terms of the *mathematical agreement* process we can say: if *we want* Cardano's formula to give the real solution to $y^3 + py + q = 0$, *we must agree* on the existence of and operations on numbers of the form $A + B\sqrt{-N}$ with $(\sqrt{-N})^2 = -N$ and $(\sqrt{-N})^3 = -N\sqrt{-N}$.

Finally the experimental sequence of activities provide some evidence that in spite of the students insisting that *"square roots of negative numbers do not exist,"* our sequence induced them to operate with them to find the roots of some polynomials proposed in the activities. The basic argument is that "they can be equaled," that is, that when the values are substituted in the equation, the result is zero (this considering that they have not tested all values obtained in the equations, only some). We consider that our sequence of activities suggests that it is possible to construct the meaning of the complex number and their operations through a process of mathematical agreement.

Bibliography

[1] Balacheff, N. (1999). Is argumentation an obstacle? Invitation to a debate. *Newsletter of proof*. March/June 1999. Retrieved from http://www.lettredelapreuve.ti/

[2] Bos, H.G. M.,(1975). "Differentials, Higher-Order Differentials and the Derivative in the Leibnizian Calculus." *Archive for the History of Exact Sciences 14*, 1–90.

[3] Dhombres, J., 1986, "Quelques aspects de l'histoire des équations fonctionnelles liés à l'évolution du concept de function." *Archive for History of Exact Sciences* 36(2) 91–181.

[4] Dunham, W., 1999. *Euler. The master of us all.* The Dolciani mathematical expositions, N.22, Washington: The Mathematical Association of America.

[5] Confrey, J. and D. Dennis, 1996, "The Creation of Continuous Exponents: A Study of the Methods and Epistemology of John Wallis." *CBMS Issues in Mathematics Education* 6, 33–60.

[6] Euler, L., 1748/1845, *Introduction a l'analyse infinitésimale*. Paris: L'Ecole Polytechnique. English version *Introduction to the analysis of the infinite* (John D. Blanton, tr.), New York, Springer, 1988, 1990.

[7] Goméz, A. and T. Pardo, 2005, "La enseñanza y el aprendizaje de los números complejos. Un estudio en el nivel universitario." *Actas del Noveno Simposio de la Sociedad Española de Educación Matemática SEIEM*, pp. 251–260.

[8] Katz, V., 1987,. "The Calculus of the Trigonometric Functions." *Historia Mathematica* 14, 311–324.

[9] Mariotti, A., 2006, "Proof and proving in mathematics education." In A. Gutiérrez, & P. Boero, (Eds.), *Handbook of research on the psychology of mathematics education: Past, present and future* (pp. 173-204). Rotterdam: Sense Publishers.

[10] Martínez-Sierra, G., 2005, "Los procesos de convención matemática como generadores de conocimiento." *Revista Latinoamericana de Investigación en Matemática Educativa* 8 (2), 195–218.

[11] ——, (2008). "From the analysis of the articulation of the trigonometric functions to the corpus of eulerian analysis to the interpretation of the conceptual breaks present in its scholar structure." *Proceedings of the HPM 2008 conference, History and Pedagogy of Mathematics*.

[12] Rosseel, H. and M. Schneider, 2003, 'Ces nombres que l'on dit "imaginaires." Petit X 63, 53–72.

[13] ——, 2004, 'Des nombres qui modélisent des transformations.' Petit X 66, 7–34.

[14] Stillwell, J., 1989, *Mathematics and its history*. New York: Springer-Verlag.

[15] Struik, D. J., 1986, *A source book in mathematics 1200–1800*. Princeton: Princeton University Press.

About the Authors

Gustavo Martinez Sierra is a researcher at the Centre for Research in Applied Science and Advanced Technology of National Polytechnic Institute (Mexico City). He graduated with a BA in mathematics from the National Polytechnic Institute (ESFM–IPN)and the Master of Science in the Department of Mathematics Education Research Center and Advanced Studies (CINVESTAV–IPN). He completed his doctoral studies in the Mathematics Education Program of the Center for Applied Scientific Research and Advanced Technology of IPN (CICATA–IPN). His professional work area is in mathematics education; his main line of research deals with the study of the the social construction of mathematical knowledge. Its activity is part of two lines of work. The first deals with the processes of construction of mathematical conceptual systems, and the second examines the practices and attitudes that students and teachers socially construct in relation to different social objects.

Rocío Antonio Antonio did his undergraduate and graduate studies in the area of mathematics education at University of Guerrero (México). In the field of research, he has been particularly interested in the study of the social construction of mathematical knowledge from the research of mathematical agreement processes as knowledge generators. His master's work was recognized with honorable mention in the award for the best master's thesis in Mathematics Education in 2009, awarded by the Mexican Mathematical Society.

4

Researching the History of Algebraic Ideas From an Educational Point of View

Luis Puig
Universidad de Valencia, Spain

4.1 Introduction

Since the early 1980s, my colleagues and I have been studying the history of algebraic ideas as a component of our research on the teaching and learning of school algebra[1]. In Filloy, Rojano and Puig [6, ch. 1, ch. 3 and ch. 10], we discuss in some detail in which sense our study of the history of algebraic ideas is made from the point of view of mathematics education. What do we mean by studying the history of mathematics from the point of view of mathematics education? First, we mean that the problems of the teaching and learning of algebra is what determines for us which texts must be sought out in history and what questions we should address to them.

Indeed, in our research on the teaching and learning of algebra we have observed that, when dealing with arithmetic-algebraic word problems or with the solution of equations, pupils use a stratified sign system, with strata that come from their previously acquired vernacular and arithmetic language, and also from the concrete models used in the teaching sequences. As a consequence, we conceive the construction by pupils of the language of symbolic algebra as the final identification, within a single language stratum, of those earlier language strata that are irreducible from one stratum to another until the more abstract language has been developed [6, p. 263, and chapter 6]. These observations, which come from our research on mathematics education, have led us to be interested in studying the sign systems prior to the establishment of the language of symbolic algebra (by Viète, Bombelli, Descartes, see the Section "A history of symbolization" of [22]). We are also interested in addressing to those historical texts questions concerning the way in which the characteristics of those sign systems affect the type of arguments used to solve arithmetic-algebraic problems.

Take, for instance, the *Trattato di Fioretti*, an *abbacus* book from the fifteenth century, edited by Gino Arrighi ([17]). We have shown that, to solve arithmetic-algebraic problems, Mazzinghi uses rules that he has to restate for the specific numbers with which it is necessary to operate, not being able to express a general rule in his sign system. Furthermore, it seems that expressions that we would write as $x + y + z = a$ and $x + y + z = b$ with $a \neq b$ are not fully identified as equivalents for the purposes of the solving procedures and strategies, which in the *abbacus* books depend strongly on the specific properties of the number a (or b) and its relationships with the other numbers that appear in all of the equations of the system in question. Take, in contrast with Mazzinghi's *abbacus* book, Jordanus de Nemore's

---

[1] Filloy [4] is the earliest presentation of our study of the history of algebraic ideas in the service of mathematics education.

De Numeris Datis, written two centuries earlier (a book we are going to study in Section 4.4). We have shown that the sign system used by Jordanus de Nemore in this book, with its idiosyncratic use of letters (which we describe in Section 4.4), is able to express general rules, and in this sense *De Numeris Datis* might be located in a more evolved stage, as it makes it possible to group problems that can be solved in the same way into large families by identifying more general forms [6, p. 74], [2]).

We do not mean that the sign systems on which the more abstract sign system of symbolic algebra is erected, present in history and in the history of each individual, are the same; nor the paths toward the sign system of symbolic algebra. Not only are the sign systems different, but so are the language games in Wittgenstein's sense, the set of pragmatic practices, rules, concepts and uses, as well as the socio-cultural context. We mean that a historical inquiry motivated by questions raised in mathematics education may give us new ideas in three components of the models[3] we use to organize our research in mathematics education: (a) elements of a formal competence model, (b) ways of understanding the performance of pupils and, therefore, of developing cognition models, and, lastly, (c) concrete teaching models. This is the second feature of our approach, which supports our claim that we are researching the history of algebraic ideas from an educational point of view.

Moreover, one of the concrete models we have used to teach problem and equation solving modeled equations by representing them by means of line segments and rectangles, and operated on them. This latter fact led us to be interested in studying texts in history in which arithmetic-algebraic problems were dealt with through the use of geometrical representations and cut and paste methods.

We are going to present here some results of our examination of cut and paste methods in texts from three moments in history: Old Babylonian algebra, al-Khwārizmī's algebra, and Jordanus de Nemore's *De Numeris Datis*. We will show how in Old Babylonian algebra cut and paste methods were developed, and they constitute an important part of the language in which arithmetic-algebraic problems were solved by calculating directly on the signs. In al-Khwārizmī's algebra, the system of signs used to represent the quantities and relations is vernacular and lacks operational power at the expression level. Cut and paste methods are used to justify the correctness of the algorithmic rules that solves the canonical forms. In Nemore's *De Numeris Datis* cut and paste methods do not appear explicitly. However, we will develop a reading of Nemore's proofs of some of the propositions of this book which uses a didactical artifact built on cut and paste methods. This reading will enable us to understand Nemore's proofs by translating them into a more concrete representation (the didactical artifact), inspiring then a teaching model for algebraic identities and second degree equations based on this capability of the didactical artifact.

4.2 Babylonian Cut and Paste Operations: A Tool for Analysis

Høyrup[4] has developed an interpretation of Babylonian algebra in which the solutions of the problems are read literally as statements about geometrical entities, and as cut and paste operations with these geometrical entities.

The statement

(1) Side and square accumulated[5], 110

which has been interpreted as the arithmetic-algebraic problem $x^2 + x = 110$, disguised in geometrical clothes, is seen by Høyrup as a statement about geometrical figures. This makes it necessary for the lines to become broad lines[6], to be provided with a projection, a virtual standard breadth of 1 unit. "Side and square accumulated" can be represented now by a square with a rectangle of width 1 pasted to it. And the calculations written in the tablet to solve the problem correspond with the actions of cutting the rectangle by the midpoint of its side of length 1, and moving around its outer half to paste it to an adjacent side of the square. The resulting figure is a gnomon of known area.

---

[2] See also [22] for a more detailed comparison between Mazzinghi's and Nemore's ways of dealing with this type of problems.

[3] See Chapter 2 of [6], where our idea of Local Theoretical Models and their components is presented.

[4] [12] presents a comprehensive account of Høyrup's interpretation. Previous key papers were [7] and [9].

[5] I am using a non rigorous version of the statements. Høyrup uses "surface" instead of "square" and "confrontation" instead of "side." He specially explains why he has coined the term "confrontation" extensively in [12, p. 13 and pp. 25–26]. Høyrup has also shown that there are different addition operations in Babylonian mathematics; "to accumulate" is one of them ([9] and [12, ch. 2]). The statement of the problem does not tell that the question is to find the length of the side of the square, which is what is found at the end of the calculations.

[6] Høyrup [11] discusses the idea of broad lines and surfaces made up of lines of width 1 unit in Leonardo's *Practica Geometriæ* and some other metrological texts. We will see this very idea in al-Khwārizmī's algebra.

4.2. Babylonian Cut and Paste Operations: A Tool for Analysis

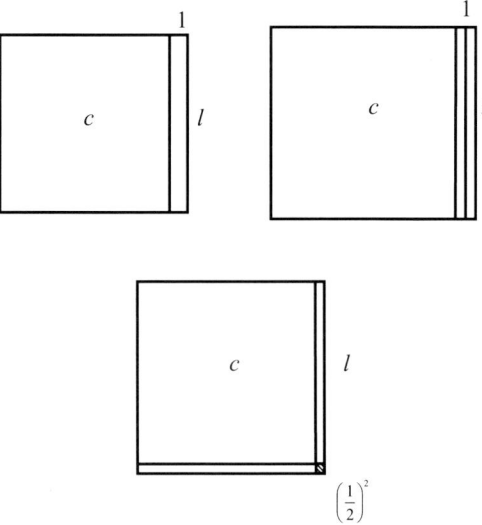

It is possible then to complete the gnomon to a square because the little square in the corner is given in area: the wide line has width 1, so its side is a half, and the area is the square of a half. As the area of the gnomon is known, the area of the big square is known, and so also its side and the side of the first square. The calculation corresponds to the formula

$$\sqrt{110 + \left(\frac{1}{2}\right)^2} - \frac{1}{2} = l$$

and to these cut, move around, and paste operations.

According to Høyrup [10], this trick appears to have been known and used by Akkadian-speaking surveyors, even if there is no direct evidence of it, but only the indirect evidence provided by the Old Babylonian scribal school, where the trick was called "the Akkadian method."

This cut and paste method is analytical by treating the unknown side and square as if they were given, and operating on them until a known configuration is reached (the completed square).

Analyticity goes beyond this in the following sequence of problems, which we present with a schematized statement, and a symbolic script.

(2) A rectangle, find length (l) and width (w), given the area (p) and difference (d) between the sides [find l, w, given p, d].

(3) A rectangle, find length and width, given the area and the sum (s) of the sides [find l, w, given p, s].

(4) A rectangle, find length and width, given the area plus difference between the sides, and sum of the sides [find l, w, given $p + d, s$].

The calculation to solve (2) can be interpreted as the result of an analysis that seeks to find the configuration of the previous problem (1). Indeed, if in this rectangle one takes away the width from the length, and splits in two the rectangle that appears, one will find the same configuration of problem (1). That makes it possible to set in motion the same actions of cutting, moving around, and pasting that are used in problem (1).

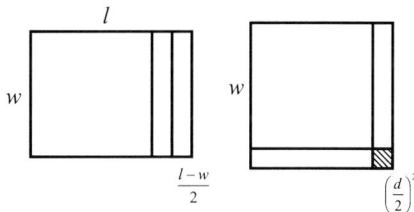

The calculation to solve (3) can be interpreted as the result of an analysis that reduces it to problem (2). If we construct the configuration of the following figure, we have the previous rectangle with a square accumulated, whose side is the width, and so we have represented the given "length plus width."

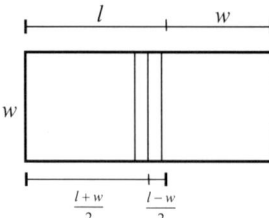

In this configuration we can search for the previous one by splitting the small rectangle in two. And then we have the sum of the sides divided by two, which is a given, and the difference divided by two, which has not been given.

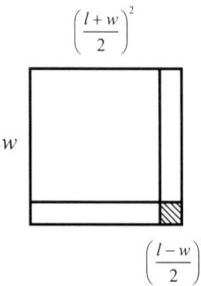

If we complete the square, now the little square is not known. But instead of trying to solve this new problem, we will reduce it to problem (2), by observing that it is possible to find the area of the little square from the area of the big one. This does not solve the problem, but gives us a way of calculating the difference between the sides. With this calculation, the condition of this problem is reduced to the conditions of the previous problem (2): a rectangle, length and width, given area and difference of sides. The analysis ends by reaching the statement of a problem one knows how to solve.

Finally, problem (4),[7] whose condition is more complex $[l, w$, given $p + d, s]$, is reduced to problem (3) by constructing a new rectangle that fulfills the conditions of problem (3) $[l, w$, given $p, s]$. The new rectangle has sides l, $w + 2$; its area is the sum of the given data $p + d$ and s, and the new sum of the sides is $s + 2$. Its area and the sum of its sides are both given, indeed.

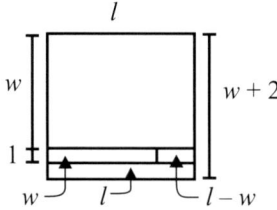

Therefore Babylonian algebra is analytical in two senses linked with the use of cut and paste operations:

- Cutting and pasting allows operating on the unknown.

- Cutting and pasting allows reducing the unknown to data (or to the unknown and data of another problem whose solution is already known).

Furthermore, the geometrical configurations of broad lines, squares and rectangles are used to translate the statement of problems that deal with other quantities. As Høyrup [12, p. 280] states, these geometrical configurations are a standard representation for problems, and their components are "functionally abstract,"[8] a segment may represent a number from a table of inverses, an area (as in BM 13901 #10), a volume (as in TMS XIX #2) or a commercial rate (as in TMS XIII). These geometrical configurations made up with broad lines, squares and rectangles, together with technical terms (length, width, "confrontation"...) are the sign system of Babylonian algebra on which one can operate directly with cut and paste operations.

---

[7] This problem appears on tablet AO 8862 #1. (See [7, pp. 309–320] and [12, pp. 162–170 (figure on p. 169)].)

[8] The expression "functionally abstract" is Høyrup's [12, p. 280]. The technical terms "length" and "width" have a *concrete* meaning, the length and width of a rectangle, but they are used to represent other quantities.

4.3 Cut and Paste Operations in Al-Khwārizmī's Algebra

To sum up, Old Babylonian algebraic problem solving is characterized by:

1. A system of signs in which problems can be represented: rectangles and broad lines.

2. A way of translating problems into this system of signs: length and width as functionally abstract terms.

3. A way of calculating directly with this system of signs: cut and paste operations.

4. A collection of configurations one knows how to solve.

4.3 Cut and Paste Operations in Al-Khwārizmī's Algebra

What is new in al-Khwārizmī's book on algebra in relation to the characteristics of Babylonian algebra we have seen is that al-Khwarizmi begins by examining what he calls the species of numbers that appear in the calculations. These species of numbers are *māl*, an amount of money, or a treasure[9], roots of the treasures, and simple numbers or *dirhams*. Al-Khwārizmi suggests having *a complete set of possibilities* of combinations of the different kinds of numbers (canonical forms). His aim is to find *an algorithmic rule* that makes it possible *to solve each* of the canonical forms, and to establish *a set of operations of calculation* with the expressions, which makes it possible *to reduce any equation* consisting of those species of numbers to one of the canonical forms. *All the possible equations would then be solvable* in his calculation. Moreover, al-Khwārizmi also establishes a method for translating any (quadratic) problem into an equation expressed in terms of those species, using another technical term, a name for the unknown, *shay'*, the thing: *all quadratic problems would then be solvable* in his calculation.

Al-Khwārizmī relies on the techniques of a tradition that comes from Babylonia, but he goes further by stating this program. Once he has a language, a complete set of possibilities, a calculation that reduces the expressions to canonical forms, and a way of translating problems to this system of signs, he only needs algorithmic rules to solve all canonical forms–and he actually has all such rules.

However, al-Khwārizmī's system of signs is made up of the vernacular language and lacks operativity at the expression level. As a result, al-Khwārizmī's is not able to develop algebraic proofs for the algorithmic rules. Babylonian cut and paste operations reappear in Al-Khwārizmī's algebra to justify the correctness of the algorithmic rules. To show this, we present here an explanation of the justification of the algorithmic rule for the fifth canonical form. In the extant manuscripts only one figure appears at the end of the justification, after the words "This is the figure." We have constructed this figure step by step following al-Khwārizmī's explanations.

When a treasure and twenty-one dirhams are equal to ten roots[10], we represent the treasure by a square whose side we do not know, the surface AD.	
To this we join a parallelogram, surface HB, the breadth of which, side HN, is equal to one of the sides of the surface AD.	
The length of the two surfaces together is equal to the line HC. We know that its length is ten in numbers; for every square has equal sides and angles, and,	

---

[9]We have decided to translate *māl* by "treasure," following Høyrup [8, n. 11], instead of translating it by "square" or x^2.

[10]The paradigmatic example of the fifth canonical form "treasure and numbers equal roots" is then equivalent to our equation $x^2 + 21 = 10x$. We are not going to translate al-Khwārizmī's sign system to symbolic algebra, nor to translate it into modern English algebraic terms. Instead we have composed this English version modifying Rosen's English translation ([25]) by checking it against the Arabic and adopting more literal translations of the technical terms and expressions. We have consulted also Gerard of Cremona's Latin translation ([14]), considered as a more faithful source than all extant Arabic manuscripts of al-Khwārizmī's original text, and the recent Rashed's ([24]) edition and French translation.

if one of its sides is multiplied by one, it is the root of the surface,	*C* ▬▬▬ *H* root of the surface
and by two, two of its roots.	*C* ▬▬▬ *H* root of the surface
As it is stated, therefore, that a treasure and twenty-one of numbers are equal to ten roots, we may conclude that the length of the line HC is equal to ten of numbers,	*C* ▬▬▬ *H* root of the surface
since the line CD represents the root of the treasure.	*D* — 10 — *C* ▬▬▬ *H* root of the treasure
We now divide the line CH into two equal parts at the point G. Then you know that the line GC is equal to the line HG, and the line GT is equal to the line CD.	*D B T N* *C A G H*
At present we extend the line GT a distance equal to the difference between the line CG and the line GT to square the surface. Then the line TK becomes equal to KM, and we have a new square, of equal sides and angles, namely the square MT.	*D B T N* *C A G H* *K M*
We know that the line TK is five; this is consequently the length also of the other sides: its surface is twenty-five, this being the product of the multiplication of half the number of the roots by themselves, five times five, that is twenty-five.	*D B T N* *C A G H* *K M*
We know that the quadrangle HB represents the twenty-one of numbers which were added to the treasure.	*D B T N* *C A G H* *K M*
We have then cut off a piece from the quadrangle HB by the line KT, which is one of the sides of the surface MT, so that only the surface TA remains.	*D B T N* *C A G H* *K M*
At present we take from the line KM the line KL, which is equal to GK.	*D B T N* *C A G R H* *K L M*
We know that the line TG is equal to ML, and the line LK, which has been cut off from KM, is equal to KG. Then the surface MR is equal to the surface TA.	*D B T N* *C A G R H* *K L M*

4.3. Cut and Paste Operations in Al-Khwārizmī's Algebra

We know that the surface HT augmented by the surface MR is equal to the surface HB, which represents the twenty-one.	
But the whole surface MT is twenty-five. If we now subtract from the surface MT, the surface HT and the surface MR, which together equal to twenty-one, there remains a small surface RK, which is twenty-five less twenty-one; this is four.	
And its root, represented by the line RG, which is equal to the line GA, is two.	
If you subtract it from the line CG, which is the moiety of the roots, then the remainder is the line AC, that is to say, three. This is the root of the first treasure.	
But if you add it to the line GC, which is the moiety of the roots, then the sum is seven, or the line RC, which is the root of a larger treasure. If you add twenty-one to this treasure, then the sum will likewise be equal to ten of its roots.	
This is the figure, [25, pp. 16–18], of the English version, and [25, pp. 11–13] of the Arabic; [14, pp. 238–239]; [24, pp. 112–117]. See also Berggren in [2, p. 545].	

As in the Babylonian cut and paste method, al-Khwārizmī looks for a gnomon to be completed into a square in the second part of his explanation. The proof is naïve in the sense that it relies on what is seen, without casting any doubt on it[11]. But the first part is more interesting: al-Khwārizmī needs to justify the representation of the algebraic terms, treasures and roots, by means of figures. To do this, he differentiates between the root of the treasure (a line without width) and the root of the surface (a broad line), and he explains how the whole rectangle represents ten roots (of the surface) because the rectangle is made up of ten broad lines. He also explains how "in numbers" the line without width HC is equal to ten, because the line without width CD represents the root of the treasure.

Al-Khwārizmī did not take for granted that his readers could understand this translation from his algebraic terms to geometrical objects as evident. Roots and treasures and lines and squares are not objects of the same nature, and it is necessary to show the links among them for the operations on the concrete level of the geometrical representation to have explanatory power.

This use of broad lines, and the distinction between them and the lines without width in the representation of algebraic terms by geometrical figures, can be traced also in Abū Kāmil's book on algebra[12] written immediately after al-Khwārizmī's. A couple of centuries later Sharīf al-Dīn al-Ṭūsī expands further the same distinction when dealing with third degree equations in his book *Treatise on equations*, where he explains the algorithmic rules for solving some cubic equations with three-dimensional cut and paste operations. To do this representation he introduces not only broad lines for the root of the surface, but also solid lines (roots), and solid treasures: "A solid treasure is a solid

---
[11] Immediately after al-Khwārizmī, Thābit ibn Qurrah rectifies this naïve character of al-Khwārizmī's proofs by referring the proofs to propositions of the second book of Euclid's *Elements* (see [16]).
[12] See [15, pp. 28–31], or [26, pp. 325–326].

whose base is a plane treasure and its height is a linear unit; the solid root of this treasure is a solid whose base is a plane root and its height a linear unit" [1, T. I, p. 16].

4.4 A Reading of Jordanus de Nemore' *De Numberis Datis*

The earliest known mathematical texts that use letters to stand for quantities are those written by Jordanus de Nemore in the 13th century, namely *De Numeris Datis* and *De Elementis Arithmetice Artis*. *De Elementis Arithmetice Artis* has been established and published in Latin with a paraphrase in English by Busard [3]. The critical edition of *De Numeris Datis* has been published with an English translation by Hughes [13], who gives 1225 as the most likely date of publication.

De Numeris Datis is written in Latin and is straightforwardly organized: three definitions at the beginning and 115 propositions distributed in four books, without any explanation of the aims of the book, neither an introduction nor transitions between books.

The propositions are presented always in the same three parts:

1) A statement asserting that if some numbers (or ratios) have been given, along with some relations between them, then some other numbers (or ratios) have also been given.

2) A series of transformations of the numbers (or ratios) and relations that either show that the numbers have indeed been given or convert them into the numbers and relations of the hypothesis of some previous proposition.

3) The calculation of an example with concrete numbers.

The propositions are then theorems on the solvability of problems. In Puig [20] we presented a detailed description of the sign system of this text, along with a translation of parts of book one, more literal than the translation by Hughes. Hughes' translation is very liberal with regard to Nemore's terminology. Even if our literality results in coarse English, his liberality makes it impossible to see some of the characteristics of Nemore's text that are essential to the description of the sign system, specially Nemore's use of letters.

This use of letters does not appear in the first proposition of the first book, that is the theorem version of Diophantos' problem I.1. Proposition I-1 of *De Numeris Datis* states that: "If a number that has been given is divided in two parts whose difference has been given, then each of the parts has been given"[13] [13, p. 57, my translation]. The argument runs as follows:

> "Since the lesser part and the difference equal the larger, the lesser with another equal to itself together with the difference make the given number. Subtracting therefore the difference from the total, what remains is twice the lesser. Halving this yields the smaller and, consequently, the greater part." [13, p. 127]

Each quantity is named using as its name Latin words with arithmetic meaning: the larger part, the lesser part, the difference, the number, the total. Nemore does not use letters in this argument, having names for all quantities he needs to refer to in the analysis.

Propositions I-3 to I-6 are closely related, and all begin also with a number that is divided in two parts[14]. If we represent the number by s, the smaller part by m, the larger part by M, the difference between the parts by d, its product by p, and the sum of its squares by s_c, these five propositions can be represented schematically by:

I-1: s and d given $\implies$ m and M given

I-3: s and p given $\implies$ m and M given

I-4: s and s_c given $\implies$ m and M given

I-5: d and p given $\implies$ s, m and M given

I-6: d and s_c given $\implies$ s, m and M given

---

[13]"Si numerus datus in duo dividatur quorum differentia data, erit utrumque eorum datum."

[14]We exclude proposition I-2 because in its statement the number is not divided into two parts, but in several parts, "If a given number is separated into as many parts as desired whose successive differences are known, then each of the parts can be found" [13, p. 127], and the argument does not reduce I-2 to I-1. We made a specific analysis of this proposition I-2 in [22, p. 215].

4.4. A Reading of Jordanus de Nemore' *De Numberis Datis*

The use of letters starts in the argument of Proposition I-3, "If a number that has been given is divided in two parts whose product has been given, then each of the parts has been given:"

(1) "Let *abc* be the number that has been given, divided in *ab* and *c*.

(2) *ab* by *c* makes *d*, given.

(3) *abc* by himself makes *e*.

(4) Let the quadruple of *d* be *f*.

(5) Taking *f* from *e* remains *g*.

(6) *g* is the square of the difference between *ab* and *c*.

(7) The square root of *g* is *b*.

(8) And *b* is the difference between *ab* and *c*.

(9) Since *b* has been given, *c* and *ab* have been given." [13, p. 58, my translation][15]

The aim of the argument is clear: what has been proved is that s and p given $\implies d$ given, hence I-3 has been reduced to I-1. The argument is analytical like the Babylonian one, which uses cut and paste transformations. Yet the details of the argument are not so clear: nothing is said to explain why the number given is represented by three letters (*abc*) instead of a single one as we would expect nowadays, and no reason is given for the assertion that (6), (7) and (8) are true.

In order to make sense of Nemore's argument as presented in the text, we introduced a representation of the quantities as segments and rectangles of a square divided *ad hoc* into parts. The number (s) is the side of the square, and the side is divided into a smaller part (m) and a larger part (M). The larger part is further divided to show the difference of the parts (d) and the relation $M = m + d$. The product of the parts (p) is represented by the rectangle shaded in the figure, and other quantities such as the square of the number, the sum of the squares of the parts, or the square of the difference of the parts are easily represented also as rectangles.

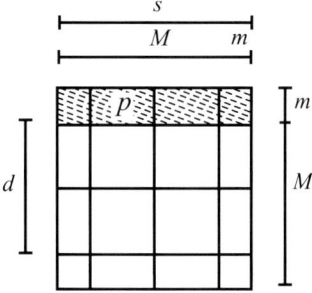

By using this device we are able to interpret Nemore's argument representing in it the letters used by Nemore (in the left column), and their meaning (in the right column).

---

[15] In Hughes' translation it is not possible to see the specific way in which Nemore uses letters, because Hughes changes the letters from the Latin text adopting in his English translation present conventions. He uses x and y for the two unknown parts in which the number has been divided, instead of *ab* and *c*, and he uses a single letter, *a*, to represent the given number instead of the juxtaposition of three letters, *abc*: "Let the given number *a* be separated into x and y..." [13, p. 128].

(1) Let *abc* be the number that has been given, divided in *ab* and *c*.	*abc* is the number (*s*); *ab* is the greater part (*M*), *c* is the smaller part (*m*)
(2) *ab* by *c* makes *d*, given.	*d* is the product of the parts (*p*)
(3) *abc* by himself makes *e*.	*e* is the square of the number (s^2)
(4) Let the quadruple of *d* be *f*.	*f* is four times the product ($4p$).
(5) Taking *f* from *e* remains *g*.	*g* is $s^2 - 4p$.

(6) g is the square of the difference between ab and c. 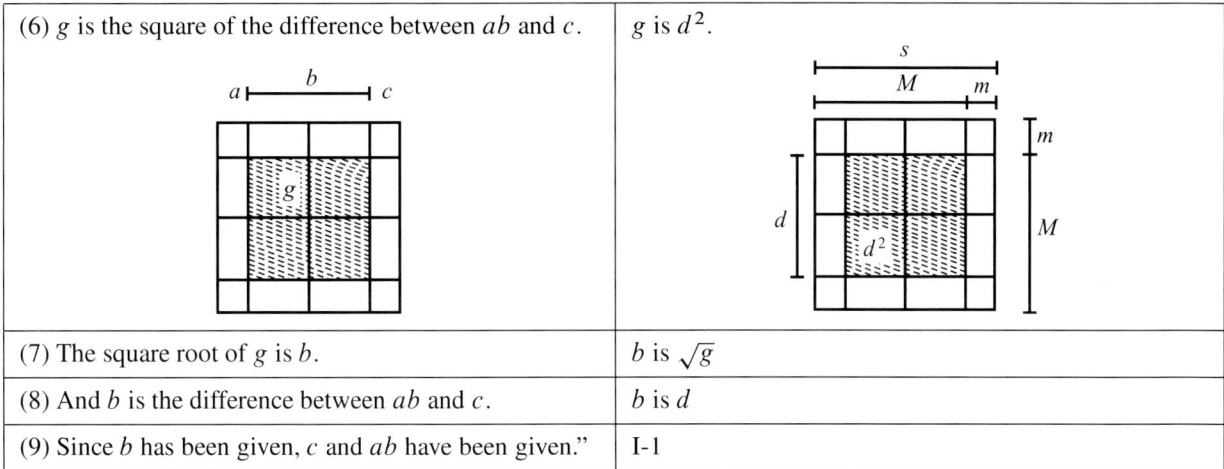	g is d^2.
(7) The square root of g is b.	b is $\sqrt{g}$
(8) And b is the difference between ab and c.	b is d
(9) Since b has been given, c and ab have been given."	I-1

The reconstruction shows how Nemore introduces from (1) to (5) a series of quantities, denoting each of them with a letter (or a juxtaposition of letters) *ad hoc*, how (6) gives a second meaning to g, and how (7) and (8) mean that b is the difference between ab and c. Hence a and c have in fact the same meaning, both referring to the smaller part. The representation in this configuration shows that the two meanings of g, as constructed from (1) to (5), that is, the square of the number minus four times the product of the parts ($s^2 - 4p$), and as stated in (6), that is, the square of the difference of the parts (d^2) refers actually to the same quantity. It explains also the use of a sign made up with three letters, abc, to represent the number that has been given.

A repeated use of this device to understand Nemore's arguments through book one, and some propositions of book four[16], led us to establish the following characteristics of Nemore's idiosyncratic use of letters (see [20], for details).

Whenever letters are used, all quantities, both known and unknown, are represented by letters; the letters are marks to denote the quantities that are built in the course of the argument and appear in alphabetical order, without any distinction between known and unknown quantities.

Besides, a quantity can be denoted by more than one letter. Each letter does not represent then a number, but the instance of the appearance of a number in the course of the argument.

There is a lack of operativity at the expression level, except for juxtaposition to mean addition; thus, when a new quantity is built using quantities already denoted by letters with an operation different from addition, the only way to denote the new quantity is to introduce a new letter to do it, there is no way of using the letters denoting the quantities involved (i.e., to denote the product of a by itself the a is useless: it is necessary to introduce another letter: a b).

This is a characteristic that Nemore's sign system shares with al-Khwārizmī's, even if al-Khwārizmī's is made of vernacular language, and Nemore's sign system includes also letters. Vernacular language lacks operativity at the expression level, as well as Nemore's idiosyncratic use of letters. Operativity at the expression level was the main characteristic of symbolic algebra, when Nesselmann introduced the distinction among rhetoric, syncopated and symbolic algebra in the middle of the nineteenth century in his book *Die algebra der griechen* [18]. In Nesselmann's own words: "We can perform an algebraic calculation from start to finish in a wholly understandable way without using a single written word" [18, p. 302]. From the internal perspective of al-Khwārizmī's or Nemore's texts this "lack of operativity" is not perceived as a lack, and they are quite able to solve the problems with their sign systems. However, what is of interest to us is the consequences that this characteristic of their sign systems had in the ways they solve the problems.

4.5 Didactical Comments

Our analysis of Nemore's sign system was used to describe the behavior of a couple of pupils who were prompted to write a message with letters to explain how they had solved a problem, equivalent in some sense to Nemore's I-1.

---

[16]This device built on cut and paste methods serves to explain the propositions that deal with simultaneous or quadratics equations, that is the propositions of book I, and some of the propositions of book IV. Book II of *De Numeris Datis* deals with proportions, book III deals with continuous proportions, and book IV combines proportions and simultaneous or quadratic equations. Our device, built on cut and paste methods, is of no use for understanding the propositions that deal with proportions.

They introduced letters in alphabetical order, one for each of the quantities they had calculated, without using the letters previously introduced by themselves to represent the new quantities they were calculating [21].

Pupils do not have to reproduce the history of algebraic ideas in their construction of their algebraic ideas. They already have the result of the historical production of mathematical ideas and sign systems embedded [23] in today's culture and language. However, some spontaneous or idiosyncratic sign production can repeat some features of productions of mathematicians from older times who were in some sense confronted with similar tasks. The study of history can give us then instruments to model pupil's performances or cognitions.

Moreover, there is a further use of our study of Nemore's book. The geometric configuration we have used as a device to understand Nemore's arguments has helped us to reveal the algebraic identities that were in play in Nemore's analytical inferences by easily representing the algebraic expressions involved in the identities as shaded rectangles, and comparing them. Indeed, the divisions of the square used to understand Nemore's arguments are selected in such a way that they easily represent the smaller part and the larger part into which a number has been divided. Alternatively, they may be used to represent the sum of two numbers, $x + y$, the difference of the numbers $x - y$, their product xy, the square of each of them, x^2 and y^2, etc. If we change Nemore's I-3 statement into the problem "to find two numbers whose sum and product are known," and we use the sign system of today school algebra instead of his idiosyncratic sign system, the sequence of shadowed figures we have used to understand Nemore's I-3 shows the algebraic identity $(x + y)^2 - 4xy = (x - y)^2$. This suggests that we could use this didactic artifact as a concrete model to teach the solving of second degree equations and algebraic identities. We sketch here how to represent in the concrete model algebraic identities, second degree equations, and the Babylonian completing the square procedure to solve second degree equations. This concrete model may be presented to pupils printed in paper, or built in a dynamic computer environment. We are currently piloting a study of such a presentation.

Algebraic identities are represented in the concrete model as sets of shaded pieces that can be obtained one from another by cutting, moving and pasting some of the pieces. By way of example, the algebraic identity "sum by difference equals difference of squares" is found by cutting the two pieces on the right of the rectangle representing the product $(x + y)(x - y)$, pasting them on the top, and identifying the resulting configuration with $x^2 - y^2$.

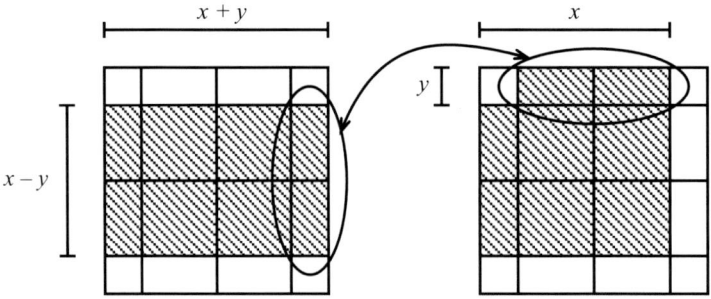

To represent second degree equations in the concrete model, the equations have to be written in one of al-Khwārizmī's canonical forms, i. e., $x^2 + bx = c$, $x^2 + c = bx$ or $bx + c = x^2$, $(b, c > 0)$, and the expressions on both sides of the equal sign have to be identified with the same rectangle. For instance, al-Khwārizmī's example of the fifth canonical form, $x^2 + 21 = 10x$, may be represented in two different forms, as shown in the following figure: x^2 is the square whose side is x, 21 (or c) is the rectangle labeled c, and so $x^2 + 21$ (or $x^2 + c$) is the rectangle shaded. One of the sides of this rectangle is x, hence the other side must be 10 (or b) in order that this rectangle may also represent $10x$.

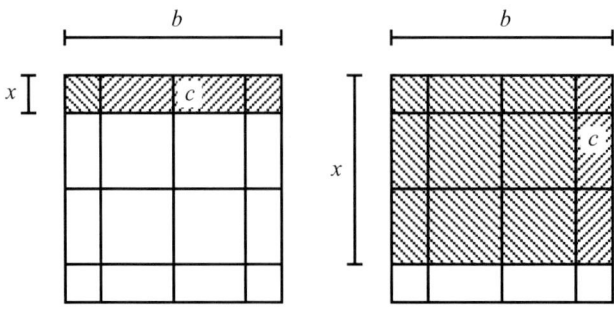

4.6. By Way of Conclusion

The Babylonian procedure of cutting and pasting to solve quadratic problems completing the square is easily represented in this figure thanks to the subdivisions of the square in halves. For instance, the first solution to the fifth canonical form may be worked out in the concrete model as shown in the following figure. One piece is cut, moved and pasted to form a gnomon, equivalent to the rectangle c, inside a square whose side is half the coefficient of x. The square which completes the gnomon is found, and then its side. Subtracting it from the side of the square which contains the gnomon gives x.

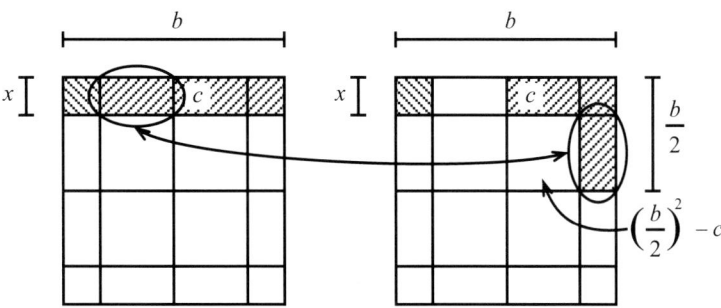

4.6 By Way of Conclusion

Cut and paste operations and the geometrical standard representation were the places in which Old Babylonian algebra operated directly. In Arab medieval algebra the construction of an algebraic language was made by using terms and expressions from the vernacular. The resulting lack of operativity did not allow the development of algebraic proofs. A version of the old Babylonian cut and paste method was used to make up for this lack of operativity by al-Khwārizmī and other medieval mathematicians to justify the algorithmic rules for solving equations by means of a more concrete model (a representation with segments and rectangles). Arabic medieval mathematicians were very careful in the explanation of the relations between the new more abstract algebraic language of treasures and roots, and the elements of the concrete model.

Nemore's sign system, with his idiosyncratic use of letters, also lacks operativity, not being able to express in his language the analytical inferences he is making to prove the propositions in *De Numeris Datis*. By using a square inspired by the cut and paste method to represent Nemore's statements, we have been able to reveal the algebraic identities that were in play in Nemore's analytical inferences. This suggests that we use this didactic artifact as a concrete model to teach the solving of second degree equations and algebraic identities. From Ancient Babylonia to present algebra classrooms: a long way for cutting and pasting operations to go.

Bibliography

[1] al-Tusi, S., 1986, *Œuvres mathématiques, Algèbre et Géométrie au XIIe siècle*. Tomes I et II. Texte établi et traduit par Roshdi Rashed, Paris: Les Belles Lettres.

[2] Berggren, J. L., 2008, "The Mathematics of Islam," in *A History of Mathematics: An Introduction. (3rd Edition)*, V. Katz (ed.) New Jersey: Pearson Education.

[3] Busard, H. L. L., 1991, Jordanus de Nemore, *De Elementis Arithmetice Artis. A medieval Treatise on Number Theory*. 2 Vols., Stuttgart: Franz Steiner Verlag.

[4] Filloy, E., 1980, "Álgebra del nivel medio y análisis epistemológico: de Bombelli a Vieta" in *Actas del V Congreso Nacional de Profesores*. México DF.

[5] Filloy, E., and T. Rojano, 1984, "La aparición del lenguaje aritmético-algebraico," *L'Educazione Matemática*, 5, 278–306.

[6] Filloy, E., T. Rojano, and L. Puig, 2008, *Educational Algebra. A Theoretical and Empirical Approach*, New York: Springer.

[7] Høyrup, J., 1990, "Algebra and Naive Geometry. An Investigation of Some Basic Aspects of Old Babylonian Mathematical Thought." *Altorientalische Forschungen*, 17, 27–69, 262–354

[8] ——, 1991, " 'Oxford' and 'Cremona': On the relations between two Versions of al-Khwarizmi's Algebra," *Filosofi og videnskabsteori påRoskilde Universitetcenter. 3. Række: Preprint og Reprints*, 1991 nr. 1.

[9] ——, 1992, "The Babylonian Cellar Text BM 85200 + VAT 6599. Retranslation and Analysis," in *Amphora. Festschrift für Hans Wussing zu seinem 65. Geburtstag*, S. S. Demidov, M. Folkerts, D. E. Rowe, & C. J. Scriba (eds.), Basel, Boston, Berlin: Birkhäuser, pp. 315–358.

[10] ——, 1994, "The Antecedents of Algebra." *Filosofi og videnskabsteori påRoskilde Universitetcenter. 3. Række: Preprint og Reprints*, 1994 nr. 1.

[11] ——, 1995, "Linee larghe. Un'ambiguità geometrica dimenticata." *Bollettino di Storia delle Scienze Matematiche*, 15, 3–14

[12] ——, 2002, *Lengths, Widths, Surfaces. A Portrait of Old Babylonian Algebra and Its Kin*, New York: Springer.

[13] Hughes, B., (ed.), 1981, Jordanus de Nemore. *De Numeris Datis*, Berkeley, CA: University of California Press.

[14] ——, 1986, "Gerard of Cremona's Translation of al-Khwārizmī's al-jabr: A Critical Edition," *Mediaeval Studies*, 48, 211–263

[15] Levey, M., (ed.), 1966, *The Algebra of Abū Kamīl, in a Commentary by Mordecai Finzi*. Hebrew text and translation, and commentary, Madison, WI: The University of Wisconsin Press.

[16] Luckey, P., 1941, "Tâbit b. Qurra über den geometrischen Richtigkeitsnachweis der Auflösung der quadratischen Gleichungen," *Sächsischen Akademie der Wissenschaften zu Leipzig. Mathematisch-physische Klasse. Berichte* 93, 93–114.

[17] Mazzinghi, M. A. di., 1967, *Trattato di Fioretti*. (Arrighi, G. Ed.) Pisa: Domus Galileana.

[18] Nesselman, G. H. F., 1842, *Versuch einer kritischen geschichte der algebra, 1. Teil. Die Algebra der Griechen*, Berlin: G. Reimer.

[19] Puig, L., 1994a, "A Study on Mathematical Sign Systems and the Method of Analysis: the Case of De Numeris Datis by Jordanus de Nemore," in *Proceedings of the First Italian-Spanish Research Symposium in Mathematics Education*, N. A. Malara, and L. Rico (eds.), Modena: Università di Modena, pp. 257–264.

[20] ——, 1994b, "El *De Numeris Datis* de Jordanus Nemorarius como sistema matemático de signos," *Mathesis*, 10, 47–92.

[21] ——, 1996. "Pupils' Prompted Production of a Medieval Mathematical Sign System," in *Proceedings of the Twentieth International Conference on the Psychology of Mathematics Education. Vol. 1*, L. Puig & Á. Gutiérrez (eds.), Valencia: PME, pp. 77–84.

[22] Puig, L. and T. Rojano, 2004, "The History of Algebra in Mathematics Education," in *The Future of the Teaching and Learning of Algebra: The 12th ICMI Study*, K. Stacey, H. Chick, & M. Kendal (eds.), Boston / Dordrecht / New York / London: Kluwer Academic Publishers, pp. 189–224.

[23] Radford, L. and L. Puig, 2007, "Syntax And Meaning as Sensuous, Visual, Historical Forms of Algebraic Thinking," *Educational Studies in Mathematics*, 66, 145–164

[24] Rashed, R. (ed.), 2007, *Al-Khwārizmī. Le commencement de l'algebre*, Paris: Librairie Scientifique et Technique Albert Blanchard.

[25] Rosen, F., 1831, *The algebra of Mohammed Ben Musa*, London: Oriental Translation Fund.

[26] Sesiano, J., 1993, "La version latine médiévale de l'Algèbre d'Abū Kamīl," in *Vestigia Mathematica. Studies in medieval and early modern mathematics in honour of H. L. L. Busard*, M. Folkerts, and J. P. Hogendijk (eds.), Amsterdam and Atlanta, GA: Editions Rodopi B. V., pp. 315–452

About the Author

Luis Puig was born in March 1948 in Valencia, Spain and graduated in Mathematics and with a Ph.D. in Psychology from the University of Valencia (Spain). He has been teaching mathematics and mathematics education at the University of Valencia since 1975. He is currently a Full Professor of Didactics of Mathematics at the Department of Didactics of Mathematics. He has authored four books on mathematics education: *Problemas aritméticos escolares*, Madrid: Ed. Síntesis, 1989 (with Fernando Cerdán); *Semiótica y matemáticas*, Valencia: Episteme, 1994; *Elementos de resolución de problemas*, Granada: Comares, 1996; and *Educational Algebra. A Theoretical and Empirical Approach*, New York: Springer Verlag, 2008 (with Eugenio Filloy and Teresa Rojano). He is also the co-author and/or editor of some 15 books of classroom materials linked to curriculum development research projects. He has also published scholarly articles and given invited scholarly lectures at national and international conferences. He is the co-editor of two series of books on Mathematics Education (Matemáticas: cultura y aprendizaje [*Mathematics: Culture and Learning*], and *Mathema*) and several series of books on Cultural Studies (*Eutopías, Eutopías Mayor, Biblioteca Otras Eutopías, La huella sonora*). He was also the co-editor of the journal *Enseñanza de las Ciencias*, from 1986 till 2006. His main areas of work have been curriculum development, heuristics, arithmetic and algebraic problem solving, and the history of algebraic ideas.

5

Equations and Imaginary Numbers: A Contribution from Renaissance Algebra

Giorgio T. Bagni
University of Udine, Italy

5.1 Editors' Introductory Comments

G.T. Bagni (1958–2009) died on 10 June 2009 in a bicycle accident, while this paper was under review. Because of his untimely death, we do not know how he would have taken into account the reviewers' comments. However, we would have suggested a somewhat different structure of the paper and the clarification of several specialized terms, not expected to be understood by many readers to whom this book is addressed. Therefore, we provide below an outline of the rationale of the paper as it has been revised according to the reviewers' comments. We are indebted to P. Boero, B. D'Amore, and L. Radford for their help; they provided several explanatory comments and clarified unclear points. Their comments have been incorporated into the text and form the basis of most footnotes.

The rationale underlying this paper is as follows:

(i) *The didactic phenomenon*: Students at school are consciously or unconsciously hesitant, or even reluctant, to accept imaginary numbers (inasmuch as they have been taught for many years that square roots of negatives are strictly prohibited or nonsensical).

(ii) *The historical context*: Imaginaries entered mathematics, not as a theoretical algebraic construct (answering the pure algebraist's 'a posteriori' question: "how is it possible to extend real numbers, so that square roots of negatives make sense?"), but as an operational tool/trick to solve cubic equations (Section 5.4).

(iii) *The empirical study*: A pilot empirical study implemented this idea in the classroom (Section 5.5):

- To raise a meaningful mathematical problem: How to solve cubic equations?
- Using the "nonsensical" tool of square roots of negatives gives a "**real**" result.
- The effectiveness of the tool orients students towards thinking that this tool may have some "real" and in any case, useful meaning.

This pilot study provided indications of positive learning outcomes, or at least, a more positive attitude of the students towards imaginaries.

(iv) Given (iii), a more systematic theoretical analysis of what happened in history and what may happen in the classroom should be done. The author makes such an attempt using concepts and ideas from a semiotic and a socio-cultural viewpoint. To this end, the theoretical tools and concepts are presented first (in Section 5.3) and then applied (mainly, in Section 5.6).

5.2 Introduction

Many authors have shown that the history of mathematics can be drawn on by teachers in the presentation of several mathematical topics to the benefit of pupils. It follows that research into the role of history of mathematics in teaching can be considered a part of research into mathematics education [12, 8, 10]. In this paper, I examine a traditional topic of High School and undergraduate mathematics (imaginary numbers) that can be approached in a historical perspective and discuss some related epistemological aspects.

The introduction of imaginary numbers is an important step in the school mathematics curriculum. However, it is not without difficulties. In Middle School, the students are often told of the impossibility of calculating the square root of negative numbers. Yet, later, the students are asked to accept the presence of a new object, $\sqrt{-1}$, named i. They are asked to use a mathematical object previously considered illicit and "wrong." No wonder that, for many students, the idea of calculating with this new object is a matter of surprise and perplexity. The habit (forced by previous educational experiences) of using only real numbers and the new possibility of using complex numbers are conflicting elements. A historical approach can be useful in order to overcome these difficulties. More particularly, our comprehension of pupils' approach to a mathematical concept can effectively take into account some epistemological aspects, for instance the development of what we can term the "semiosic chain," i.e., the signifying relationships between signs, which we will consider in Section 5.6.

5.3 Theoretical Framework

When we consider a sign, we make reference to an object, and in the case of mathematical objects, to a concept. However my approach does not deal only with "concepts." Font, Godino and D'Amore [9, p. 14] state that although

> to understand representation in terms of a semiotic function, as a relation between an expression and a content established by 'someone,' has the advantage of not segregating the object from its representation, [...] in the onto-semiotic approach [...], the type of relations between expression and content can be varied, not only be representational, *e.g.*, 'is associated with;' 'is part of;' 'is the cause of/reason for.' This way of understanding the semiotic function gives us great flexibility, not to restrict ourselves to understanding 'representation' as being only an object (generally linguistic) that is in place of another, which is usually the way in which representation seems to us mainly to be understood in mathematics education.

Furthermore, in the line of inquiry that I am following here, a central role is given to the cultural way in which objects of knowledge become signified in mathematical activity. In the socio-cultural approach outlined by Radford [17, p. 32], "mathematical knowledge is more than merely concomitant with its cultural environment [...]; the configuration and the content of mathematical knowledge is properly and intimately defined by the culture in which it develops and in which it is subsumed." In tune with the socio-cultural approach, the ontosemiotic approach to mathematics cognition also assumes socio-epistemic relativity for mathematical knowledge, since

> "knowledge is considered to be indissolubly linked to the activity in which the subject is involved and is dependent on the cultural institution and the social context of which it forms part." [9, p. 9].

My inquiry into Bombelli's use of imaginary numbers draws on these representational and sociocultural considerations but is also linked to some considerations about semiotic aspects, based upon a Peircean approach. According to Peirce we cannot "think without signs," and signs consist of three interrelated parts: an object, a proper sign (*representamen*), and an interpretant[1]. Peirce considered either the immediate object represented by a sign, or the dynamic object, progressively originated in the semiosic process. As a matter of fact, an interpretant can be considered as a new sign. The limit of this process is the *ultimate logical interpretant* and is not a real sign, which would induce a new interpretant. It is a *habit-change* ("meaning by a habit-change a modification of a person's tendencies toward action, resulting from previous experiences or from previous exertions of his will or acts, or from a complexus of both kinds of cause:" [15, §5.475]).

---

[1] *Editors' footnote*: In his later works, Peirce uses the term sign both for the triad "object, sign, interpretant" and the representamen.

5.3. Theoretical Framework

The sign determines an interpretant by some features of the way the sign signifies its object to generate and shape our understanding. Peirce associates signs with cognition, and objects "determine" their signs, so the cognitive nature of the object influences the nature of the sign (in this work "mathematical objects" will be considered as "objectualized procedures" [22, 11]). If the constraints of successful signification require that the sign reflects some qualitative features of the object, then the sign is an *icon*; if they require that the sign utilizes some physical connection between it and its object, then the sign is an *index*; if they require that the sign utilizes some convention or law that connects it with its object, then the sign is a *symbol*.

According to Peirce, the formulas of our modern algebra are icons, i.e. signs which are mappings of that which they represent [15, §2.279]. Nevertheless pure icons, according to Peirce himself (1931-1958, §1.157), only appear in thinking, if ever. Pure icons, pure indexes, and pure symbols are not actual signs. In fact, every sign "contains" all the components of Peircean classification, although one of them is predominant. It is worth noting that a sign in itself is not an icon, index or symbol. From the educational viewpoint, the identification of signs is not a question of classifying a sign (as *e.g.* an icon), but it is a question of showing their cognitive import in the students' ongoing understanding of the "rules of the game" [18, 3]).

Peirce underlined the importance of iconicity and noticed that

> deduction consists in constructing an icon or diagram the relations of whose parts shall present a complete analogy with those of the parts of the object of reasoning, of experimenting upon this image in the imagination, and of observing the result so as to discover unnoticed and hidden relations among the parts. [15, §3.363]

In fact, Peirce distinguished three different kinds of icons: images, metaphor, and diagrams. According to Radford [20], since the epistemological role of "diagrammatic thinking" rests on making apparent some hidden relations, it relates to actions of objectification, and a diagram can be considered a *semiotic means of objectification*. As I deal here with the formation of a mathematical object, let us remember the importance of Sfard's view on concept formation and the "constant three step pattern" she suggests:

> a constant three step pattern can be identified in the successive transitions from operational to structural conceptions: first there must be a process performed on the already familiar objects, then the idea of turning this process into a more compact, self contained whole should emerge, and finally an ability to view this new entity as a permanent object in its own right must be acquired. These three steps will be called interiorization, condensation and reification. [23, pp. 64–65]

I shall take into account these steps in order to describe some stages in the analysis of the semiosic chain, and we shall use the terms introduced by Sfard [22]. However we must also take into account an important issue: the relationship between psychological and historical developments. Radford [17] reminds us that in some constructivist and neo-constructivist works on mathematics education we find an implicit or explicit reference to the Piaget and Garcia [16] relationship between psychological and historical developments. (They suggested that the relationship between psychological and historical developments must be seen in terms of mechanisms mediating the transitions between the intraoperational, the interoperational and the transoperational stages). Radford quotes Sfard [24, p. 15]:

> indeed, there are good reasons to expect that, when scrutinized, the phylogeny and ontogeny of mathematics will reveal more than marginal similarities. At least, this is what follows from the constructivist view according to which learning consists in the reconstruction of knowledge. [...] It is probably because of the inherent properties of knowledge itself, because of the nature of the relationship between its different levels, that similar recurrent phenomena can be traced throughout its historical development and its individual reconstruction.

However, as Radford [17] argues, recent developments in the sociology of knowledge are challenging the presupposition of a mind free of social and cultural forces. He refers to Wertsch [26] and Otte and quotes the latter saying that "the development of knowledge does not take place within the framework of natural evolution but within the framework of sociocultural development. [...] Knowledge is necessarily social knowledge." Otte [14, p. 309]. In the same paper, Radford [17] points out that fifty years ago Werner (1948) argued that there are essential differences in phylogenesis and ontogenesis that make them *incomparable*; in fact, the development of the child is encompassed by an interaction with the world of adults. According to Werner, what a phylogenesis-ontogenesis comparison may bring about is a merely formal connection [25, pp. 7–28]. After mentioning the recurrent failure of constructivist approaches in including culture in a decisive manner in theoretical accounts of knowledge formation, Radford [17, p. 29] concludes: "the socio-cultural perspective suggests that the effects of culture and society are fundamental to the way in which we come to know." In the next Section we will consider the case of complex numbers according to this perspective.

5.4 History of Mathematics and Imaginary Numbers

The cultural movement of the Renaissance in Europe had a deep impact on mathematics and involved a major shift in mathematical ideas. The study of mathematics began to be considered important because of its association with trade and commerce. In part because of these practical needs, the main direction of mathematics was toward algebraic methods (see [19]). In fact, beginning in the fourteenth century the abacists had to teach the merchants both the decimal place-value system and the algorithms for using it. During the fourteenth and fifteenth centuries they extended those methods by introducing abbreviations and symbolism, while developing new methods for dealing with algebraic problems.

The invention of the printing press also had an important effect on mathematics and mathematics education. Both ancient mathematical texts and the works of modern mathematicians were printed. The diffusion of the printing press on the one hand ensured a quick, widespread circulation of new techniques, while on the other further enhanced the introduction of abbreviations and symbolisms and their selection according to typographical needs (see [7, p. 384] for an example).

In what follows, I briefly consider the resolution of cubic equations. I will not dwell on the dispute about the paternity of the resolution of cubic equations, as this is a well known topic. G. Cardan (1501–1576) published his *Ars Magna* in 1545, one of the first Latin works devoted to algebra. It was in this work that the first computation with complex numbers appeared (although Cardan did not completely understand it). In 1572, R. Bombelli (1526–1573) published his masterwork, *Algebra*, in which he recapitulated Cardan's solution of cubic equations and showed that sometimes their resolution makes it necessary to consider complex numbers. Bombelli was the first mathematician to give rules for addition and multiplication of complex numbers and showed that correct real solutions could be obtained from Cardan's method, even in the "irreducible" case. Bombelli used new symbolism (*e.g., R.q.* for square root, *R.c.* for cubic root, and symbols for powers[2]), and Leibniz cited him as an *outstanding master of the analytical art*.

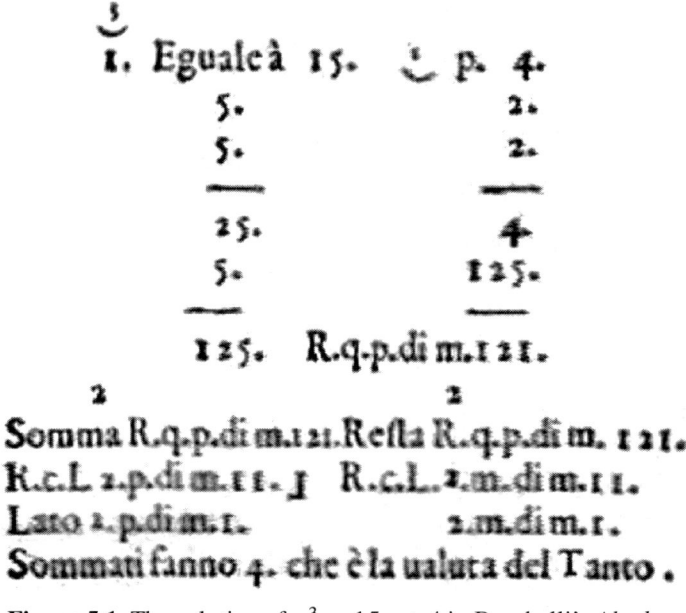

Figure 5.1. The solution of $x^3 = 15x + 4$ in Bombelli's *Algebra*.

---

[2]*Editors' footnote*:He used a horizontal concave circular arc, with the exponent indicated over it (see Figure 5.1).

5.4. History of Mathematics and Imaginary Numbers

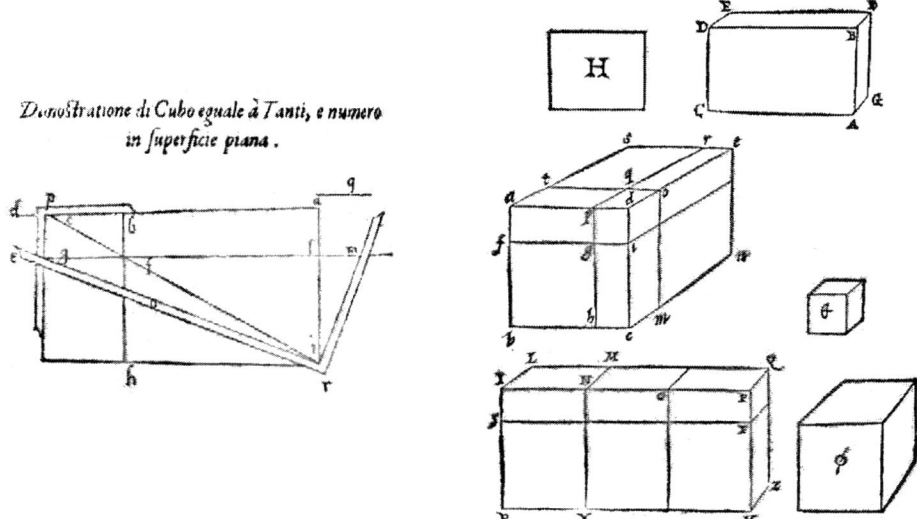

Figure 5.2. Bombelli's geometric constructions: *Proof of Cube equal to Unknowns, and number in plane Surface*

Let us consider an example. In Bombelli's *Algebra* we do not find the use of the symbol i for $\sqrt{-1}$. Bombelli wrote $p\,d\,m$ and $m\,d\,m$ for i and $-i$.[3] The solution of the equation $x^3 = 15x + 4$ leads to the sum of radicals[4], where $2 + 11i = (2 + i)^3$ and $2 - 11i = (2 - i)^3$. So a real solution of the original equation is $x = (2 + i) + (2 - i) = 4$. This problem appears on p. 294 of Bombelli's *Algebra*, as shown in Figure 5.1.

Its transcription is as follows[5]:

$x^3 = 15x + 4$ $\qquad\qquad\qquad\qquad\qquad\qquad\qquad\qquad$ x^3 Equale à [equals to] $15x$ p.4
$(4/2)^2 - (15/3)^3 = 2^2 - 5^3 = 4 - 125 = -121$
$\sqrt{(-121)}$ $\qquad\qquad\qquad\qquad\qquad\qquad\qquad\qquad\qquad\qquad$ R.q.p.di m. 121
$\qquad\qquad\qquad\qquad\qquad\qquad$ Somma [*Sum*] 2 R.q.p.di m.121 resta [*remains*] R.q.p.di m.121
$\sqrt[3]{2 + 11i} + \sqrt[3]{2 - 11i}$ $\qquad\qquad\qquad\qquad\qquad\qquad\qquad$ R.c.L 2 p.di m. 11⌋ R.c.L 2 p.di m. 11⌋
$(2 + i) + (2 - i)$ $\qquad\qquad\qquad\qquad\qquad\qquad\qquad\qquad$ Lato[6] [*side*] 2 p.di m.1 2 m.di m.1
4 $\qquad\qquad\qquad\qquad\qquad\qquad\qquad$ Sommati [*when added*] fanno [*make*] 4, che è la valuta del
$\qquad\qquad\qquad\qquad\qquad\qquad\qquad\qquad\qquad\qquad$ Tanto [which is the value of the Unknown]

Figure 5.2 shows the geometrical constructions (both two-and three-dimensional) that Bombelli proposed on pp. 296, 298 of *Algebra* in order to confirm his procedure (see details in: [5]).

In Figure 5.3 we can see the original "rules" as listed on p. 169 of *Algebra*.

In view of footnote 3 we have:

Più means *plus*, with the present meaning of $+1$; *Meno* means *minus*, with the present meaning of -1; *Più di meno* is Bombelli's terminology for $+i$; *Meno di meno* is Bombelli's terminology for $-i$; *Via* means *multiplied by*, with the present meaning of ; *Fa* means [*it*] *does* or (more literally) [*it*] *makes*, which could be represented by $=$;

---

[3]*Editors' footnote*: Bombelli wrote that "As one cannot call it either plus or minus, I shall call it 'più di meno' when it is added and 'meno di meno' when it is subtracted." This terminology stands for "più radice di meno" and "meno radice di meno," i.e. in modern notation $+\sqrt{-1}$ and $-\sqrt{-1}$ respectively ('più' means plus; 'meno' means minus; 'radice' means root), and Bombelli used the abbreviation p.dm, m.dm or p.dim, m.dim [7, p. 127].

[4]*Editors' footnote*: The solution of $x^3 = px + q = 0$ is obtained as the sum of the cubic roots of the roots of an associated quadratic equation, with discriminant $4[(q/2)^2 - (p/3)^3]$. In this example, it involves $\sqrt{(-121)} = 11i$.

[5]*Editors' footnote*: See Figure 5.1 explained in footnote 3. Note also, Bombelli's abbreviations are explained in footnote 4. Note also that Bombelli almost always used $R.c.L$ (or $R.q.L$) to indicate a cubic (or square) root of more than one terms, followed by a mirrored L, i.e., ⌋, to indicate the end of the expression that is radicated; (R stands for *radix* and L for *legata*; see [7, Vol I., p. 124]).

[6]*Editors' footnote*: It is clear that Bombelli conceives the cubic root as the side of a cube, hence the wording in his text.

> Più uia più di meno, fà più di meno.
> Meno uia più di meno, fà meno di meno.
> Più uia meno di meno, fà meno di meno.
> Meno uia meno di meno, fà più di meno.
> Più di meno uia più di meno, fà meno.
> Più di meno uia men di meno, fà più.
> Meno di meno uia più di meno, fà più.
> Meno di meno uia men di meno fà meno.

Figure 5.3. Bombelli's rules of calculation with imaginary numbers.

Hence, Figure 5.3 is translated into modern notation as

$$(+1) \cdot (+i) = +i \qquad (-1) \cdot (+i) = -i$$
$$(+1) \cdot (-i) = -i \qquad (-1) \cdot (-i) = +i$$
$$(+i) \cdot (+i) = -1 \qquad (+i) \cdot (-i) = +1$$
$$(-i) \cdot (+i) = +1 \qquad (-i) \cdot (-i) = -1$$

The linguistic aspect is important: for example, we have translated the term "fa" with the symbol "=," but the modern equality symbol can be referred to as a relation "in two directions," while the term "fa" means that the result of the multiplication in the first member is written in the second member.

Moreover, in *Algebra* we find (p. 70):

> "Più via più fa più.   Meno via meno fa più.
> Più via meno fa meno.   Meno via più fa meno."

which can be transcribed in the following way:

$$(+1) \cdot (+1) = +1 \qquad (-1) \cdot (+1) = -1$$
$$(+1) \cdot (-1) = -1 \qquad (-1) \cdot (-1) = +1$$

So we can write the following *Cayley table*:

x	$+1$	-1	$+i$	$-i$
$+1$	$+1$	-1	$+i$	$-i$
-1	-1	$+1$	$-i$	$+i$
$+i$	$+i$	$-i$	-1	$+1$
$-i$	$-i$	$+i$	$+1$	-1

From a modern viewpoint, the set of elements $\{+1, -1, +i, -i\}$ along with the multiplication rules can be interpreted as the multiplicative group $(\{+1; -1; +i; -i\}; \cdot)$ of the fourth roots of unity, a finite Abelian group. (It is well known that it is a cyclic group and it can be generated either by i or by $-i$. Of course Bombelli did not notice this property, and he did not notice this for its cyclic subgroup generated by -1.)

So in *Algebra* we cannot find a modern introduction either to complex numbers or to the formal notion of a group: Bombelli just indicated some mathematical objects in order to solve cubic equations. These ideas were not immediately accepted following the publication of Cardan's and Bombelli's works. Nevertheless the *formal* introduction of i in *Algebra* is important and modern [6, pp. 91–92]. We notice, according to Sfard [22, p. 12], that complex numbers were introduced simply as an *operational* concept:

> Cardan's prescriptions for solving equations of the third and fourth order, published in 1545, involved [...] even finding roots of what are today called negative numbers. Despite the widespread use of these algorithms, however, mathematicians refused to accept their by-products [...] The symbol [square root of -1 was] initially considered nothing more than an abbreviation for certain 'meaningless' numerical operations. It came to designate a fully fledged mathematical object only after mathematicians got accustomed to these strange but useful kinds of computation. [22, p. 12]

We can consider the idea of group in a similar way. It would be incorrect to ascribe to Bombelli an explicit awareness of the group concept, three centuries before Galois and Dedekind; however, we can state that he implicitly introduced—in action—one of the most important concepts of mathematics. So a question is the following: is it possible and useful to introduce the group concept by a pre-axiomatic first treatment? [2].

5.5 Imaginary Numbers from History to Mathematics Education

In this Section we present the idea that the introduction of imaginary numbers through the intermediate steps of resolution of cubic equations (as computation symbolic devices) could facilitate the students' approach to these new entities. The results of an empirical research project seem to support this idea.

The introduction of imaginary numbers, historically, did not take place in the context of quadratic equations, as in $x^2 = -1$. It took place in the resolution of cubic equations. Sometimes their resolution does not take place entirely in the set of real numbers, but one of their solutions is always real. A substitution of $x = 4$ in the equation $x^3 = 15x + 4$ considered by Bombelli is possible ($4^3 = 15 \cdot 4 + 4$) in the set of real numbers. In the quadratic equation $x^2 = -1$ the results themselves are not real, so their acceptance needs the knowledge of imaginary numbers. Although the focus of this paper is not on the discussion of experimental data, let us briefly summarize the results of a research study [1]. In a first stage we examined 97 third-year High School students (Italian *Liceo scientifico*, students 16–17-years-old) and fourth-year students (17–18-years-old). In all classes, at the moment of the test, pupils knew the resolution of quadratic and of cubic equations, but they did not know imaginary numbers:

- about the quadratic equation $x^2 + 1 = 0 \longrightarrow x = \pm i$ only 2% stated acceptance of the solution (92% denied it; 6% did not answer).

- immediately after this test, the resolution of the cubic equation $x^3 - 15x - 4 = 0 \longrightarrow x = \sqrt[3]{2 + 11i} + \sqrt[3]{2 - 11i} \longrightarrow x = (2 + i) + (2 - i) = 4$ was accepted by 54% of the pupils (35% denied it; 11% did not answer).

So imaginary numbers *in the process* of the solution of an equation, but *not* in its result, are frequently accepted by pupils (the *didactical contract*[7] ascribes great importance to this result). Under the same conditions, a test was then proposed to another group of 52 students of the same age, where the equations were presented in the reverse order:

- 41% accepted the solution of the cubic equation (25% rejected it and 34% did not answer).

- immediately after that, the solution of the quadratic equation was accepted by 18% of the students, with only 66% rejecting it (16% did not answer).

These data suggest that teaching a subject by taking account of some basic facts in historical development may help students to acquire a better understanding of it. Nevertheless this statement is still unspecific and requires a detailed analysis. What happened when Bombelli introduced his procedure and described the role of pdm and mdm? Is this historical introduction really useful in order to understand students' learning? As previously noticed, the diffusion in the Renaissance of the decimal place-value system and of the algorithms for using it, and in particular the introduction of suitable abbreviations and symbolisms, were important elements to be considered with reference to the development of new methods for dealing with algebraic problems. The invention of the printing press is another important aspect (see above). However these historical facts are rather general, and they are not enough to comprehend the specific features of the historical introduction of a new mathematical object. Above all, we must take into account that historical facts cannot be uncritically considered with reference to the behavior of our pupils today [17].

In order to understand the educational importance of the historical formation of a mathematical object, we need a fine-grained cultural analysis of the considered mathematical practice; thus we need to frame the emergence of the new entity. This will lead us to point out some features of the semiosic chain.

5.6 The Semiosic Chain

Let us turn back to the educational aspect. Of course, as noted above, the didactical contract ascribes great importance to the result, but this is not enough to account for the conceptualization of pupils. Now we shall consider some features of students' approach, making reference to Peirce's *unlimited semiosis*. As noted above, every step of the interpretative process produces a new "interpretant n" that can be considered the "sign $n + 1$" linked to the object (considered in the sense of an objectified procedure). However we must ask ourselves: what about the very first sign to be associated to our object? Can we find an element from which the semiosic chain originates?

---

[7] *Editors' footnote*: The didactical contract is the brilliant idea introduced by Guy Brousseau; it refers to the reciprocal expectations that emerge between teacher and pupil in classroom situations, expectations that stem from the interpretations regarding the modalities of the two agents.

Our mathematical "object" (a procedure to solve an equation) would be represented by a first "sign": in fact, "absence" itself can be considered as a sign. Peirce [15, §5.480] made reference to "a strong, but more or less vague, sense of need" leading to "the *first logical interpretants* of the phenomena that suggest them, and which, as suggesting them, are signs, of which they are the (really conjectural) interpretants." So we suppose that this kind of absence can be the starting point of the semiosic process. Educationally speaking, this is influenced by important elements, e.g., the theory in which we are working, the persons (students, teacher), the social and cultural context. Of course by this we do not mean that there is one and only one starting point of the mentioned kind for every "mathematical object." Nevertheless this starting point can be described as a *habit* linked to the absence of a procedure (or, better, a procedure to be objectified). So the situation is characterized by intuitive sensations and by the influence of social, cultural, traditional elements. Later, with the emergence of formal aspects, our object will become more "rigorous."[8]

According to an ontosemiotic approach, knowledge is linked to the activity in which the subject is involved and it depends on the cultural institution and the context [9, 17]. In the case considered, pupils have the perception of an absence, referred to the strategy to be followed, namely the procedure to be objectified. Historical references give them the opportunity to consider a situation, and the context is characterized by the "game to be played" (the solution of an equation) at the very beginning of our experience. We cannot make reference to a semiotic function related to an object to be represented. The "object" will be considered just later, on the basis of the solving strategy. A real strategy is actually absent, and only a "potential object" is connected to the possibility of finding an effective procedure in order to play the (single) game considered (see Figure 5.4).

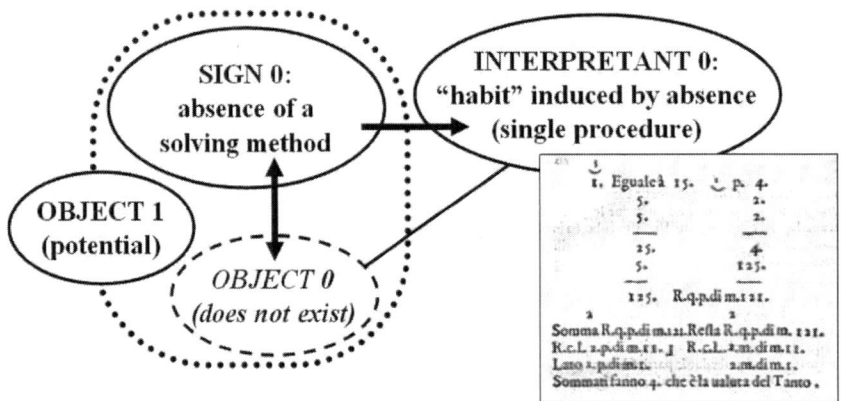

Figure 5.4. The beginning of a semiosic chain[9]

In Bombelli's original work, a form of iconicity can be pointed out in the written disposition of the numbers employed. This can be relevant to students' first approach (further research can be devoted to this issue). Educationally speaking, in this stage the effectiveness of the procedure is fundamental. There is not a real mathematical object to be considered, nevertheless pupils have a "game to be played," so the possibility to provide a first "structure" to the strategy (*e.g.*, the consideration of some standard actions) makes possible its becoming a procedure to be objectified (see Figure 5.5).

Both from the historical viewpoint (see the aforementioned geometrical constructions of Bombelli) and from the educational viewpoint (with reference to the substitution of the result, $x = 4$, in $x^3 - 15x - 4 = 0$ so $4^3 - 15 \cdot 4 - 4 = 0$), a first kind of objectification can be pointed out. In Sfard's words, pupils are in the stages of interiorization and condensation. Nevertheless the path considered does not allow us to state that our pupils have reached a complete reification. In Figure 5.6 the interpretant 2 is related to an objectified procedure, and it is referred to the "rules" listed by Bombelli.

---

[8]Making reference, of course, to the conception of rigor in a historical and cultural context; Bombelli's rigor and that of a modern mathematician are different.

[9]*Editors' footnote*: Here, the "potential object" is the object that emerges as a possible interpretation of the situation brought into play; in the aforementioned example, the solution strategies. The "single procedure" refers to a "preliminary procedure," to the absence and the sense of need that give rise to successive procedures: the procedure to be objectified, the objectified procedure etc.

[10]*Editors' footnote*: The procedure to be objectified could be interpreted as a fixed pattern (using Radford's terminology) that has to be recognized (objectified) both at a personal and cultural level through suitable semiotic means, thereby acquiring a cultural and historical experience.

5.6. The Semiosic Chain

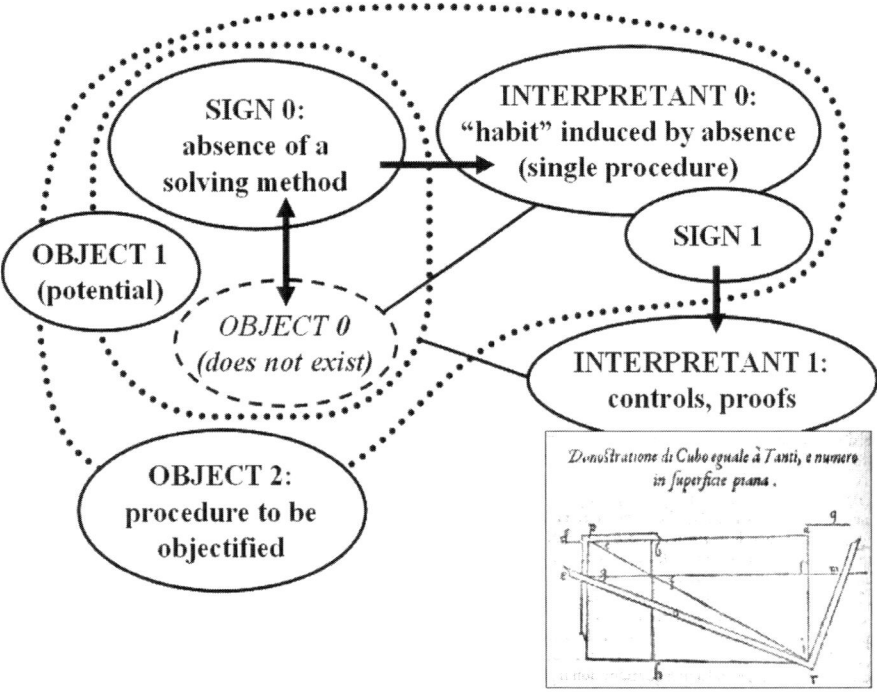

Figure 5.5. Next step in the semiosic chain[10]

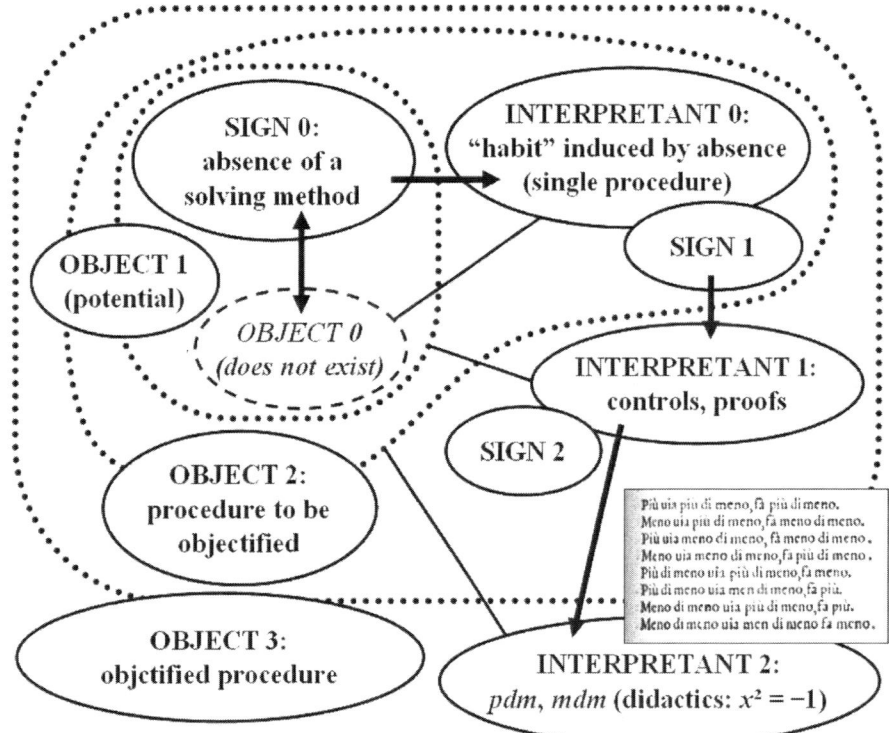

Figure 5.6. The interpretant 2^{11} is related to an objectified procedure[12].

---

[11]*Editors' footnote*: Although Peirce considers the sign that derives from the interpretant of the previous step of the semiosic chain both as representamen and referring to the triad as a whole (see Section 5.3), here the author considers the sign as representamen; hence, the interpretants lie outside the dotted closed curve in Figures 5.4–5.6.

[12]*Editors' footnote*: The interpretation addresses all information brought into play; in particular, the objectification based on the process that is being followed. Therefore, the procedure itself is objectified. Object 1, 2 & 3 and the semiosic chain can be traced back to Sfard's interiorization,

The presence of the "rules" for *pdm* and *mdm* (p. 169 of *Algebra*) can emphasize the symbolic feature of the sign employed. Later, the strategy will become an autonomous object and its transparency (in the sense of [13]) will be important from the educational point of view. It will not be linked to a single situation and it will be applied to different cases.

According to Font, Godino and D'Amore ([9], p. 14), "what there is, is a complex system of practices in which each one of the different object/representation pairs (without segregation) permits a subset of practices of the set of practices that are considered as the meaning of the object." The starting point of the semiosic chain cannot be considered in the sense of semiotic function. More properly, it can be considered as a first practice that will be followed by other practices in order to constitute the meaning of the object.

5.7 Final Reflections

As regards educational uses of historical references, if we operate on teaching to improve its quality by *thought transference*, some reactions, especially those in pupils' minds, are just inferred. We can propose an example in the historical sphere, in order that students will "learn" in this sphere, but so that the knowledge achieved will not be confined to the historical sphere, evolution to different spheres is necessary. The main limitation of the notion of mathematics education as *thought transference* lies in the uncertainty about real effects upon the learning of teachers' choices. We make no claims for the teaching of mathematics as regards the nature and the meaning of mathematical objects. Here several problems are opened, involving some fundamental philosophical questions.

An important point to be considered can be summarized as follows: is the analysis of the semiosic chain important cognitively or epistemologically speaking? Of course social practices are relevant in order to give sense to a new mathematical object, as noticed above. We do not agree with the outmoded paradigm of an isolated solitary thinker busy with the internal mental dynamics of his own symbols. Nevertheless, in our opinion, the mathematical practices mentioned and their epistemological analysis are useful in order to comprehend the conceptualization of a mathematical object. More particularly, our comprehension of pupils' approaches to a mathematical concept can take into account effectively the initial development of the semiosic chain. In the described process we can point out some interesting elements, some hints that lead us to a better comprehension of the early stage of pupils' approach to a mathematical idea, *e.g.*, the important role of the cue provided by the research of a strategy for the solution of a cubic equation. By that we do not support Piaget and Garcia [16] in their recapitulation notion, according to which there is a relationship between psychological and historical developments. In our opinion this theoretical structure cannot be pointed out just by taking into account that "similar recurrent phenomena can be traced throughout its historical development and its individual reconstruction" [24, p. 15]. Modern cognitive developments are embedded in social and cultural contexts that are different from those of the past [17].

To conclude, the importance of an ontosemiotic approach to representations can be highlighted by a Peircean perspective giving sense to the starting point of the semiosic chain, and, as previously stated, the analysis of this stage of the chain can help us to comprehend both our pupils' modes of learning and the essence of mathematical objects themselves.

Yet the question remains: what is the role of culture in the way Bombelli dared to trespass the lands of old knowledge and walk into new territory? What does culture have to do with Bombelli's inventing gesture? As previously seen, the semiosic chain is based on the idea of a method. And method here is based on the idea of calculation and the symbols in which calculations and objects are expressed. This idea was not particular to Renaissance algebra. More generally, Renaissance mathematics could indeed be characterized to an important extent as a methodical inquiry–a new form of inquiry where the focus is not on objects (as it was for the Greek *episteme*) but on methods and operations on symbolic operations and objects [4]. As Radford [19] argues, such a shift in mathematics, particularly in the new algebra of Cardan, Bombelli, and other Renaissance mathematicians should be linked to the rise and spreading of manufacture and a new systematic and methodic dimension in human actions. More specifically, Radford claims, such a shift was encompassed by the appearance of a world in which machines and new forms of labor transformed human experience,

---

condensation and reification (cf. the text below Figure 5.5). The objectified procedure can be identified with reification. It is a procedure, a fixed pattern that has an independent existence in the culture that subsumes it. The objectified procedure both emerges from, and is referred by semiotic means. The ontosemiotic and cultural-semiotic approaches, addressed by the author, call the objectified procedure, as institutional and cultural object, respectively.

5.7. Final Reflections

introducing a systemic dimension that acquired the form of a metaphor of efficiency, not only in the mathematical and technical domains, but also in aesthetics and other spheres of life [19, p. 512]. Here is the vital contribution of culture.

Culture creates the cultural–epistemological conditions for new knowledge to emerge. As Radford and Empey [21, p. 250] put it, "mathematical objects are not pre-existing entities but rather conceptual objects generated in the course of human activity." And, as these authors argue, "mathematics is much more than just a form of knowledge production—an exercise in theorization. If it is true that individuals create mathematics, it is no less true that, in turn, mathematics affects the way individuals are, live and think about themselves and others."

Bibliography

[1] Bagni, G.T., 2000a, Introducing complex numbers: an experiment. In J. Fauvel & J. van Maanen (eds.), *History in mathematics education. The ICMI Study*. Dordrecht: Kluwer, 264–265.

[2] ——, 2000b. The role of the history of mathematics in mathematics education: reflections and examples. In I. Schwank (ed.), *Proceedings of CERME-1*. Osnabrueck: Forschungsinstitut fuer Mathematikdidaktik, vol. II, 220–231.

[3] ——, 2006, Some cognitive difficulties related to the representations of two major concepts of Set Theory. *Educational Studies in Mathematics* 62, 3, 259–280.

[4] ——, 2009, Bombelli's Algebra (1572) and a new mathematical object. *For the Learning of Mathematics*, 29, 2, 30–32.

[5] Bombelli, R., 1966, *L'Algebra*. U. Forti & E. Bortolotti (eds.). Milano: Feltrinelli.

[6] Bourbaki, N., 1960, *Eléments d'histoire des mathematiques*, Hermann, Paris.

[7] Cajori, F. (1928-29): *History of mathematical notations*. La Salle, Il.: The Open Court Publishing Company.

[8] Fauvel, J. and J. van Maanen, (eds.), 2000, *History in mathematics education. The ICMI Study*. Dordrecht: Kluwer.

[9] Font, V., Godino, J.D. and B. D'Amore, 2007, An ontosemiotic approach to representations in mathematics education. *For the Learning of Mathematics* 27, 2, 9–15.

[10] Furinghetti, F., L. Radford, and V. Katz, (Eds.)(2007). The History of Mathematics Education: Theory and Practice. Special Issue. *Educational Studies in Mathematics*, 66(2), 107–271.

[11] Giusti, E., 1999, *Ipotesi sulla natura degli oggetti matematici*. Torino: Bollati Boringhieri.

[12] Jahnke, H.N., 1996, Mathematikgeschichte fuer Lehrer, Gründe und Beispiele, *Mathematische Semesterberichte* 43, 1, 21–46.

[13] Meira, L., 1998, Making sense of instructional devices: the emergence of transparency in mathematical activity. *Journal for Research in Mathematics Education* 29, 2, 121–142.

[14] Otte, M., 1994, Historiographical trends in the social history of mathematics and science. In K. Gavroglu & Al. (eds.). *Trends in the historiography of sciences*. Dordrecht: Kluwer, 295–315.

[15] Peirce, C.S., 1931–1958, *Collected papers*. Cambridge: Harvard University Press.

[16] Piaget, J. and R. Garcia, 1989, *Psychogenesis and the history of science*. New York: Columbia University Press.

[17] Radford, L., 1997, On psychology, historical epistemology and the teaching of mathematics: towards a socio-cultural history of mathematics. *For the Learning of Mathematics* 17, 1, 26–33.

[18] ——, 2000, Signs and meanings in the students' emergent algebraic thinking: a semiotic analysis. *Educational Studies in Mathematics* 42, 3, 237–268.

[19] ——, 2006, The cultural-epistomological conditions of the emergence of algebraic symbolism. In F. Furinghetti, S. Kaijser & C. Tzanakis (Eds.), *Proceedings of the 2004 Conference of the International Study Group on the Relations between the History and Pedagogy of Mathematics & ESU 4—Revised edition* (pp. 509–524). Uppsala, Sweden.

[20] ——, 2008, Diagrammatic thinking: Notes on Peirce's semiotics and epistemology. PNA: *Revista de Investigación en Didáctica de la Matemática*, 3(1), 1–18 (www.pna.es/)

[21] Radford, L. and H. Empey, 2007, Culture, knowledge and the self: mathematics and the formation of new social sensibilities in the Renaissance and Medieval Islam. *Revista Brasileira de História da Matemática*. Festschrift Ubiratan D'Ambrosio, Especial 1, 231–254.

[22] Sfard, A., 1991, On the dual nature of mathematical conceptions: reflections on processes and objects as different sides of the same coin. *Educational Studies in Mathematics* 22, 1–36.

[23] ——, 1992, Operational origins of mathematical objects and the quandary of reification-the case of function. In G. Harel & E. Dubinsky (eds.). *The concept of function: aspects of epistemology and pedagogy*, MAA Notes 25. Washington: MAA, 59–84.

[24] ——, 1995, The development of algebra: confronting historical and psychological perspectives. *Journal of Mathematical Behavior* 14, 15–39.

[25] Werner, H., 1948, *Comparative psychology of mental development*. New York: International Universities Press.

[26] Wertsch, J.V., 1991, *Voices of the mind. A sociocultural approach to mediate action*. Cambridge: Harvard University Press.

About the Author

Giorgio Tomaso Bagni graduated in Mathematics at the University of Padova (Italy). He was a university researcher in History, Epistemology and Didactics of Mathematics at the Department of Mathematics, University of Rome "La Sapienza," and afterwards at the Department of Mathematics and Computer Science, University of Udine.

Thanks to his wide and deep mathematical culture he was able to teach different mathematical subjects to mathematics and science students. He also taught several courses for prospective teachers. He delivered invited talks in national and international meetings. He strongly engaged in collaboration with other researchers, in Italy and abroad. He chaired working groups in national and international conferences. He attended HPM conferences.

He authored about 300 publications, mainly concerning the use of the history of mathematics in mathematics teaching, history of mathematics, and epistemology. He also studied problems related to the teaching and learning of algebra. He was killed tragically in a biking accident in 2009.

6

The Multiplicity of Viewpoints in Elementary Function Theory: Historical and Didactical Perspectives

Renaud Chorlay
IREM de l'Université Paris, France

6.1 Introduction

From 2002 to 2006, the "history of mathematics" group of Paris 7 IREM[1] contributed to a research project funded by the *Institut National de la Recherche Pédagogique* (INRP). We chose to work on the multiplicity of viewpoints on functions. In spite of the fact that some didactical and some historical research work was available on this topic, we felt the relevant connections still needed to be pointed to and explored. We also made use of fresh historical research work, namely R. Chorlay's doctoral dissertation on the emergence of the concepts of "local" and "global" in mathematics [7, 8].

We borrowed from didactical works the notion of viewpoint (as opposed to theoretical frame and semiotic register [12] and the distinction between four viewpoints in mathematical Analysis: point-wise, infinitesimal, local and global. Didactical work focused either on issues of cognitive flexibility (versatility)—the ability to change viewpoints, frames or semiotic registers in problem-solving—and its growing importance in higher education (Advanced Mathematical Thinking), or on curricular discontinuities: a point-wise/global dialectic when the concept of function is first encountered, then the infinitesimal and local viewpoints come into play with calculus, then an all-encompassing abstract theoretical frame in higher education. We focused our more historical investigation on a series of *hot spots* in which the four viewpoints interact and, eventually, were made explicit in the 19th century: proofs of the mean value theorem; proofs of the link between the sign of f' and the variations of f; differentiation of the three notions of maximum, local maximum and upper bound; emergence of the domain concept. From a curricular viewpoint, these topics cover material that students study from the beginning of high-school (*e.g.*, maximum) all the way to undergraduate college Analysis.

In this paper, we shall present some of the results of this investigation. After a general and a-historical introduction of the four viewpoints we will briefly present a short case-study. We will then introduce a more general explanatory

---

[1] IREM : Institut de Recherche sur l'Enseignement des Mathématiques. The e-mail address for the history of mathematics group is iremmath@yahoo.fr. The members of the Paris group involved in this specific project were : Philippe Brin, Renaud Chorlay and Anne Michel-Pajus. Professional webpage for Chorlay: www.rehseis.univ-paris-diderot.fr/spip.php?article35

framework, distinguishing between a "world of quantity" and a "world of sets." We will eventually outline didactical perspectives.

6.2 The Interplay Between Four Viewpoints

Let us give a few examples in order to illustrate the intricacy of the interplay between the four viewpoints in rather simple statements in function theory. Let us first consider "function f is positive at 2," the property it states is clearly point-wise. Things get more tricky with "function f is differentiable at 2": here is stated an infinitesimal property, which is also a local property of function f; in addition, from a purely syntactical viewpoint, the statement looks perfectly point wise ("when $x = 2$"). Now "function f has a local (or relative) maximum when $x = 2$" is of a local nature, yet stated in a point-wise fashion, and often (but not necessarily, since the property is perfectly well defined for non-differentiable functions) related to the infinitesimal behavior of the function. If one is to give an example of a global property, "function f is bounded on $[0, 1]$" can do: the fact that the property is global is reflected in the fact that a domain relative to which the property holds appears explicitly in the statement (though it could be left implicit if a domain had been set once and for all). This domain-component of the statement helps differentiate global statements from the three other kinds of statements, in which the domain is usually left implicit. It should be noted straight away that, in this context, the term "viewpoint" doesn't primarily denote a subjective property: in spite of the fact that the word "viewpoint" places the emphasis on cognitive processes, our four viewpoints refer to mathematical properties of mathematical statements and objects.

The basic element is the fact that, to say it roughly, functions are objects of a relational nature, they "do something somewhere." The syntactic structures which reflects this relational nature is:

$$\text{Function } f \text{ is } [property] \text{ on } [domain],$$

a syntactic structure that (French) students are required to use systematically from their first encounter with the function concept (at age 15). One could consider characterizing the four viewpoints by classifying domain-types: the property is point-wise if it can be defined on a single-point domain, it is local if its proper definition requires neighborhoods of a point. It is hard to go much further in this direction, for two essential reasons. The first one is specific to the infinitesimal level: it cannot be characterized through a specific domain type, unless one introduces tools of unreasonable sophistication such as tangent spaces (if only the first order is concerned) or "infinitesimal neighborhoods" of the kind differential geometry or modern algebraic geometry consider. The other reason is specific to the dialectic between local and global. A property is not global because it deals with a certain *type* of domain, it is global because the domain involved plays a specific *role* in the statement of the property. This was explicitly remarked by the first (at least to our knowledge) mathematician who tried to explain the meaning of "local" and "global" in a mathematical treatise, in 1901. In the article on the theory of functions of complex variables which he wrote for Felix Klein's *Encyclopaedia of Mathematical Sciences*, American mathematician Willian Fogg Osgood gave the following criterion (we paraphrase the German original): in function theory, the behavior of a function is local if it refers to the neighborhood of a point (or a subset), it is global if it refers to a domain which had been set right from the start, as opposed to domains whose extent is determined afterwards to fit the requirements of the problem [?, p. 12]. The wording may, at first, seem a little obscure, but it stresses the fact that what matters is the role that domains play in the syntactic structure of the statement. What Osgood had in mind was the difference between a local and a global inversion theorem: in the global theorem, the conclusion holds for the very domain which had been set at the start; in the local case, the conclusion holds for some domain which is usually a sub-domain of the domain we started with. A more elementary example can be found in the distinction between a global maximum (a maximum over the whole domain you started with) and a local maximum (for which some unspecified yet specifiable domain is referred to). This distinction between "given domains" and "specifiable domains," or, to put it differently, between "domains given right from the start" and "domains determined afterwards" is reflected in the order of quantifiers in the formal statement. Unfortunately, the illuminating Osgood distinction does not say it all, at least in the local case. The global nature of a property may be completely characterized by the role of the domain, but it is not so for local properties: a neighborhood is also a *type* of domain; the local/global dialectics may be captured on the syntactic level, but the local part has deeper, non syntactic, topological roots.

It seems there is no easy way out; yet our job is to design paths for students to gain some insight into this, whether at high-school or university level. Of course, one could argue that these intricacies are to be avoided completely. We think, and there is empirical evidence to support this claim, that complete avoidance of these problems has a high cost in the long term in terms of cognitive flexibility (in problem solving) and ability to adapt to evolving theoretical frameworks; we think some degree of awareness of this interplay is necessary for students to be able to do more than routine calculation in Analysis, and that teachers and those who train them cannot shun the topic altogether.

Before reflecting on a historical case-study, let us try to show how reasonable classroom work could help raise awareness. We will consider three pairs of statements; in each case, students could be asked if they are true or false; in the latter case, they could be required to exhibit (graphically, for instance) some counter-example.

Statement 1: *if f is continuous at 2 and positive at 2, then f is positive in the neighborhood of 2.*

Statement 2: *if f is continuous at 2 and positive at 2, then f is positive at* 2.00001.

A graphical counter-example to statement 2 should help point to the fact that no given size can be assigned to the domain over which the conclusion is valid, in spite of the fact that there is such a domain. According to the teacher's goal for the discussion, statement 1 could either be considered to be the right one (as opposed to false statement 2), or too vague to be *the* right one. The first statement could then be amended by introducing "*on a neighborhood of 2*" (in which sentence "neighborhood" is a definable domain type and not only a metaphor from daily life) or quantified statements such as "*one can find a positive number A such that f is positive in $[2-A, 2+A]$.*"

Statement 3: *if $f(2) = g(2)$ then $f'(2) = g'(2)$*

Statement 4: *if $f'(2) = g'(2)$ then $f(2) = g(2)$*

Both statements are false and are classical mistakes, but they point to two different kinds of errors. If one tries to characterize these errors, the first one comes from an improper understanding of the relative scopes of the hypothesis (which is point-wise) and the conclusion (which is infinitesimal, thus local); the second one can be ascribed to the intrinsic loss of information that derivation entails, a loss of information that students often fail to see in spite of the fact that they learn that a primitive (over an interval) of a continuous function is given only up to an additive constant. Whether the explicit wording of the reasons for which the statements are false should be a goal for classroom work is left to the teacher's choice. For instance, it could be considered irrelevant in high-school but highly relevant at the beginning of university Calculus or in a teacher-training session. In any case, high school students should be able to identify these statements as false, and draw counter-examples.

Statement 5: *a continuous function is bounded.*

Statement 6: *an everywhere locally increasing function is an increasing function.*

Both statements deal with the passage from local hypotheses (at every point though) to global conclusions. They should help point to the fact that, in order for global conclusions to hold, some knowledge about the nature of the domain is required. For instance, the first statement is valid if the domain is a closed and bounded (i.e. compact) interval, and student should be able to come up with counter-examples, even on bounded intervals (consider for example the tangent function tan on $(-\pi/2, \pi/2)$). As for statement 6, it is valid for intervals (for reasons of connectedness) but not for disconnected domains, as $-1/x$ on $\mathbf{R}\setminus\{0\}$ clearly shows. Both can help stress the importance of the domain part in the "function f is [*property*] on [*domain*]" structure, a part whose importance usually fails to strike students at any level of the educational system.

6.3 A Historical Case-Study of f' and Variations of f According to Cauchy

We conducted several case-studies, some of which we will leave aside here—for instance the study of the history of the implicit function theorem [5], or that on the evolution of the meaning of "maximum" [6, 7]. For lack of space, we will only present here some elements from one case-study, on the theorem linking the sign of the derivative and the variations of the primitive.

This theorem is one of the first ones that students encounter in Calculus and, in (French) high-schools, it tends to become the main application of the notion of derivative (approximation aspects play a lesser role in the current curriculum). Heuristic arguments for this theorem are given at high-school level, but no proof; the proof that has become standard since the late 19th century (which depends on the mean value theorem, whose proof, in turn, depends on a maximum argument which depends on topological properties of **R**) is usually given at the very beginning of College Calculus, and is one of the first occasions to experience the wealth of links between the four viewpoints. As far as History is concerned, we studied the historical emergence of this proof-scheme in the works of Bonnet, Serret, Dini and Jordan, to name a few. We also studied a former generation of proofs, namely those of Ampère and Lagrange.

Yet another proof-scheme can be found in Cauchy's 1823 treatise on differential calculus. Here is a translation from the French :

> "Problem. Assuming that the function $y = f(x)$ is continuous relative to x in the neighborhood of a specific value $x = x_0$, one asks whether the function increases or decreases from this value, as the variable itself is made to increase or decrease.
>
> Solution. Let Δx, Δy denote the infinitely small and simultaneous increments of variables x and y. The $\Delta y/\Delta x$ ratio has limit $dy/dx = y'$. It has to be inferred that, for very small numerical values of Δx and for a specific value x_0 of variable x, the ratio $\Delta y/\Delta x$ is positive if the corresponding value of y' is positive and finite. [...] This being settled, let us assume that the function $y = f(x)$ remains continuous between two given limits $x = x_0$ and $x = X$. If the variable x is made to increase by imperceptible degrees from the first limit to the second one, the function y shall increase every time its derivative, while being finite, has a positive value."[4][2]

The proof architecture is quite clear: the behavior of the function is studied in the neighborhood of a given point (and Cauchy's inference is correct), then the conclusion is extended to intervals of continuity. A few things are worth stressing.

First, what Cauchy proves in the first part of the proof is not that the function is locally increasing, as counter-example $x + 10x^2 \sin \frac{1}{x}$ shows when $x = 0$: the derivative at $x = 0$ is positive, yet the function is monotonic in no interval around 0. Of course Cauchy never claimed he had proved such a thing, but this way of reading it is naturally induced by our definition of increasing and decreasing functions[3].

By the way, and this is the second point that is worth stressing, in no part of Cauchy's treatises does one find a definition for increasing functions, in sharp contrast to current curricula which (in France) include a definition of what it is for a function to increase: a function, defined on a domain D, is an increasing function if, whenever a and b are any two elements of D such that $a \leq b$, then $f(a) \leq f(b)$; increasing functions are "order preserving" functions, a notion in which only the point-wise and the global viewpoints are involved, but which requires arbitrary pairs of points to be considered. The latter definition is mathematically elementary, but proves difficult for most students. More often than not, this *definition* is not included in the *concept image*[4] of increasing functions, a concept image which is sufficient to tell increasing functions from decreasing functions when a graph or a table is exhibited. In his proof, Cauchy does not rely on this (or any) definition, but rather on the following *cognitive root*[5]: a function is increasing for a specific

---

[2]*"Problème. La fonction $y = f(x)$ étant supposée continue par rapport à x dans le voisinage de la valeur particulière $x = x_0$, on demande si, à partir de cette valeur, la fonction croît ou diminue, tandis que l'on fait croître ou diminuer la variable elle-même.*

Solution. Soient Δx, Δy les accroissements infiniment petits et simultanés des variables x et y. Le rapport $\Delta y/\Delta x$ aura pour limite $dy/dx = y'$. On doit en conclure que, pour de très petites valeurs numériques de Δx et pour une valeur particulière x_0 de la variable x, le rapport $\Delta y/\Delta y$ sera positif si la valeur correspondante des y' est positive et finie. (...) Ces principes étant admis, concevons que la fonction $y = f(x)$ demeure continue entre deux limites données $x = x_0, x = X$. Si l'on fait croître la variable x par degrés insensibles depuis la première limite jusqu'à la seconde, la fonction y ira en croissant toute les fois que sa dérivée étant finie aura une valeur positive."

[3]It should be emphasized that, in this passage, Cauchy does *not* confuse the local with the infinitesimal. On the whole, his proof scheme is sound and a little rewriting is all it takes to turn it into a proof that meets our requirements of rigor. Cauchy's proof-scheme weaves the infinitesimal, local and global viewpoints in a rigorous way, yet in a way that is quite different from that which is found is the now standard proof (which starts by using the global existence theorem for an absolute maximum for a continuous function defined on a closed interval, then derives an infinitesimal consequence).

[4]We found the distinction between "concept definition" and "concept image," which Tall and Vinner formulated in 1981, to be a pretty adequate tool. "Concept definition" refers to the formal definition; Quoting Tall and Vinner, "We shall use the term concept image to describe the total cognitive structure that is associated with the concept, which includes all the mental pictures and associated properties and processes. It is built up over the years through experiences of all kind, changing as the individual meets new stimuli and matures." [21, p. 152].

[5]Again, we borrowed the notion of "cognitive root" from Tall's work in cognitive psychology. The term is self-explanatory: "A cognitive root is an anchoring concept which the learner finds easy to comprehend, yet forms a basis on which a theory may be built. (...) Local straightness proves to be a cognitive root for the calculus." [19]. We need not discuss here whether or not the word "concept" should have been used in this definition. The "local straightness" example is given for the sake of clarity, not because we endorse it.

value $x = x_0$ of the variable if sufficiently small increments Δy and Δx have the same signs; a cognitive root whose embodied nature is striking. The formal definition which one can draw from this cognitive root may be less easy to handle than the order-preservation one, but the link between the two notions may be worth eliciting.

The third point that should be stressed is the role of the domain. In the current curriculum, the statement of the theorem goes: "If a function f is defined and differentiable on an interval I, and if its derivative is positive on I, then f is an increasing function on I;" the nature of the domain is an essential part of the statement (for the theorem is false for non-connected domains) but usually overlooked by students. Things are quite different in Cauchy. First, the pre-definition cognitive root on which he relies implicitly entails connectedness of the domains over which the final conclusion can hold. Second, the fact that the conclusion holds for intervals is not entirely captured by the conjuring up of "limits" x_0 and X. One must recall that Cauchy's approach to continuity is not entirely ours. For instance, we teach students that the reciprocal function $1/x$ is defined and continuous on $\mathbf{R}^*$; saying that it is discontinuous at 0 is just a common mistake: this function has no property at all at 0, since it lies outside the domain of definition. In Cauchy's treatises the notion of domain of definition never appears, and the behavior of the reciprocal function at 0 is exactly what he calls "being discontinuous." In the proof above, the connectedness of the domain on which the conclusion holds is partly hidden (for us) behind the continuity hypotheses "let's assume function $y = f(x)$ remains continuous between two given limits $x = x_0$ and $x = X$."

This example shows quite clearly that our "case-studies" are of a theoretical nature: we studied historical texts with epistemological and didactical questions in mind; we didn't use these texts either in the classroom, or in teacher training[6]. However, as a second phase of this project, field-work on the teaching of function variation in high-school has started. An outline of this new phase was presented at CERME 6[7].

6.4 Ways of World-Making: The Case of Functions

As mentioned earlier, the historical case-study sketched in the previous paragraph is one of several. They helped us characterize different ways of "working with and speaking about" functions, different ways of (function) world-making. We shall sketch the outline of two such ways of world-making, the "world of quantity"(WOQ) and the "world of sets" (WOS). One of the goals is to get a *positive* grasp of the pre-Weierstrassian framework. By "positive" we do not mean a comparison of the respective values of both frameworks; we mean to provide a description of what it does and how it is structured which does not rely entirely on negative descriptive elements: lists of shortcomings, of implicit assumptions, forms of syntactic vagueness etc. In short, the "world of quantity" is not only something that fails to be the "world of sets." This part of our work partly rests on Moritz Epple's analysis of the end of the world of quantity [13], and parallels Gert Schubring's work [18][8].

A first feature of the world of quantity (or "world of magnitude") is what we call the *universally local approach*: everything is known about a function when its behavior is known *everywhere* (that is, for each and every specific value of the variable). In the WOS, the point-wise viewpoint is the fundamental one, the starting point: for a specific value of the variable, the function has a *value*. In the WOQ a function has a behavior and not only a value. A striking example is that in the 19th century functions do not *take on* the value 0, they *vanish*. The fundamental viewpoint is not the point-wise one, it is a mixture of local and infinitesimal viewpoints; what is described is not only a value but an event: functions do something/something happens to them. In the WOQ, mathematical statements and proofs are written in what we called the *narrative style*, as opposed to the *static*, explicitly quantified Weierstrassian style.

This rather abstract description in terms of "universally local approach" and "descriptive style" emerged from several case-studies. For instance, it helps describe Cauchy's notion of increasing function: an increasing function is a function which, at every point (universally local approach), displays some typical behavior; in turn, this typical behavior can be captured either formally (by the condition that $\Delta x.\Delta y > 0$) or narratively (as the independent variable

---

[6]Both things being what the IREM people usually do.

[7]Chorlay, R., 2009, *From Historical Analysis to Classroom Work: Function Variation and Long-Term Development of Functional Thinking*. CERME 6 (6th Conference of European Research in Mathematics Education), Lyon, Jan. 28th–Feb. 1st 2009. To appear in the proceedings. In this paper, we also discuss various obstacles that may account for students' difficulties in understanding of the notion of variation (algebra of inequalities, universal quantifiers etc.).

[8]Schubring's impressive study and ours overlap only partially, both in terms of topic (emphasis on numbers on the one hand, on the notion of domain on the other hand) and in terms of periods investigated (since we studied only 19th century material, with a focus on the second half of the century).

x increases, the dependent variable y also increases). A quick glance at another case-study can also help flesh this out. The problem of finding the maximum (or minimum) of a quantity that depends on another quantity is known to be one of the historical sources of calculus (Fermat, Leibniz, L'Hôpital). As far as the notion of maximum is concerned, we now distinguish between two different notions: one of global maximum (the greatest of all values taken on by a function on its domain, if there be such a value), and one of local maximum. The connections between the two are numerous but not straightforward. For instance, a local maximum is not usually a global maximum (consider $y = -x^3 + 3x$ at 1); conversely, the function defined on [0, 1] by $y = x$ clearly has a global maximum at 1, but it is a tricky question whether or not it also has a local maximum. In the current French curriculum, the notion of global maximum is introduced first, since it relies on the point-wise and global viewpoints only; the local notion is introduced later, and is presented as a derived notion (on open intervals, local maxima are global maxima of restrictions of the function to open subintervals). Not only is this second notion a derived one, but it is also seen as a more complex one, since it involves the local viewpoint and, more often than not, the infinitesimal viewpoint (with the vanishing of the derivative as a necessary condition). A study of 19th century textbooks on function theory shows that, until the second half of that century, the notion of local maximum was considered to be central, the global (or absolute) notion being either secondary and derived, or not mentioned at all. This focus on local extrema is in keeping with the universally local approach which we consider typical of the WOQ. Furthermore, the notion of local extremum was usually defined in narrative terms. A striking example can be found in Lacroix's *Traité élémentaire de calcul différentiel et intégral*, written at the turn of the 19th century (and reissued many times, well into the 19th century); the definition reads:

> "The value which occurs in the passing from increasing to decreasing, being greater than those immediately preceding and following it, is called a maximum" [15, p. 96][9]

It is quite likely that most students today would find this "definition" more understandable than the formal definition mentioned above[10].

An outline of the comparison between the WOQ and the WOS requires that some more abstract features be mentioned. First, this notion of behavior has a formal equivalent in the WOS, if one uses the (somewhat) sophisticated notion of function germ; the notion of germs-ring and of quotient rings can even enable us to state infinitesimal properties in a clean-cut, set-theoretic fashion[11]. Second, in the universally local approach, local and infinitesimal properties are not distinguished. This has deep historical reasons, which we studied in some detail in [7, chapter 6]. Third, in the universally local approach, some global aspects elude grasp since the reference to *all* points or *all* values is of a distributive nature: in this context, an important class of global properties, namely uniform properties (such as uniform continuity for a function or uniform convergence for a sequence of functions) seem to be out of reach. In this respect, the advent of the WOS in the work of Weierstrass has (at least) three related components: (1) considering the purely point-wise viewpoint to be the most fundamental one (since it is the only one fitted for "arbitrary" functions)[12] [10], (2) distinguishing between point-wise and uniform properties (for continuity of functions, convergence of function sequences, etc.) (3) distinguishing between "infinitely near points" (infinitesimal neighborhood[13]) and "sufficiently near points" (topological neighborhood). In this respect, the systematic use of explicit quantifiers appears as a tool, an essential tool indeed, but a tool nonetheless.

---

[9]"*La valeur qui a lieu dans le passage de l'accroissement au décroissement, étant plus grande que celles qui la précèdent et la suivent immédiatement, s'appelle un maximum.*" [15, p. 96].

[10]This case study on the history of the various notions of maximum is presented in much more detail in [6]; it relies on systematic comparisons of texts by Lacroix, Cauchy, Serret, Weierstrass, Cantor, Heine, du Bois-Reymond, Darboux and Osgood.

[11]Namely: if R is a local ring, with the (only) maximal ideal $\mathbf{m}$, R can be seen as the generalization of the ring of admissible functions defined in the neighborhood of a point P (on an algebraic variety for instance); in a sense, R captures the local situation at point P. The passage to the quotient field $R \longrightarrow R/\mathbf{m}$ captures the point-wise viewpoint, as it generalizes the evaluation of a function at a point. The passage to the quotient ring $R \longrightarrow R/\mathbf{m}^2$ captures the infinitesimal viewpoint (of the first order), and the module $(\mathbf{m}/\mathbf{m}^2)^*$ can be seen as the tangent space at P, or the first order infinitesimal neighborhood of P. In order to capture in this algebraic setting the infinitesimal neighborhoods of all orders, one can consider the $\mathbf{m}$-adic completion of the local ring R. This algebraic interpretation of the notions of *point-wise*, *local* and *infinitesimal* viewpoints can be found, for instance, in the preface of Bourbaki's *Algèbre commutative* [3].

[12]On the interaction between the growing role of "arbitrary" functions in elementary analysis, and the articulation of the point-wise viewpoint, important primary sources are Heine and Darboux; for a general presentation, see [10].

[13]For those who are not familiar with the notions of local and infinitesimal viewpoints in commutative algebra and algebraic geometry on which we relied above to flesh out the distinction between neighborhood and infinitesimal neighborhood, different instantiations can be found in differential geometry. On a smooth surface (embedded in 3-d Euclidean space), a (topological) neighborhood of a point P is the intersection of the surface with an open ball with center P; the tangent plane at P can be seen as the (first order) infinitesimal neighborhood of P.

In defense of the WOQ, it can be argued that it expresses global properties rather efficiently in terms of *systems of singularities*, an epistemic scheme that proved highly seminal in the hands of Riemann and Poincaré; it led to (algebraic) topology and to the qualitative theory of differential equations. In this respect, Riemann is the perfect example of the *versatile* thinker, switching between WOQ and proto-WOS in his works on complex or real function theory (respectively).

To contrast sharply the WOQ against the WOS, one can use the purely set-theoretic description of a function as a map between sets:

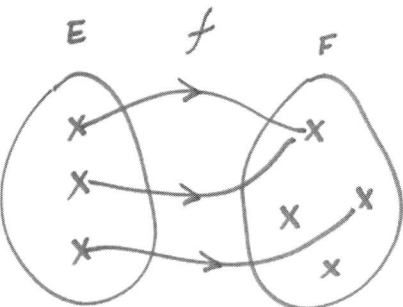

In this formal scheme, two types of "objects" come into play: sets on the one hand—the domain of definition E and the target set F, whose explicit statement is mandatory; another object, the map, which is of a second-order, relational nature: the map is the set of arrows (i.e. a part of the cartesian product $E \times F$). At this level of generality, the only relevant viewpoints are point-wise and global. This system of arrows may be arbitrary up to a certain point, and sets E and F play asymmetrical roles: every element of E has to have one (universality) and only one (single-valuedness) counterpart in F, while some elements of F may have no counterpart (i.e. the image-set may not coincide with the target set) or more than one counterpart in E. In the WOS, a function is just a map between number sets. Many of these elementary but structural features of the WOS are in sharp contrast to the WOQ: in the WOQ no statement of sets is necessary and the roles of what could be considered as definition and target sets are completely symmetrical, in particular since many-valued functions are the rule. If today's functions are just special maps, the multi-valued function of the 19th century can be seen, in retrospect, as general relations whose maximal domain is to be determined.

The notion of domain is now considered to be a defining element of the notion of function (i.e. of function as map), which is in striking contrast to the WOQ take on function. For instance, no notion of domain is necessary to define local maxima; on the contrary, no notion of global maximum can be defined if no domain is set from the start.

6.5 From Historical Research to Didactic Engineering[14] and Research in Didactics

We mentioned in the first part of this paper various statements (1–6) that can provide starting points for the designing of classroom sessions, whether in secondary or higher education: the exhibition of counter-examples may be adequate at high-school level; in higher education, discussion of those statements could trigger off the search for precise definitions and proof-ideas. They could also help to show students how to manage proof-tasks when no formulas are given for the functions that are being studied: this new type of task is quite specific to higher education (at least in France) and proves quite unsettling for most students. From a research viewpoint, we touch on questions that have been studied from a purely didactical perspective, by Aline Robert for instance [2] (see also [17]). Adjectives such as "local" or "global" belong to what A. Robert calls the *meta*-level: they are instances of a form of mathematical knowledge that says something *about* mathematics; they help sort mathematical statements and definitions in high-level categories; they enable you to "find your way around" in an ever-growing, ever more complex mathematical environment. This

---

[14] In the French school of didactics of mathematics, "didactic engineering" (*ingéniérie didactique*) denotes the design of teaching sequences; or, more precisely, the designing process in which the theoretical (whether epistemological or didactical) hypotheses are stated as explicitly as possible, and the results of field-work analyzed for future improvement of the sequence.

knowledge about mathematics helps you spot potential difficulties (e.g. "in this problem we are to go from local to global, I know this is usually quite tricky"), identify the right theoretical tools (e.g. to choose between Taylor-Lagrange and Taylor-Maclaurin formulas, that is between a global and an infinitesimal formula) or conjure up problems of a similar kind whose solution you remember. Whether or not this meta-level knowledge is to be made explicit for the students, or even taught as such, is a debated issue.

We mentioned in Section 6.3 of the essay how we found some classical notions from the psychology of mathematical learning—such as that of cognitive root, or the difference between concept image and concept definition—helpful to link historical work and teaching issues[15]. As far as the notion of functional variation is concerned we decided to venture in the world of didactic engineering, and field-work is under way.

The distinction between two formal models, WOQ and WOS, helped us point to a general problem in the current introduction of the function concept. In France, students encounter from the very beginning (age 15) the abstract notion of "arbitrary function on a given domain of definition." This notion is a mixture of elements which, as far as history is concerned, emerged in rather distinct contexts. On the one hand, the notion of arbitrary function emerged in the debate on the foundations of real Analysis; no *natural* domain can be ascribed to an arbitrary function; domain restrictions and extensions are completely trivial in this context. On the other hand, questions about domains emerged in the case of highly non-arbitrary functions; for instance, to some convergent power-series (usually of a complex variable), one can assign two natural domains: its region of convergence, and the domain of holomorphy of the unique holomorphic functions that it represents (domains of meromorphy can also be considered). More generally, questions of domain extension become relevant only when functions with specific properties are concerned. In fact, empirical work would certainly show that this "arbitrary function on a given domain of definition" is usually not included in the function concept-image in high-school, partly for ecological reasons but maybe also for the reason we point to here.

Finally, the historical perspective may help to guide didactical choices, in particular by distinguishing more clearly between (mathematical) necessity and convention. "Elementary" and seemingly "natural"—mathematically natural, not psychologically natural—notions such as those of function (as arbitrary, one valued correspondence between sets), domain of definition, maximum etc. are partly *conventional*; which by no means entails that they are arbitrary! They appear to us as "elementary" and "natural" because mathematicians left us a body of mathematics in which massive conceptual restructuring had taken place, especially in the 19th century. It is not mathematically necessary to assume that a function is single-valued; for rigorous mathematics to be written, it is not necessary that domains of definitions be stated from the outset: choosing between either conventions will change the priority between notions, the syntactic rules for writing rigorous mathematics, alter the meaning of a theorem. By their very nature, such conventional parts of our function-world cannot be expected to emerge from classroom work as the best solution to a well-designed problem; there lies an intrinsic limit for didactic engineering. Conventions are neither true or false; the shift of conventions is a slow and high-level process in which mathematicians react to the global state of mathematics and not to a specific problem.

6.6 Conclusion

By way of conclusion, we would like to stress the interest of a macro-historical approach that aims at identifying ideal and coherent "worlds" such as the world of quantity and the world of sets. This is neither historical work in the strict sense—in terms of the standards of the community of historians of mathematics—nor directly didactical work, but we think it helps *build bridges* without resorting to dubious "ontogeny recapitulates phylogeny" arguments. In particular, the classical Bachelard-Piaget notion of *obstacle* can be interpreted in a less teleological way ("as mathematics developed, mathematicians overcame obstacles which students are, in turn, faced with..."). The issue of the multiplicity of coherent worlds—none of them being *the* right one (of which all others are superseded archaic forms)—helps us to see obstacles as translation problems that are bound to come up when changing worlds. Such translations are difficult for at least two different reasons. First, in such a translation task both semiotic *and* conceptual aspects are deeply interconnected; second, because of the large scale coherence of such "worlds," there is no such thing as a strictly autonomous, purely local move. This type of history-based investigation should throw light on both versatility issues and long term curricular discontinuity problems.

---

[15] We do not claim, however, that we used this notions to study the mathematical practice of top-ranking mathematicians (*e.g.*, Cauchy). For an analysis of mathematical practice, we tend to prefer other conceptual tools; see, for instance, [10].

Bibliography

[1] Artigue, M., 1992, "Functions from an Algebraic and Graphic Point of View: Cognitive Difficulties and Teaching Practices", in *The Concept of Function: Aspects of Epistemology and Pedagogy*, E. Dubinsky & G. Harel (eds.), Washington DC: The Mathematical Association of America, pp. 109–132.

[2] Baron, M. and A. Roberts (eds.), 1993, *Métaconnaissances en IA, en EIAO et en Didactique des Mathématiques*, Cahier DIDIREM spécial. Paris : IREM Paris 7

[3] Bourbaki, N., 1989, *Elements of Mathematics, Commutative Algebra (Chapters 1–7)*. New-York: Springer.

[4] Cauchy, A.-L., 1882–1974, *Œuvres Complètes* tome 16. Paris : Gauthiers-Villars.

[5] Chorlay, R., 2003, "Fonctions Implicites: de la Notion au Théorème", *Mnémosyne* n° 18, p.15–58.

[6] ——, 2007(a), "La multiplicité des points de vue en Analyse élémentaire comme construit historique", in E. Barbin & D. Bénard (eds.) *Histoire et enseignement des mathématiques : rigueurs, erreurs, raisonnements*. Lyon : INRP, p.203–227.

[7] ——, 2007(b), *L'émergence du couple local / global dans les théories géométriques, de Bernhard Riemann à la théorie des faisceaux* (1851–1953), Thèse de l'Université Paris 7.

[8] ——, 2010, "From Problems to Structures: The Cousin Problems and the Emergence of the Sheaf Concept", *Archive for History of Exact Sciences*, 64, 1–73.

[9] ——, 2011, ""Local—Global": The First Twenty Years", *Archive for History of Exact Sciences*, 65 (1), 1–66.

[10] ——, forthcoming(b), "Questions of Generality as Probes into Nineteenth Century Analysis", in *Handbook on Generality in Mathematics and the Sciences*, by Chemla, K., Chorlay, R., Rabouis, D. (eds.), forthcoming. Available online at www.rehseis.univ-paris-diderot.fr/spip.php?article35.

[11] Darboux, G., 1875. "Mémoire sur les fonctions discontinues", *Annales Scientifiques de l'E.N.S. (2ème série)* 5, 57–112.

[12] Didirem, 2002, *Actes de la Journée en Hommage à Régine Douady* (14 juin 2001). Paris : IREM Paris 7.

[13] Epple, M., 2003, "The End of the Science of Quantity: Foundations of Analysis, 1860–1910", in H-N. Jahnke (ed.) *A History of Analysis*. New-York : AMS, 2003, pp.291–324.

[14] Jahnke, H.-N. (ed.), 2003, *A History of Analysis*. New-York : AMS.

[15] Lacroix, S.-F., 1867, *Traité Elémentaire de Calcul Différentiel et Intégral* (tome I). Paris: Gauthiers-Villars.

[16] Osgood, W. F., 1901, "Analysis der komplexen Grssen", in *Encyclopädie der Mathematischen Wissenschaften* II.2, Leipzig: Teubner, 1901-1921, pp.1–114.

[17] Praslon, F., 1994, *Analyse de l'aspect Meta dans un enseignement de Deug A concernant le concept de dérivée. Etude des effets sur l'apprentissage* (Mémoire de D.E.A.). Paris : Université Paris 7.

[18] Schubring, G., 2005, *Conflicts between Generalization, Rigor and Intuition. Number Concepts Underlying the Development of Analysis in 17–19 Century France and Germany*. New-York: Springer, 2005.

[19] Tall, D., 1989, "Concept Images, Generic Organizers, Computers & Curriculum Change", *For the Learning of Mathematics*, 1989, 9(3), 37–42.

[20] ——, (ed.), 1991, *Advanced Mathematical Thinking*. Boston : Kluwer, 1991.

[21] Tall, D. and S. Vinner, 1981, "Concept Image and Concept Definition in mathematics, with particular Reference to Limits and Continuity", *Educational Studies in Mathematics*, 1981, 12(2), 151–169.

[22] Volkert, Klaus, 1987, "Die Geschichte der pathologischen Funktionen", *Archive for History of Exact Sciences* 37(1), 193–232.

About the Author

Renaud Chorlay works as a historian of mathematics (modern period), at the SPHERE research team (CNRS-Université Denis Diderot). His latest papers deal with the design of the sheaf structure in the context of problem-solving (1883–1953) and with the emergence of the meta-concepts of "local" and "global," at the turn of the twentieth century (1898–1918) (*Arch. Hist. Exact Sci.* 64(1) and 65(1)). After ten years as a high-school teacher, he now works at a teacher-training university in Paris (IUFM-Paris IV). As a member of the IREM and HPM communities, he works on the various connections between history of mathematics on the one hand, and teaching and teacher-training of the other hand. He published an example of the use of an original source in the classroom, based on a little known Leibnizian text on probability theory. Through epistemological reflections on the history of analysis in the nineteenth-century, he endeavors to tackle issues related to curriculum design and Advanced Mathematical Thinking.

7

From History to Research in Mathematics Education: Socio-Epistemological Elements for Trigonometric Functions

Gabriela Buendia Abalos and Gisela Montiel Espinosa
Centro de Investigación en Ciencia Avanzada y Tecnología Aplicada del IPN, Mexico

7.1 Introduction

We propose to investigate the construction of the trigonometric function from a perspective that challenges what the educational system 'teaches' and, consequently, how the student learns; this implies questioning not only how we teach, but what we teach. By means of the theoretical perspective of *Socioepistemology* [5, 6, 2] we intend to recognize the *uses* and *significations* associated with the trigonometric functions in a particular historical setting.

This theoretical perspective problematizes the knowledge confronting the mathematics of the educational system with the *uses* of knowledge in different settings, like historical, professional, or even school settings when experiencing non-traditional pedagogical proposals. The purpose of this paper is to recognize those *significations* that belong to knowledge but usually become diluted, altered or lost when setting up a school discourse.

There is a change from the didactical approaches concerned with the design of accessible presentations for school mathematics content or strongly centered on cognitive aspects to recognizing socio-cultural settings as part of the explanation of didactical phenomena. In this last paradigm, the socio-epistemological research framework focuses its attention on the normative role of *social practices* in the construction of mathematical knowledge: that which makes human individuals and groups do what they do [10].

This allows us to extend the didactical or cognitive explanations, dominant in the literature, about didactical phenomena related to the trigonometric functions. In this respect, before proposing educational innovations, we seek to recognize in the historical significations and uses, elements that help to explain didactical phenomena and then, based on epistemologies of practices, re-design the school mathematical discourse.

7.2 Socio-Epistemological Theoretical Approach

The socio-epistemological approach to research in mathematics education has as its fundamental task examining the situated knowledge that addresses particular socio-cultural circumstances and settings, characterizing knowledge as the outcome of epistemological and social factors [4]. In this context, we focus on the social subject, one who acts and thinks in interaction [20]; this social subject belongs to specific settings and while facing particular problems, builds mathematical knowledge using and giving significations to that knowledge. That is, what guides the social thinking towards scientific thinking is not the mathematical knowledge, but from social thought and mathematical activity there is a path of development of scientific thinking.

The theoretical approach assumes social practices as the actions of a social group that has its own meanings and intentions, is located in a historical context or a current one, acts in accordance with prevailing ideologies, and uses mathematics as a tool for building knowledge [9]. It is not intended that students get a utilitarian sense of the mathematical concepts to use them inside or outside of school. Our intention is that students can achieve a school mathematics knowledge which is truly functional, that is a knowledge which is integrated into their own lives, transforming and giving them new meanings [7].

Under this view, a change in language is needed, one that emphasizes the change from the object to the practice. From now on, we are going to consider the construction of trigonometric functionality through the significations associated with the trigonometric functions, based on a certain epistemology of practice that we will try to develop.

7.3 Didactic Difficulties with the Trigonometric Functions

In order to problematize how the trigonometric functions have been taught and from there, to discuss the significations obtained from a particular historical setting and to propose the construction of trigonometric functionality, we present the main difficulties research has reported in dealing with the trigonometric functions. We present also some of the didactic-cognitive explanations from recent research.

Grabovskij and Kotel'Nikov [17] recognized that the teaching of the trigonometric functions (with a numeric argument) in high-school faces the difficulty of using geometric ideas of the student to understand them, rather than analytical methods. The authors, in response, propose the modeling of physical phenomena as a means for the recognition of the conceptual origin of the trigonometric quantity and of the properties of trigonometric functions.

This conflict between the geometric and analytic aspects in trigonometry is expressed more clearly in the research reported by De Kee, Mura and Dionne ([13]), who studied the comprehension of sine and cosine in the context of the right triangle and trigonometric circle. The authors found that students do not distinguish between the sine (or cosine) as a trigonometric ratio and the sine (cosine) as a trigonometric function. In the context of the right triangle, they consider the sine and the cosine as a process of dividing the lengths of two sides of the triangle, while in the unit circle, sine and cosine refer to the point whose coordinates are in the trigonometric circle; trigonometric functions were remembered as wavy curves. Students who participated in the research had worked in the context of the trigonometric circle around the time of the study; however, they showed more familiarity with the triangle with which they had worked long before.

This research allowed authors to question the role of the trigonometric circle as a didactic strategy:

> ...if we think that the circular function is only a teaching tool designed to make the construction of trigonometric functions more visual and more "concrete," this finding is puzzling. We must recognize that this approach materializes the definition of trigonometric functions by paying the price of making it more complicated. (p. 25)

Maldonado [23] designed a questionnaire based on the institutional content of three Mexican educational systems to explore students' conceptions about the trigonometric functions. Her design draws three school stages: the approach of the trigonometric ratio, the relationship between angles measured in degrees and radians, and the understanding of the properties. In this experience, students did not evidence understanding of the sine and cosine as mathematical functions. That is, students continue to use the trigonometric ratio and its units of measurements—degrees—as the instrument for the resolution of that which is related to the sine, cosine, tangent and their inverses. For example, in Figure 7.1 we show the answer of a student to a task that consisted of graphing $y = \sin x$ and $y = \cos x$ and telling the values of x that makes $\sin x = \cos x$. Notice that the x-axis is graduated in degrees.

Maldonado identified the relation (equivalence) of the radian and the real number as the central point to construct and understand the functional notion of the trigonometric ratios and concludes that when this is not explicit in the school mathematical discourse, the school treatment of the trigonometric aspect as ratio or as function will be completely indistinct to the student.

7.3. Didactic Difficulties with the Trigonometric Functions

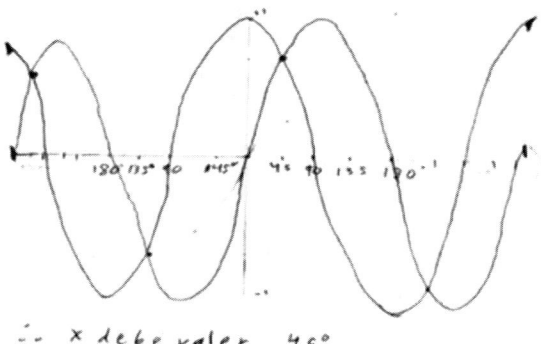

Figure 7.1. The answer is "*x* must be 45°."

Finally, we found in Weber [27, 26] a proposal for teaching trigonometric functions as a *procept*[1]. He reports that college students showed ability to approximate the values of basic trigonometric expressions, to identify properties of trigonometric functions, and to justify why they have such properties. In doing this, the method of the unit circle is favored: the intersection of the unit circle and the line segment with an angle of inclination (measured in degrees) is located, and the corresponding value on the *x*-axis defines the cosine while the corresponding value on the *y*-axis defines the sine (Figure 7.2).

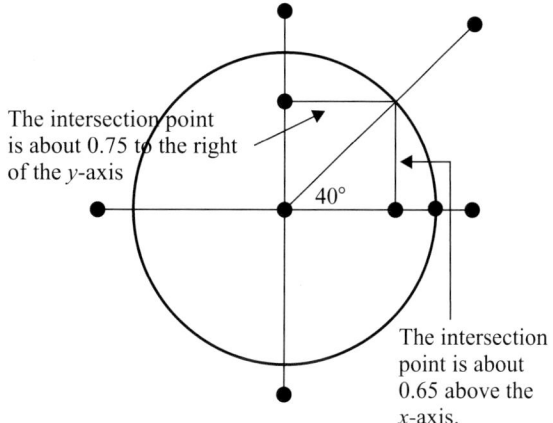

Figure 7.2. The unit circle as a favored method (Source: [26])

The instructional design of this investigation did not lead to graphing the trigonometric functions with the classical form of wavy curves, so did not allow us to make inferences about the transition from the geometric to the analytical settings.

Weber [27] also reported that students who did not work through the design showed a strong tendency to 'need' a triangle labeled with values in their sides and angles to calculate the sines and cosines requested, and therefore found it difficult to argue about the properties of functions. From his theoretical framework, the lack of a building process of the circle (which he calls a geometric process) and not handling the function as procept are what makes the performance of these students so difficult.

This brief review of previous research shows mainly the ambiguous use of the measurement unit for angles and an over reliance on the unit circle as *the* didactic strategy that will get the student into the meaningful use of trigonometric functions. We are now in position to discuss the significations that a particular historical setting can provide to our proposal for trigonometric functionality.

---

[1]*Procept* is an amalgam between a process that produces a mathematical object and a symbol used to represent both the process and object ([19], in [27])

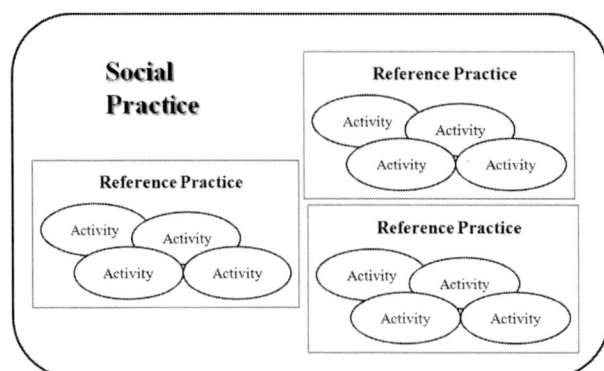

Figure 7.3. Model of practices.

7.4 Socio-Epistemological Review: Towards Trigonometric Functionality

To establish a basis of signification for trigonometric functionality, we conducted a socio-epistemological research project to analyze the ideas identified in history that explain the construction of a specific mathematical topic, from its gestation until it becomes erudite mathematical knowledge. But this analysis is done considering not only those mathematical results obtained by human beings, but also those circumstances surrounding what they did [2]. That is, a socio-epistemological review takes into account the social-cultural-historical setting where the humans moved while dealing (proposing or developing) with mathematical issues that eventually we—in a present time—will know as institutionalized mathematical knowledge. We propose to extract significations from this review that will explain what we mean by trigonometric functionality.

What results from this socio-epistemological review from history is an epistemology of practices; this is an explanation of the social nature of mathematical knowledge in terms of human practices. Montiel [25] proposes a practices model (Figure 7.3) to explain that these practices have different epistemological status in explaining the construction of the mathematical knowledge:

> ...the *activity* is what we observed in both individuals and human groups; the articulated set of intentional activities which follows a specific purpose has been denominated as *reference practice*. Finally, the *social practice* is assumed as the normative practice of the reference one and therefore, of the associated group of activities too (p. 126).

Altogether, activities, reference practices, and social practices are what explain *why humans did what they did* in the historical setting, recognizing the relation of these practices with the generation of meaningful mathematical knowledge.

In the next sections, we present some results from the socio-epistemological review of the trigonometric functions. Since the introduction of the trigonometric functions into Calculus was done by Euler (18th century), this historic fact will be taken as a breakpoint to study the knowledge construction and the production context for trigonometric functionality.

7.4.1 The transcendental trigonometric quantity

We can find contributions to trigonometry from astronomical studies and from construction techniques in Greek geometry, which deals with immobile objects, as a kind of universe of ideas ([11]). In this sense a conception of motion in the study of celestial phenomena was nonexistent.

For the construction of models that could represent celestial reality, the orbits were thought of as resembling geometric figures, and the relationships among the celestial bodies were thought of as arcs and chords of circles and as triangles. The scale and proportion were kept to fit the empirical data. As a result, the trigonometric quantities arose to estimate the length of the sides of a plane or spherical triangle inserted in a circle or sphere, particularly that which turns out to be the subtended chord in relation to the central angle.

We have named this production context, where the transcendental trigonometric quantity emerged, a *static-proportional context*; it does not yet have a functional character. To obtain the latter, a new concept was necessary, one where a mathematizable conception of motion had to be developed. Thus, the measurement of the semi-chord in relation to the central angle, as the quantity that arises from the circle, must be seen as a curve or trajectory on the plane.

7.4.2 Mathematization of physics

Mathematization, such as the manipulation done to explore the structure of reality, was identified by De Guzmán [12] as an initial perception of certain similarities in objects that can be perceived by the senses, perception that leads us to abstract what they have in common and, in a rational, symbolic elaboration, to address the underlying structure of such perceptions more clearly. Under this view, it becomes natural to understand *mathematics as a language of science or as its universal method of representation*. This is applicable to diverse scientific disciplines that have an instrumental relationship with mathematics in which the latter subject intervenes as a purely technical instrument to manipulate the quantitative aspects of the problem. Nevertheless, a sufficiently defined separation exists between the conceptual arsenal inherent in these scientific domains and the mathematical techniques utilized [22]. Of a very different nature is the relationship between mathematics and physics that Lévy-Leblond calls a constitutive relationship [22, p. 73].

For this author, physics interiorizes mathematics: physics not only uses mathematics to express the physical idea, but is part of its thinking. As a result, physical concepts are inextricably associated with one or several mathematical concepts. This relationship is much more evident in the historic moment in which some physical and mathematical concepts were constructed. From this constitutive view we speak of the mathematization of physics to situate the construction of the functional nature of the transcendental trigonometric quantity.

Before Euler, the geometric aspects of the sine and cosine were the object of study and not their analytical properties. Katz [21] points out that the trigonometric functions could be avoided because there was no *reasonable use* for them. Perhaps it was the *new uses* of trigonometric quantities that stripped them of their geometric nature; from being lines in a circle, they became quantities that describe certain phenomena, in particular periodic movements. We think that the reasonable use mentioned by Katz is of a completely socio-cultural nature since it arises out of and develops within specific tasks related to the *mathematization of oscillatory movement*.

The uses of trigonometric functions emerge via differential equations when Euler addresses the problem of harmonic oscillatory movement. Elements can be found in this author's work that characterize these uses and show the construction and development of knowledge within the organization of human groups.

7.4.3 The uses of trigonometric function in Euler

In 1739, Euler presented the work *De novo genere oscillationum* about movements with a common property: oscillation. Among them he identifies the vibrating chord, the sound waves produced by a bell, waves of water and ocean flows (or currents). Nowadays we call this harmonic oscillatory movement. In that study some important changes came to light such as the change of focus from time to movement of objects, from the periodicity of time to the periodicity of movement of objects. Euler's work, like that of his contemporaries, was influenced by the dominant paradigm of the eighteenth century, the mathematization of movement.

Concurrently with this treatment, and on proposing a solution to the vibrating chord problem[2], Euler stated that the property of periodicity that the function that solves the problem supposedly possessed was restrictive, because that kind of solution did not take into account algebraic functions and some transcendental curves. His proposal admitted a parabola $f(x) = hx(a - x)$ as the initial form of the chord since it can be *made* periodic by reflecting the arcs in relation to the straight lines $x = \pm na$ (Figure 7.4).

Grattan-Guinness ([18]) believes that Euler was suggesting a broadening of the theory of functions, by returning to geometry. This broadening would change the definition of discontinuous function. When Euler introduced his theory of "discontinuous" functions defined by means of "continuous" segments, a revolutionary step needed to be taken in order to assign an interval of definition corresponding to each completely independent segment of algebraic form. This

---

[2] During the 18th century, the analysis of the vibrating chord was one of the relevant investigations carried out. Published works on the topic defined the terms for reflection on the definitions given to several of the concepts addressed until then, among them that of function, its properties and representations. The problem of the chord can be described in the following way: suppose that a flexible chord is pulled taught and its ends are fixed at 0 and a in the abscissa. Then the chord is slackened until it takes the form of a curve $y = f(x)$ and is then let go. The question is: What is the kind of movement described by the curve? One of the first proposed solutions is that the function $f(x)$ has to be odd and periodic.

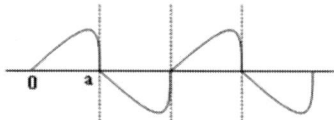

Figure 7.4. Euler's proposal for periodic functions ([1])

is what would have been involved in a return to geometry as well as in assigning a "geometric periodicity" to $f(x)$. However, at the time, the idea was not followed up.

There is controversy in how periodicity was being interpreted. In the eighteenth century, attributes such as periodicity were intrinsically related to an analytical expression, without any reference to a domain that was independent of it; the analytical expression was the function itself. Therefore, we can speak of an "algebraic periodicity" such as that of the trigonometric functions valid on the entire x-axis, or a "geometric periodicity" as in current piecewise-defined functions, where periodic functions can be created from algebraic or transcendental functions [2].

So, in Euler's proposal, periodicity is associated with the combinations of certain functions. Although in today's educational system, periodicity is used to qualify a function (in particular, the trigonometric functions), Euler used periodicity to qualify a *certain behavior* and not a certain function. This is a more meaningful use for the periodicity property.

Some years later, Euler published *Introductio in analysin infinitorum* (1748), a scientific work that answered the need for a systematic analysis of the current knowledge of the mathematical community about functions. It is after this work that analysis no longer dealt only with the properties of curves, but also with the properties of functions [14]. At this time of formalization, properties like periodicity were associated with trigonometric functions.

Thus, to propose a meaningful basis for trigonometric functionality which recognizes the social nature of mathematics, we have to distinguish the historical moments when a concept is used from those moments when a systematic and ordered presentation of the tools, notions and concepts turns out and is recognized to be necessary; *Introductio in analysin infinitorum* would appear to have such a dual nature. In this work, we find a simultaneous use of the measurement of an angle in degrees and radians, and how Euler used the periodic behavior while constructing new—systematic and ordered—analytic function knowledge. In the next Section, we present some extracts from Euler's work to point out those moments, from which we can extract meanings for trigonometric functionality.

7.5 A Socio-Epistemological Approach to the *Introductio in Analysin Infinitorium*

7.5.1 A new transcendental quantity and its properties

Euler begins chapter VIII of vol. 1 of the *Introductio*, entitled *Des quantitatibus trascendentibus ex Circulo ortis*, by considering the sines and cosines of circular arcs, a genre of quantities with the same transcendental status as logarithmic and exponential quantities.

In this chapter, Euler uses arcs and their lengths with the measurement of the angle in degrees simultaneously. As we can see in Figures 7.5 and 7.6, in this chapter the measurement of the angle in degrees coexists with the measurement of the length of the arc in terms of π, depending on what the author is trying to develop. This is a kind of articulation of the geometric context in which the said quantities originated and the operations with sines, cosines and tangents, articulation in which effectiveness is gained.

The properties of these new trigonometric quantities which will lead to trigonometric functionality are extracted from the circle where they originated. For example, in Figure 7.7 Euler points out explicitly the range of the sine and cosine functions as a result of considering their characteristic values in the circle, concluding that all the values are contained between the limits $+1$ and -1.

In general, Euler uses this chapter to revisit theorems, formulas, equations and calculations from many other previous works. In doing so, he uses the periodic behavior. For example, he generalizes the properties shown in the table of Figure 7.8, to obtain the table of Figure 7.9.

7.5. A Socio-Epistemological Approach to the *Introductio in Analysin Infinitorium*

Figure 7.5. (Page 99)

Figure 7.6. (Page 103)

Figure 7.7. The limits for sine and cosine. (Page 93)

Figure 7.8. (Page 94)

Figure 7.9. (Page 95)

He took all the multiples expressed in the form $\frac{(4n+k)}{2}\pi + x$ for $k = 1, 2, 3, 4$. The role of the fraction $k/2$ is to change the starting point of the circle, while the expression $4n/2 = 2n$ represents the period of repetition of the circle. So, periodicity was meaningfully used as a property that qualified the behavior of the new function.

7.5.2 Transcendental expressions become algebraic expressions

In chapter IX of this volume, *De investigatione Factorum trinomialium*, Euler presents for the first time the development of the sine and cosine as an infinite product (Figure 7.10).

158. Si x fiat quantitas imaginaria, formulæ hæ exponentiales in Sinum & Cosinum cujuſpiam Arcus realis abeunt. Sit enim $x = z\sqrt{-1}$; erit $\frac{e^{z\sqrt{-1}} - e^{-z\sqrt{-1}}}{2\sqrt{-1}} =$ *ſin.* $z = z - \frac{z^3}{1.2.3} + \frac{z^5}{1.2.3.4.5} - \frac{z^7}{1.2.3\ldots7} + \&c.$, quæ adeo expreſſio hos habet Factores numero infinitos $z(1 - \frac{zz}{\pi\pi})(1 - \frac{zz}{4\pi\pi})(1 - \frac{zz}{9\pi\pi})(1 - \frac{zz}{16\pi\pi})(1 - \frac{zz}{25\pi\pi})$ &c., ſeu erit *ſin.* $z = z(1 - \frac{z}{\pi})(1 + \frac{z}{\pi})(1 - \frac{z}{2\pi})(1 + \frac{z}{2\pi})(1 - \frac{z}{3\pi})(1 + \frac{z}{3\pi})$ &c.. Quoties ergo Arcus z ita eſt comparatus, ut quiſpiam Factor evaneſcat, quod fit ſi $z = 0$, $z = \pm\pi$; $z = \pm 2\pi$, & generaliter ſi $z = \pm k\pi$, denotante k numerum quemcunque integrum, ſimul Sinus

Hence, each time the arc z is such that a factor disappears, which happens if $z = 0$, $z = \pm\pi$, $z = \pm 2\pi$ and in general, $z = k\pi$, k denoting any integer, then the sine of that arc must be = 0, which is evident.

Figure 7.10. (Page 120)

The product is obtained as the result of well-structured arguments, but Euler's main emphasis is that the zeros of the sine function are $\pm k\pi$ with k as any natural number. That is, he considered the power series as "infinite polynomials" that obey the same rules of algebra as ordinary finite polynomials: he turned transcendental expressions into algebraic expressions.

7.5.3 The behavior of the trigonometric curve

Finally, on dealing with transcendental curves, periodicity is identified as a property of the new function. In chapter XXII (vol. II) of the *Introductio*, entitled *On Trascendental Curves*, Euler constructs a curve $\frac{y}{a} = \arcsin\frac{x}{c}$ pointing

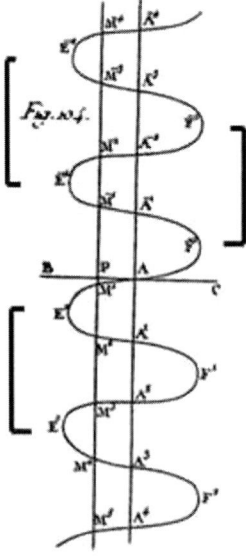

Figure 7.11.

out the infinite number of arcs of a circle whose sine is $\frac{x}{c}$, and where the ordinate y is a multivalued function. The y-axis and any other vertical parallel line intersect the curve on an infinite number of points (Figure 7.11).

He characterizes the arcsine curve by pointing out those sections that are equal. He mentions that the intervals $E^1E^2, E^2E^3, E^1E^{-1}, E^{-1}E^{-2}$; and $F^1F^2, F^1F^{-1}, F^{-1}F^{-2}$ are all equal to $2a\pi$.

It is an argument that, making use of the curve behavior, characterizes the periodic property of the sine function.

7.6 A Socio-Epistemology Based on Uses and Significations Developed in a Historic Setting

By analyzing a certain issue in mathematics teaching and proposing a socio-epistemology of the mathematical knowledge involved, we can be in a position to understand in depth that which is being taught and to discuss how it is taught and the consequences of these teaching choices. The current school mathematics discourse for trigonometric knowledge is structured around the sequence *trigonometry*⟶ *trigonometric circle*⟶ *trigonometric function*, usually thought as necessary and transparent to the student. We are now able to re-think what is taught and what is learned taking into account the socio-epistemological elements extracted from the historical review we have done.

7.6.1 The unit measure

While current conflicts are identified in school mathematics due to the ambiguous handling of the unit of measurement for the variable of the trigonometric function, in contrast, we see that Euler uses both units (degrees and radians) without difficulty owing to the kinds of tools that were available to calculate the trigonometric quantities and the kinds of tasks in which he is involved.

The currently available instruments to the student for obtaining the value of any trigonometric quantity are the trigonometric tables (used less and less in school), the calculator or, for relatively simple cases, the right triangle. Among them, the only instrument that can be re-built in the classroom is the right triangle, and so it is in this static context associated with geometry that the quantity arises for the student. Paradoxically, although trigonometric ratios are used to calculate the values, the proportional nature underlying these objects is not recognized.

Euler does not argue explicitly for the use of one or another unit of measurement. Nevertheless, on the one hand he has previously used trigonometric quantities for the modeling of oscillatory movements. On the other hand, the way he deals with operations[3] (such as additions and subtractions in Figure 7.5) in the calculations is what makes him use radians instead of degrees. So, the use of the radian measurement, over the degree measurement, is a necessity imposed by the setting, the context or the kind of problems that are being resolved and not by the school discourse, as in today's curricula. The necessity to homogenize equations, a classic issue in physics, can be recognized when a student is asked to work with functions like $y = \sin x + x$; if the decision is to express x in degrees, then a conflict in doing the sum will be evident. That is, the use and *operability* of the trigonometric quantities, within specific settings, contexts and problems, are factors that determine the use of the unit of measurement.

7.6.2 Periodic and bounded behavior

The properties of periodicity and bounded range that characterize the trigonometric functions must be seen more as qualities of the trigonometric behavior. Buendía [2] proposes that, according to the socio-epistemogical review she has done for the periodic property, these qualities can be identified meaningfully on a repetitive graph function by setting up tasks of prediction. Recalling Euler's work, the periodic aspect of functions is recognized by the contemporary need to systematize the analysis of movements of objects in which prediction plays a predominant role. In short, prediction, as an argument, compares different types of repetition and allows for the construction of the notion of periodicity by means of the redefinition of the form and type of repetition of the motion graph. This leads us to the development of practices such as prediction, a fundamental task of physical sciences, and thus to favor the establishment of laws that govern behavior of systems in the mathematical environment [6]. The behavior of the function and the practice of prediction become elements that shape our basis of significations for trigonometric functionality.

---

[3] What we are going to call *operability*

We can recognize that, for example, the periodic aspect is not just related to the fulfillment or proof of the equality $\sin(x + 2\pi) = \sin(x)$, but to the characterization of its repetitive nature, as in Euler's work. While periodicity is not limited to the trigonometric function sine or cosine, *trigonometric functionality* is unavoidably established based on its periodic aspect. The bounded behavior $[-1, 1]$ will be related to the kind of phenomenon and its conditions.

7.6.3 Other trigonometric significations

The work with Euler's infinite polynomials leaves out the (conflictive) discussion of degree/radian as a unit of measurement and the, also conflictive, handling of the terminology *sine*, *cosine* that on occasions the student uses as variables, such as when wanting to simplify:

$$y = \frac{\sin x}{x} = \frac{\sin \cancel{x}}{\cancel{x}} = sin$$

The Eulerian infinite polynomial allows a clear way for the transition and the link between algebraic and graphic representations of trigonometric behavior, for example associating each factor to the zero in the graph (Figure 7.12). The numerical values of the function can be approximated by simple arithmetic and not just by entering values in the calculator as is usually done at school.

$$\sin z = \cdots \left(1 + \frac{z}{3\pi}\right)\left(1 + \frac{z}{2\pi}\right)\left(1 + \frac{z}{\pi}\right) z \left(1 - \frac{z}{\pi}\right)\left(1 - \frac{z}{2\pi}\right)\left(1 - \frac{z}{3\pi}\right) \cdots$$

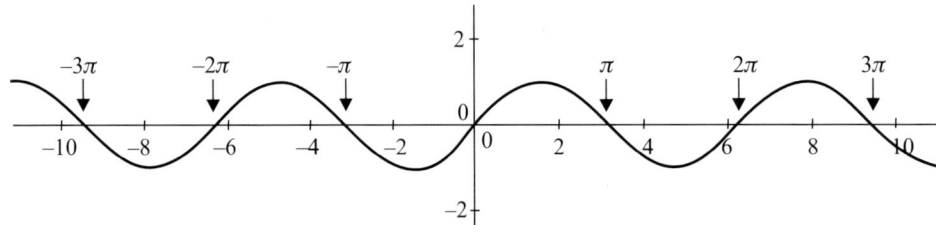

Figure 7.12. Sine graph with zeros identified from the polynomial expression.

However, placing this product in school discourse has the cost of bringing around the notion of infinity before calculus. Historically for Euler, this was not a special difficulty, and in the case of high school students this treatment can be used to introduce the notions of the infinitely large and infinitely small.

7.6.4 Trigonometric functionality

In the first place, we have to consider the context of origin of the trigonometric function: a *dynamic context*, where the study of the motion of objects in oscillatory phenomena sheds light on their properties (bounded and periodic behavior) and the way the unit of measurement is used to represent them. The use of a unit of measurement (angles/radians) is totally contextual and the recognition of a property, as periodicity, has more to do with recognizing the behavior of the mathematical object (the graph, for example) than with knowing how to apply the periodic formula to it.

A socio-epistemological base has been created, an explanation of the construction of trigonometric functions that takes into account the particular uses of mathematical knowledge and the significations that humans obtain while using it. Recalling Montiel's model of practices, prediction can be seen as a social practice that norms the construction of trigonometric functionality; the mathematization of oscillatory movement can be understood as a reference practice that articulates activities such as measuring, modeling, calculating (Figure 7.13). In this setting, several uses of the trigonometric function and significations for the periodic or bounded aspect can be developed meaningfully. The set of those practices, uses and significations confirm trigonometric functionality.

We can also speak of the *development of a functional trigonometric thought* when the learner is capable of recognizing, from periodic-bounded behavior, a useful tool to develop such tasks as predicting or modeling. And, finally, the trigonometric quantity can be distinguished and characterized from other algebraic or transcendental quantities by its variation and successive variations (how it changes and how its changes change).

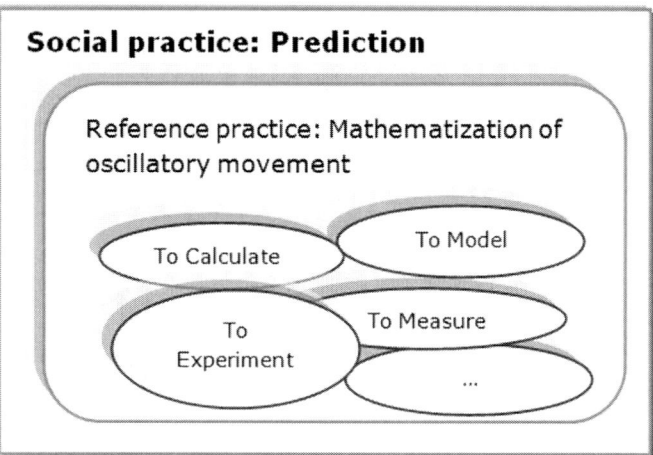

Figure 7.13. Model of practices for trigonometric functionality.

7.7 Problem Situations

We present in this Section situations that revisit diverse elements of the socio-epistemological proposal to illustrate its viability. Our purpose is to give evidence about the role of the social practice, the reference practices, and the type of problems and the context in the use of mathematical tools while constructing trigonometric functionality. Nevertheless, it is important to point out that these problem-situations require modification so they can be incorporated into the mathematics classroom as part of the curriculum.

7.7.1 A problem-situation to identify the unit of measurement

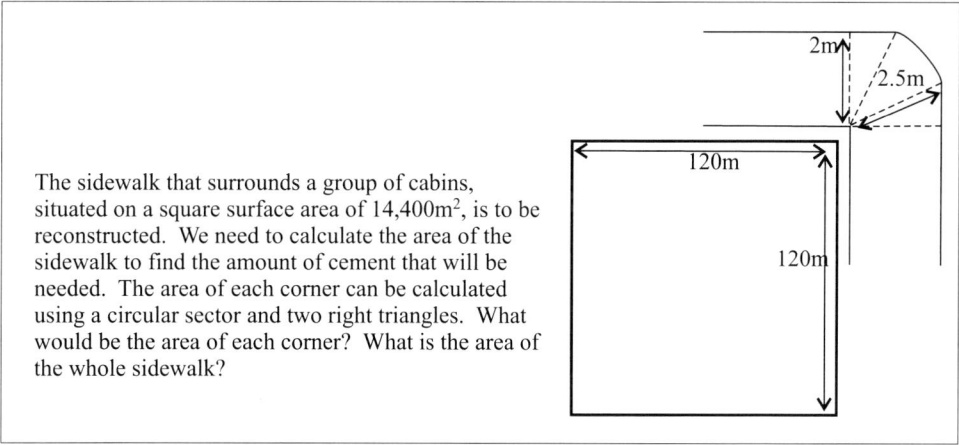

Figure 7.14.

In a non-school environment[4] which allowed a certain freedom to express ideas and beliefs, a group of eighteen mathematics teachers considered the problem-situation of Figure 7.14. This can be solved by calculating first the simplest areas: four rectangles of 120m in length and 2m in width and 8 right triangles of 2m base and 1.5m height, the last measurement being obtained from the Pythagorean Theorem. The main issue is the area of the circular sector and this is the task we want to discuss here.

One way of facing this task is like the answer shown in Figure 7.15 in which the teacher considers each corner as consisting of two right triangles and one circular sector. Since the concurrent angles form a right angle, she uses the expression $2\beta + \alpha = 90°$, where β is the concurrent angle in the right triangle that can be calculated, given that she

---

[4]During a Workshop in the 19th Reunión Latinoamericana de Matemática Educativa (Relme) with teachers from several countries in Latin America and working at different educational levels: from high school to college.

knows the values of one leg and the hypotenuse. So, she first obtains β from $\cos \beta$, and she uses it to find the unknown side of the triangle (marked with x). Finally, with the expression $2\beta + \alpha = 90°$, she obtains α, but in radians!

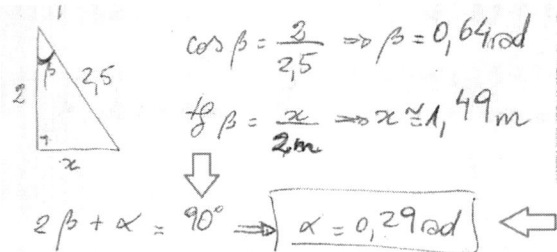

Figure 7.15. Obtaining the angle of the circular sector.

Notice that somehow she uses simultaneously both units of measure (degrees and radians), as in Euler's work; this does not generate conflicts while solving the problem-situation. Degrees are being used as a *referent*, as in $2\beta + \alpha = 90°$, and radians are used to operate with and to homogenize the equations as we can see in the next part of her solution (Figure 7.16).

To find the area of the circular sector, the teacher considers the proportional relation $\alpha : 2\pi :: $ **area of the sector** : **total area** (the angle of the sector is to 2π, as the area of the sector is to the area of a whole circle), so for the area A of the circular sector, she gets, wrongly, $A = \alpha \pi r^2$, by forgetting to divide by 2π. From the original figure $r = 2.5$, so $A = 5.684 \text{m}^2$.

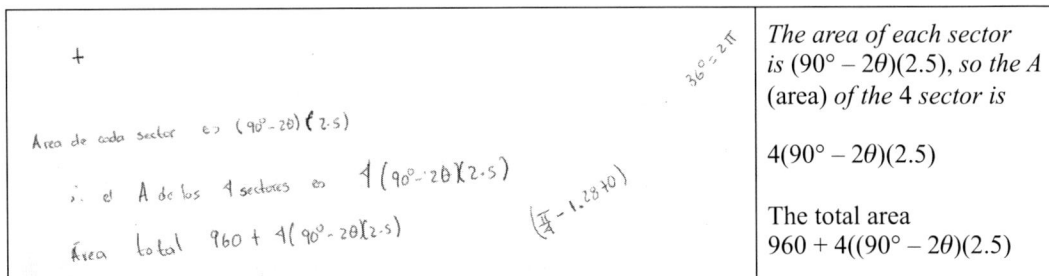

Figure 7.16. Operating with radius.

Trying to follow the same idea, another teacher continues to use angle α in degrees, $\alpha = 90° - 2\theta$, then she tries to express the multiplication by $r = 2.5$, in order to calculate the four circular sectors, but she could not continue with the exercise, because she obtained an expression, $960 + 4(90° - 2\theta)(2.5)$, in which a multiplication between meters and degrees had to be done in an arbitrary way (Figure 7.17).

	The area of each sector is $(90° - 2\theta)(2.5)$, so the A (area) of the 4 sector is
	$4(90° - 2\theta)(2.5)$
	The total area $960 + 4((90° - 2\theta)(2.5)$

Figure 7.17. When the teacher tries to operate with degrees, he could not go further.

Interestingly, in the debate that ensued with and among the teachers, there were no arguments to uphold the decision to use the angle in degrees. The choice of the mathematical tools used to solve the problem is influenced by the context: a triangle drawing without the explicit measurement of one of its sides causes the teachers to use the trigonometric ratio. In this geometric context the teachers rely more on the use of the degree as the unit of measurement, although they have no real justification. And it is the need to accomplish certain operations that can causes a change in the use of unit of measurement.

7.7.2 A problem situation for characterizing the functionality of the sine and cosine: the periodic aspect and the bounded behavior

The following situation was initially proposed to show the role of prediction in the meaningful construction of the periodic property ([3]) and now strengthens the socio-epistemological proposal about trigonometric functionality. In this

7.7. Problem Situations

problem-situation, students deal with objects in motion by analyzing time-distance graphs (continuous/discontinuous, sine form / lineal elements). The activities asked were: (1) Describe the movement shown in each graph; (2) Predict the position of the object when $t = 321$; (3) Determine, which ones are periodic graphs (Figure 7.18):

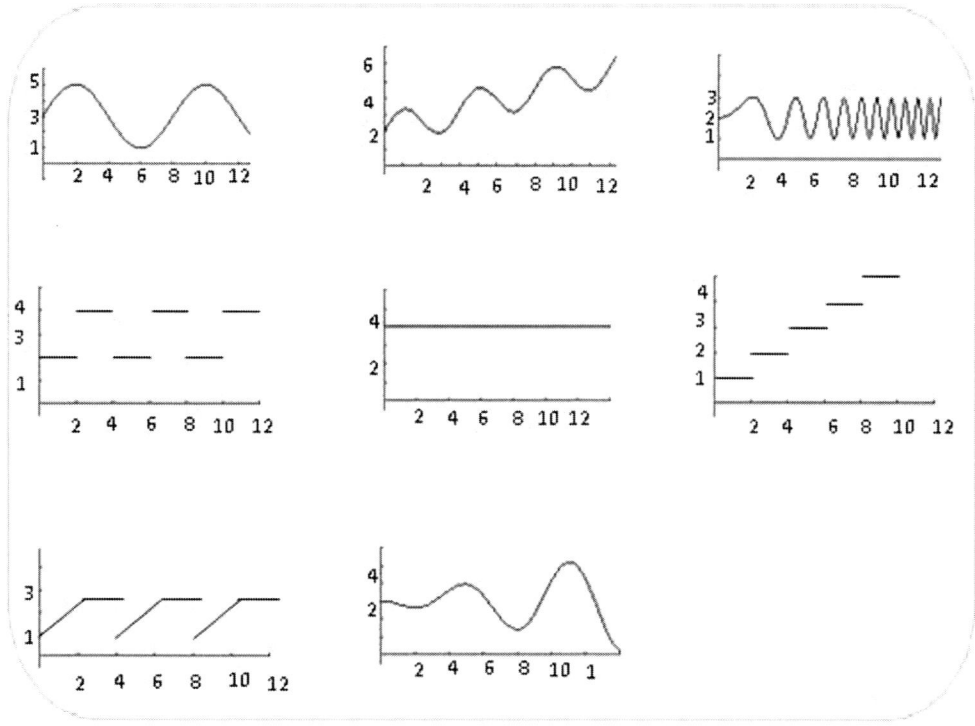

Figure 7.18. Time-distance graphs

The practice of prediction allows us to bring into play and to contrast meanings about the kind of repetition that each movement presents. The socio-epistemology of the periodic aspect has been carefully analyzed by Buendia in [3]. In particular, she considers how the initial meaning about periodicity as *anything that repeats itself* evolves to the question of *how does it repeat*. Prediction favors this evolution because the first step in prediction is the identification of a period of repetition[5]. Then, its use enables the calculation of the required prediction. As a result, different predictive procedures are developed whose context and tools (graphic, analytic, numeric, among others) are totally situational and developed in interactive contexts.

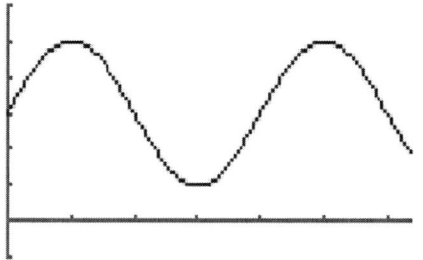

Juan Carlos (JC), teacher in charge of coordinating an engineering program; Researcher (R)
JC: *The object is coming and going; from one side to another. For example, my activity would be to go and see in a given time a teacher, over there, then I come back and then, mmm, is the same time I do to get there and back once more because I forgot to tell him some instructions.*
R: *To the same teacher?*
JC: *Yes, and in the same place.*

Figure 7.19.

Predicting is not just for obtaining a numeric answer (a position, in this case); it is a practice proposed through several activities that allows us to re-define properties such as the periodicity and the bounded behavior of a function. These acquire meanings in the exercise of that practice and not as the application of their respective analytical forms, so we are facing two significations that students construct in relation to trigonometric functionality.

---

[5]These periods of repetition have a more general connotation than their value associated with trigonometric functions (2π): it is a unit of analysis which, given a certain piece of relevant information, enables the prediction of the future state of the system in question.

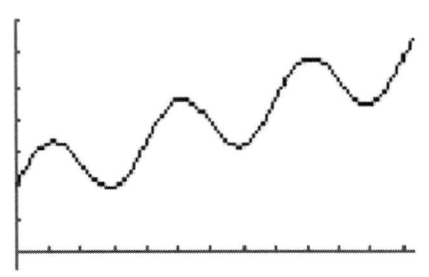

Juan Carlos (JC); Angel (A), teacher in an engineering program, Researcher (R)
JC: *Here, I am going to some other place; maybe I am returning. Well, maybe not; well, maybe yes but is like if distance is increasing and as if I were going to another place.*
Like going to the principal's office and then to Linking-Building[6], so the distances are different.
R: *And, do you return?*
JC: *mmm, no, I don't think so.*
A: *The graphic shows what he is saying about going and returning, here* (pointing to the first descendent part) *is when he comes back from his first task but, mmm, first, there is a certain distance but then there would be another distance from his first task. There is a variation in his distance, on his return; that is, he does not return to his first position.*

Figure 7.20.

As an example, we show some responses obtained in experimental settings.

In Figures 7.19 and 7.20, we can see a dialogue between two teachers and the researcher. In the first example, Juan Carlos identifies the coming and going as forgotten tasks that he has to accomplish in a given time; in the second example, he correctly identified—by referring to the position of different buildings that he usually visits—the distance variation but he could not finish expressing his idea. Angel helped him by mentioning that he never returns to his first position: Juan Carlos' forgetfulness would cause him to come and go several times but farther and farther! We can see that the interpretation of characteristics as periodic behavior and bounded range are totally contextual ones.

In the next example, Enrique and Andrés—teachers and master degree students—were asked to classify, before and after the activity of prediction, the graphs by their differences or similarities. The results presented correspond to the classification done after predicting (Figure 7.21):

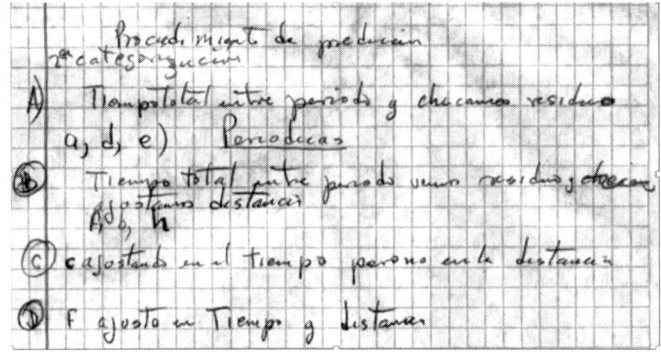

Prediction procedure. 2a. characterization
A) *Total time divided by the period and then we check the residue: a, d, e (referring to the graphs)* <u>Periodic ones</u>.
B) *Total time divided by the period, we check the residue and we adjust the distance: f, b, h*
C) *c: adjusting the time but not the distance*
D) *f: adjusting the time and the distance*

Figure 7.21. After predicting.

For Enrique and Andrés, the periodic graphs were the ones in which they used the same predicting procedure. They noticed that what happens in each axis (time or distance) influences the kind of procedure that they must follow. Perhaps, the decision to label the first group as periodic graphs was influenced by the presence of the typical sine graph, but they used the behavior of the graph as an argument and not only a thoughtless or poor reference to "*is periodic because is like the sine.*"

7.8 Final Reflection

The pair History–Education referring to mathematics has found a number of successful manifestations in school. Seen as a source of interesting problems that trigger great mathematical ideas, the history of mathematics can enrich the knowledge we impart in today's classrooms; it enables us to present a more *human* knowledge. The use of primary

---

[6]He is referring to different buildings in the university campus. We were working in classroom-buildings. The three mentioned buildings were more or less situated in a straight line. If the classroom-building is thought as the origin, the next building is the Linking-building, and the Principal's office-building.

historic sources allows us also to recognize the socio-cultural factors in the development of mathematics, such as its potential to solve societal problems and its impact on the cultural development of human groups.

In this paper, we propose another pair that does not place history directly in the classroom, since we already recognize the transformation that mathematical knowledge undergoes when taken into a classroom situation. Our proposal of the pair History–Mathematics Education considers that the first contributes to the second bringing elements related to (1) the explanation of phenomena in the classroom and (2) the construction of epistemologies of practice that condition and give rise to mathematical knowledge. Thus, *Socioepistemology* develops research strategies through history to determine those circumstances that illustrate why we deal with school mathematics today as we do.

To understand how a concept was used in a certain historic period, what problems it addressed and what obstacles limited its scope, we must avoid delving into the past to track down only the roots of modern knowledge. Maltese [24] realizes that to face this task, the object under investigation must be placed in the framework of a certain era, seeking its influence on previous scientific practice and even beyond that, uncovering the procedures and algorithms to decipher the way it was used.

The history of trigonometric functions, besides illustrating a historic period of scientific revolution for humanity that is interesting in its own right and that contributes to the mathematical culture of teachers and students, has given us socio-cultural and epistemological elements to explain certain classroom phenomena, and to establish a basis of signification for trigonometric functionality. This is the kind of basis on which the socio-epistemological approach will employ its didactical resources toward the re-design of school mathematics discourse.

Bibliography

[1] Antolín, A., 1981, *De Euler a Fourier: Crisis y Abandono del Concepto Clásico de Función*. Master degree thesis, México: UNAM.

[2] Buendía, G. and F. Cordero, 2005, "Prediction and the Periodical Aspect as Generators of Knowledge in a Social Practices Framework", *Educational Studies in Mathematics* 58, 299-333

[3] Buendía, G., 2006, "Una Socioepistemología del aspecto periódico de las funciones", *Revista Latinoamericana de Matemática Educativa* 9, 227–251.

[4] Cantoral, R. and R. Farfán, 2004, "La sensibilité à la contradiction: logarithmes de nombres négatifs et origine de la variable complexe". *Recherches en Didactique des Mathématiques* 24, 137–168.

[5] ——, 2003, "Mathematics Education: A vision of its evolution". *Educational Studies in Mathematics* 53, 255–270.

[6] Cantoral, R. and M. Ferrari, 2004, "Uno studio socioepistemologico sulla predizione", *La matematica e la sua didattica* 2, 33–70.

[7] Cordero, F., 2006, "El uso de las gráficas en el discurso del cálculo escolar. Una visión socioepistemológica", in *Investigaciones sobre enseñanza y aprendizaje de las matemáticas. Un reporte latinoamericano*, Cantoral, R. Covián, O., Farfán, R., Lezama, J., Romo, A. (eds), México: Comité Latinoamericano de Matemática Educativa y Díaz de Santos, pp. 265–286.

[8] ——, 2005, "El rol de algunas categorías del conocimiento matemático en educación superior. Una socioepistemología de la integral" *Revista Latinoamericana de Matemática Educativa* 8, 265–286.

[9] ——, 2001, "La distinción entre construcciones del cálculo. Una epistemología a través de la actividad humana", *Revista Latinoamericana de Investigación en Matemática Educativa* 4,103–128.

[10] Covián, O., 2005, *El papel del conocimiento matemático en la construcción de la vivienda tradicional: El caso de la Cultura Maya*. Tesis de maestría. México: Centro de Investigación y Estudios Avanzados del IPN Cinvestav.

[11] De Gandt, F., 1999, "Matemáticas y realidad física en el siglo XVII (de la velocidad de Galileo a las fluxiones de Newton), in *Pensar la Matemática*, Guénard, F. y Lelievre, G. (Eds.), Barcelona, España: Tusquets Editores, pp. 41–68

[12] De Guzmán, M., 1996, *El rincón de la pizarra*. Madrid: Pirámide.

[13] De Kee, S., R. Mura, and J. Dionne, 1996, "La compréhension des notions de sinus et de cosinus chez des élves du secondaire", *For the Learning of Mathematics* 16, 19–22.

[14] Dunham, W., 2001, *Euler: El maestro de todos los matemáticos*. España: Nivola.

[15] Ferrari, M., 2001, *Una visión socioepistemológica. Estudio de la función logaritmo*. Master degree thesis. México: Centro de Investigación y Estudios Avanzados del IPN

[16] Gascón, J, 1998, "Evolución de la didáctica de las matemáticas como disciplina científica", *Recherches en Didactique des Mathématiques* 18, pp. 7–34.

[17] Grabovskij, M. and P. Kotel'Nikov, 1971, "The use of kinematic models in the study of trigonometric functions", *Educational Studies in Mathematics* 3, 147–160.

[18] Grattan-Guinness, I., 1970, *The Development of the Foundations of Mathematical Analysis from Euler to Riemann*. Cambridge, Mass: The MIT Press.

[19] Gray, E. and D. Tall, 1994, "Duality, ambiguity, and flexibility: A proceptual view of elementary arithmetic", *Journal for Research in Mathematics Education* 26, 114-141.

[20] Guimelli, C., 1999, *La pensée sociale*, Paris: Presses Universitaires de France.

[21] Katz, V., 1987, "The Calculus of the Trigonometric Functions", *Historia Mathematica* 14, 311–324.

[22] Lévy-Leblond, J., 1999, "Física y Matemáticas", in *Pensar la Matemática*, Guénard, F. y Lelievre, G. (Eds.), Barcelona, España: Tusquets Editores, pp. 69–84

[23] Maldonado, E., 2005, *Un análisis didáctico de la función trigonométrica*, Masters degree thesis. Cinvestav-IPN, México: Centro de Investigación y Estudios Avanzados del IPN.

[24] Maltese, G., 2000, "On the relativity of motion in Euler's science", *Archive for history of exact sciences* 54, 319–348.

[25] Montiel, G., 2005, *Estudio socioepistemológico de la función trigonométrica*. Tesis de Doctorado no publicada. México: Centro de Investigación en Ciencia Aplicada y Tecnología Avanzada del IPN. Available at www.matedu.cicata.ipn.mx/tesis03.html

[26] Weber, K., 2008, "Teaching trigonometric functions: Lessons learned from research", *Mathematics Teacher* 102, 144–150.

[27] ———, 2005, "Student's understanding of trigonometric functions", *Mathematics Education Research Journal* 7, 91–112.

About the Authors

Gabriela Buendia Abalos and **Gisela Montiel Espinosa** are researchers in Mathematics Education. They work at the Mathematics Education Program in CICATA, a research center belonging to the National Polytechnic Institute of Mexico. The master and doctoral courses they teach are online, so they work with graduate students that at the same time are teachers from all over Mexico and Latin America. Their main research interest is the social construction of school mathematics considering mathematical knowledge as a social, cultural and historical product of humanity, so their interest in history is as a source to generate bases of meanings for school mathematics.

8

Harmonies in Nature: A Diaglogue Between Mathematics and Physics

Man-Keung Siu
University of Hong Kong, China

8.1 Why is an Enrichment Course on Mathematics-Physics Designed?

In school it is a customary practice to teach mathematics and physics as two separate subjects. In fact, mathematics is taught throughout the school years from primary school to secondary school, while physics, as a full subject on its own, usually starts in senior secondary school. This usual practice of teaching mathematics and physics as two separate subjects has its grounds. To go deep into either subject one needs to spend at least a certain amount of class hours, and to really understand physics one needs to have a sufficiently prepared background in mathematics. However, such a practice deprives students of the opportunity to see how the two subjects are intimately interwoven. Indeed, in past history there was no clear-cut distinction between a scientist, not to mention so specific as a physicist, and a mathematician.

Guided by this thought we try to design an enrichment course for school pupils in senior secondary school, who are about to embark on their undergraduate study in two to three years' time, that tries to integrate the two subjects with a historical perspective. Conducting it as an enrichment course, we are free from an examination-oriented teaching-learning environment and have much more flexibility with the content. Admittedly, this is not exactly the same as the normal classroom situation with the constraint imposed by an official syllabus and the pressure exerted by a public examination. However, just like building a mathematical model, we like to explore what happens if we can have a bit more freedom to do things in a way we feel is nearer to our ideal.

Albert Einstein and Leopold Infeld sum up the situation succinctly, "In the whole history of science from Greek philosophy to modern physics there have been constant attempts to reduce the apparent complexity of natural phenomena to some simple fundamental ideas and relations. This is the underlying principle of all natural philosophy." [3]. Such a process makes demand on one's curiosity and imagination, but at the same time requires disciplined and critical thinking. Precision in mathematics as well as in words is called for. Galileo Galilei already referred to mathematics as the language of science in his *Il Saggiatore* (*The Assayer*) of 1623, "Philosophy is written in this grand book — I mean the universe — which stands continually open to our gaze, but it cannot be understood unless one first learns to comprehend the language and interpret the characters in which it is written. It is written in the language of mathematics, and its characters are triangles, circles, and other geometric figures, without which it is humanly impossible to understand a single word of it; without these, one is wandering about in a dark labyrinth."

By promoting this view Galileo made a significant step forward in switching the focus from trying to answer "why" to trying to answer "how (much)", that is, from a qualitative aspect to a quantitative aspect. In the Eastern world a similar sentiment was expressed by many authors of ancient classics that may sound like bordering on the mystical side. One such typical example is found in the preface of *Sun Zi Suan Jing* (*Master Sun's Mathematical Manual*) in the 4th century, "Master Sun says: Mathematics governs the length and breadth of the heavens and the earth; affects the lives of all creatures; forms the alpha and omega of the five constant virtues; acts as the parents for yin and yang; establishes the symbols for the stars and the constellations; manifests the dimensions of the three luminous bodies; maintains the balance of the five phases; regulates the beginning and the end of the four seasons; formulates the origin of myriad things; and determines the principles of the six arts."

The conviction in seeing beauty and order in Nature was long-standing. Plato's association of the five regular polyhedra to the theory of four elements in *Timaeus* (c.4th century B.C.) is an illustrative example. Over a millennium later, Johannes Kepler tried to fit in the motion of the six known planets (Saturn, Jupiter, Mars, Earth, Venus, Mercury) in his days with the five regular polyhedra in *Mysterium Cosmographicum* of 1596. By calculating the radii of inscribed and circumscribed spheres of the five regular polyhedra nestled in the order of a cube, a tetrahedron, a dodecahedron, an icosahedron and an octahedron, he obtained results that agreed with observed data to within 5% accuracy! He also thought that he had explained why there were six planets and not more! Now we realize the lack of physical ground in his theory, beautiful as it may seem. Still, it is a remarkable attempt to associate mathematics with physics, and indeed it led to something fruitful in the subsequent work of Kepler.

Well into the modern era the explanatory power of mathematics on Nature is still seen by many to be mystical but fortunate. Eugene Paul Wigner, 1963 Nobel Laureate in physics, refers to it as "the unreasonable effectiveness of mathematics in the natural sciences". Heinrich Rudolf Hertz even said (referring to the Maxwell's equations which predicted the presence of electromagnetic wave that he detected in the laboratory in 1888.), "One cannot escape the feeling that these mathematical formulas have an independent existence of their own, that they are wiser than we are, wiser even than their discoverers, that we get more out of them than was originally put into them." Robert Mills, an eminent physicists of the Yang-Mills gauge theory fame, says, "You can't hope to understand the [physics / math] until you've understood the [math/physics]." [11]. This dictum that emphasizes a two-way relationship between mathematics and physics furnishes the guideline for our enrichment course.

8.2 How is Such a Course Run?

The enrichment course, with its title same as that of this paper, ran for ten sessions each taking up three hours on a weekend (outside of the normal school hours). It had been run four times, in the spring of 2006 to 2009, in collaboration with a colleague at the Department of Physics in my university. Much as we wish to offer a truly integrated course, other constraints and factors (individual expertise, affordable time of preparation, inadequacy on our part, lack of experience in this new venture) force some sort of division of labour so that each one of us took up about half of the course. However, we still tried to maintain a spirit of integration in having a balanced emphasis on the mathematics and the physics in a suitable manner. In this paper I will naturally tell more about the part I took up, which involved the first two sessions, two intermittent sessions and the final session.

The underlying theme of the course is the role and evolution of mathematics, mainly geometry and calculus, with related topics in linear algebra, in an attempt to understand the physical world, from the era of Isaac Newton to that of James Clerk Maxwell and beyond it to that of Albert Einstein. In other words, it tries to tell the story of triumph in mathematics and physics in the past four centuries. The physics provides both the source of motivation and the applications of a number of important topics in mathematics. Along the way both ideas and methods are stressed, to be learnt in an interactive manner through discussion in tutorials and group work on homework assignments. A rough sketch of the content of the course is summarized in Table 8.1. Considering the level of the course, it is to be expected that topics near to the end are treated only after a fashion, mainly for broadening the vista of the students rather than for teaching them the technical details.

8.3 A Sketch of the Content of the Course

Each session of the enrichment course consists of a lecture in the first hour followed by a tutorial. The lecture serves to highlight some keypoints and outline the development of the topic. What is covered is selective in the sense that the

8.3. A Sketch of the Content of the Course

Time period	Physics	Mathematics (mainly)
4th century B.C.	Physical view of Aristotle	Euclidean geometry
many centuries in between		geometry (area / volume) algebra (equations)
17th century	physical view of Copernicus, Kepler, Galileo, Newton, ...	vectors in $\mathbb{R}^2$ and $\mathbb{R}^3$, calculus in one variable (functions, including polynomial, rational, trigonometric, logarithmic and exponential)
18th century	wave and particle	differential equations, Fourier analysis, complex numbers
19th century	theory of electromagnetism (Maxwell's equations)	vector calculus, Stokes' Theorem (Fundamental Theorem of Calculus)
20th century	theory of special and general relativity, quantum mechanics	probability theory, non-Euclidean geometries of spacetime

Table 8.1.

material illustrates some theme rather than provides a comprehensive account. Interested students are advised to read up on their own relevant references suggested in each session. (A selected sample of such books can be found in the list of references, some of which are more suitable for the teacher than the student [1, 2, 3, 4, 7, 9, 10, 11, 12, 13, 15, 16].) The course is seen as a means to arouse, to foster and to maintain the enthusiasm of students in mathematics and physics more than as a means to equip them with a load of knowledge.

To keep within the prescribed length of the paper I would not give a full account of the content but select certain parts, particularly the beginning part that sets the tone of the course, with supplementary commentary, to illustrate the intent of the enrichment course. The intent is to highlight the beautiful (some would say uncanny!) and intimate relationship between mathematics and physics, in many cases even mathematical ideas that have lain quietly in waiting for many years (sometimes more than a thousand years!) that enhance theoretical understanding of physical phenomena. In fact the relationship is two-way so that the two subjects benefit mutually from each other in their development. In Section 8.4 some sample problems in tutorials are appended in the hope of better illustrating this intention.

The course begins with a discussion on the Aristotelian view of the physical world that came to be known since the 4th century B.C.. All terrestrial matters, which are held to be different from heavenly matters, are believed to contain a mixture of the four elements in various compositions. Each of the four elements is believed to occupy a natural place in the terrestrial region, in the order of earth (lowest), water, air, fire (uppermost). Left to itself, the natural motion of an object is to go towards its natural position, depending on the composition and the initial position. Hence, a stone (earth) falls to the ground but a flame (fire) goes up in the air. A natural motion has a cause. It is believed that the weight of a stone is the cause for its free falling motion. According to the Aristotelian view, a heavier stone will fall faster than a lighter one. Any motion that is not a natural motion is called a violent motion, believed to be caused by a force.

We next bring in the physical world view that Galileo propounded in the first part of the 17th century. In particular, he demolished the theory that a heavier object falls faster by mathematical reasoning (thought-experiment) in *Discorsi e dimonstrazioni matematiche intorno a due nuove scienze* (*Discourses and Mathematical Demonstrations Concerning Two New Sciences*) of 1638. Suppose object A_1 has a larger weight W_1 than the weight W_2 of object A_2. Tie the objects A_1 and A_2 together to form an object of weight $W_1 + W_2$. The more rapid one will be partly retarded by the slower; the slower one will be somewhat hastened by the swifter. Hence, the united object will fall slower than A_1 alone but faster than A_2 alone. However, the united object, being heavier than A_1, should fall faster than A_1 alone. This is a contradiction! [6, p. 446]. A commonly told story says that Galileo dropped two balls of different weights from the top of the Tower of Pisa to arrive at his conclusion. There is no historical evidence that he actually did that. The significant point does not lie so much in whether Galileo actually carried out the experiment but in his arrival at the conclusion by pure reasoning. Together with pure reasoning, Galileo was known for his emphasis on observations and experiments as well, notably his experiments with an inclined plane. By observing that a ball rolling down an inclined plane will travel up another inclined plane joined to the first one at the bottom until it reaches the same height, he saw that the ball will travel a greater distance if the second inclined plane is placed less steep than the first one, the greater if the second inclined plane is less steep. From thence a thought-experiment comes in again. If the second inclined plane is actually placed in a horizontal position, the ball will travel forever without stopping. "Furthermore we may remark that any velocity once imparted to a moving body will be rigidly maintained as long as the external causes of acceleration or retardation are removed, a condition which is found only on horizontal planes. ...it follows that motion along a horizontal plane is perpetual..." [6, p. 564]. This motivated him to announce his famous law of inertia, which becomes the first law of motion in Newton's *Philosophiae naturalis principia mathematicas* (*Mathematical Principles of Natural Philosophy*) of 1687: "Every body perseveres in its state of rest, or of uniform motion in a right line, unless it is compelled to change that state by forces impressed thereon." [6, p.743]. This fundamental modification on the Aristotelian view (in a sense actually more natural according to daily experience!) that a force acting on an object is exemplified not by the speed of its motion but by the change in speed (acceleration), led to a quantitative description of this relationship in Newton's second law of motion (which yields the famous formula $F = ma$). It turned a new page in the development of physics. We follow with a discussion on the work of Johannes Kepler in calculating the orbit of Mars based on the meticulously kept observed data of Tycho Brahe [8]. On the one hand the story displays a beautiful interplay between theory and experiment. On the other hand Kepler's laws on planetary motion provide a nice lead to a discussion on Newton's law of universal gravitation.

We next discuss the theory of wave motion along with the mathematics, culminating in the theory of electromagnetism and Maxwell's equations. Mathematics owed to physics a great debt in that a large part of mathematical analysis that was developed in the 18th and 19th centuries have to do with the Vibrating String Problem. We talk about the all-important notions of function and of equation. Together with the discussion on vector calculus and the generalized Fundamental Theorem of Calculus, there is much more material than enough to take up the second part of the course. The unification of electricity, magnetism and light through the electromagnetic wave is a natural lead into the final third of the course, which is spent on a sketch of the theory of relativity and on quantum mechanics. Some probability theory is introduced to let students appreciate the stochastic aspect that is not usually encountered in the usual school curriculum. The close relationship between geometry and physics is stressed in the final episode on the theory of general relativity. In a letter to Arnold Sommerfeld dated October 29, 1912 (collected in A. Hermann, *Einstein/Sommerfeld Briefwechsel*, Schwabe Verlag, Stuttgart, 1968, p. 26) Albert Einstein wrote, "I am now exclusively occupied with the problem of gravitation, and hope, with the help of a local mathematician friend, to overcome all the difficulties. One thing is certain, however, that never in my life have I been quite so tormented. A great respect for mathematicians has been instilled within me, the subtler aspects of which, in my stupidity, I regarded until now as pure luxury. Against this problem, the original problem of the theory of relativity is child's play." The 'mathematician friend' refers to Einstein's school friend Marcel Grossmann, and the mathematics refers to Riemannian geometry and tensor calculus. The story on the work of Carl Friedrich Gauss and Georg Friedrich Bernhard Riemann in revealing the essence of curvature which lies at the root of the controversy over the Fifth Postulate in Euclid's *Elements* (but which had been masked for more than two thousand years when the attention of mathematicians was directed into a different direction) and its relation to Einstein's idea on gravitation theory is fascinating for both mathematics and physics. No wonder Riemann concluded his famous 1854 lecture titled *Über die Hypothesen welche der Geometrie zu Grunde liegen* (*On the hypotheses which lie at the foundation of geometry* (an English translation can be found in [17, pp. 411-425]) with: "This path leads out into the domain of another science, into the realm of physics, into which the nature of this present occasion forbids us to penetrate."

8.4 Some Sample Problems in Tutorials

In this course more than half of the time in each session is spent as a tutorial, which is regarded as an integral part of the learning experience. Students work in small groups with guidance or hint provided on the side by the teacher and a team of (four) teaching assistants. At the end of each session there is a guided discussion with presentations by students. A more detailed record of the solution is put on the web afterward for those who are interested to probe further. Some sample problems in the tutorials are given below to convey a flavor of the workshop.

8.4. Some Sample Problems in Tutorials

Question 1. A, B, C, D move on straight lines on a plane with constant speeds. (The speed of each chap may be different from that of another.) It is known that each of A and B meets the other three chaps at **distinct** points. Must C and D meet? Under what condition will the answer be 'yes' (or 'no')? [The question was once given in an examination at Oxford University.]

Discussion: C and D will (respectively will not) meet if they do not move (respectively move) in the same or opposite directions. The catch is a commonly mistaken first reaction to draw a picture with two straight lines emanating from a common point M_{AB} (the point where A and B meet) and two more straight lines, one intersecting the first line at M_{AC} and the second line at M_{BC}, the other one intersecting the first line at M_{AD} and the second line at M_{BD}. It seems that the answer comes out obviously from the picture until one realizes that a geometric intersecting point needs not be a physical intersecting point! This problem is set as the first problem in the first tutorial to lead the class onto the important notion of spacetime, which will feature prominently in the theory of relativity. Viewed in this context, no calculation is needed at all!

Question 2. Suppose you only know how to calculate the area of a rectangle—our ancestors started with that. Explain how you would calculate the area of a triangle by approximating it with many many rectangles of very small width. This answer, by itself, does not sound too exciting. You can obtain it by other means, for instance by dissection—our ancestors did just that! However, what is exciting is the underlying principle that can be adapted to calculate the area of regions of other shapes. Try to carry out a similar procedure for a parabolic segment. (Find the area under the curve given by $y = kx^2$ from $x = 0$ to $x = a$. What happens if you are asked to find the area under the curve $y = kx^3$? $y = kx^4$? $\cdots$? Later you will see how a result enables us to solve this kind of problem in a uniform manner.)

Discussion: This problem is set at the beginning of the course to introduce some ideas and methods devised by ancient Greeks and ancient Chinese on problems in quadrature, to be contrasted with the power of calculus developed during the 17th and 18th centuries, culminating in the Fundamental Theorem of Calculus with its generalized form (Stokes' Theorem) established through the development of the theory of electromagnetism in the 19th century. For this particular problem some clever formulae on the sum of consecutive rth power of integers $1^r + 2^r + 3^r + \cdots + N^r$ are needed. That kind of calculation is not totally foreign to the experience of school pupils and yet offers some challenge beyond what they are accustomed to, which is therefore of the level of difficulty the workshop is gauged at. After struggling with specific but seemingly ad hoc 'tricks' of this sort, students would appreciate better the power afforded by the Fundamental Theorem of Calculus when they learn it later.

Question 3(a) By computing the sum
$$1 + z + z^2 + \cdots + z^n$$
where $z = e^{i\theta}$, and using Euler's formula
$$e^{i\theta} = \cos\theta + i\sin\theta,$$
find a simple expression for
$$1 + \cos\theta + \cos 2\theta + \cdots + \cos n\theta$$
and
$$\sin\theta + \sin 2\theta + \cdots + \sin n\theta.$$

(b) Apply the result in (a) to calculate the area under the curve $y = \sin x$ on $[0, \pi]$ from scratch in the way you did for $y = x^2$ in the first tutorial. Do the same for $y = \cos x$ on $[0, \pi]$. (How do you normally calculate this area in your class at school?)

Discussion: Besides introducing a most beautiful formula in mathematics, this problem further strengthens students' appreciation of the Fundamental Theorem of Calculus. In the course of explaining Euler's formula students are led into the realm of complex numbers, to the 'twin' functions of logarithm and exponentiation.

Question 4(a) Pierre Simon Laplace (1749–1827) once said, "By shortening the labors, the invention of logarithms doubled the life of the astronomer." To appreciate this quotation, let us work on an multiplication problem (81276×96343) like people did before the invention of logarithm. The method, known as "prosthaphaeresis", is based on the addition formula of trigonometric functions.

(i) If $2\cos A = 0.81276$ and $\cos B = 0.96343$, find A and B.

(ii) Calculate $A + B$, $A - B$, and hence calculate $\cos(A+B)$, $\cos(A-B)$.

(iii) Calculate $\cos(A+B) + \cos(A-B)$, which is equal to $2\cos A \cos B$, and hence find out what 81276×96343 is.

(b) Compare Napier's logarithm with the natural logarithm you learn in school.

(c) Making use of the idea Leonhard Euler (1707–1783) explained in Chapter XXII of his *Vollständige Anleitung zur Algebra* (1770), compute the common logarithm of 5, log 5, in the following steps:

(i) As 5 lies between 1 and 10, so log 5 lies between 0 and 1. Take the average of 0 and 1, which is $1/2$. Compute $10^{1/2}$, which is the square root of 10, say a_1.

(ii) Decide whether 5 falls into $[1, a_1]$ or $[a_1, 10]$. Hence decide whether log 5 falls into $[0, 1/2]$ or $[1/2, 1]$. It turns out log 5 falls into $[1/2, 1]$. Take the average of $1/2$ and 1, which is $3/4$. Compute $10^{3/4}$, which is the square root of 10 multiplied by the square root of $10^{1/2}$, say a_2.

(iii) Decide whether 5 falls into $[a_1, a_2]$ or $[a_2, 10]$. Hence decide whether log 5 falls into $[1/2, 3/4]$ or $[3/4, 1]$. It turns out log 5 falls into $[1/2, 3/4]$. Take the average of $1/2$ and $3/4$, which is $5/8$. Compute $10^{5/8}$, which is the square root of $10^{1/2}$ multiplied by the square root of $10^{3/4}$, say a_3.

(iv) Continue with the algorithm until you reach a value of log 5 accurate to three decimal places.

Discussion: Note the similar underlying idea of converting multiplication to addition in "prosthapharesis" and in logarithm. That allows the class to see how John Napier and later Henry Briggs devised their logarithm in the early 17th century. The bisection algorithm explained in **(c)**, though seemingly cumbersome from a modern viewpoint, is nonetheless very natural and simple, reducing the calculation to only finding square root. It provides an opportunity to go into the computation of square root by the ancients, first propounded in detail in the ancient Chinese classics *Jiu Zhang Suan Shu (Nine Chapters on the Mathematical Art)* compiled between 100 B.C. and 100 A.D. For the generation of youngsters who grow up with calculators and computers, this kind of 'old' techniques may add a bit of amazement as well as deeper comprehension.

Question 5. In an $x - t$ spacetime diagram drawn by an observer S who regards himself as stationary, draw the world-line for S and the world-line for an observer S' moving with uniform velocity v (relative to S). At $t = 0$ both S and S' are at the origin O. Both S and S' observe a light signal sent out from O at $t = 0$, reflected back by a mirror at a point P, then received by S' at Q. Which point on the world-line for S' will S' regard as an event **simultaneous** with the reflection of the light signal at P? Call this point P'. Show that the slope of the line $P'P$ is equal to v/c^2, where c is the speed of light (units omitted). [The physical interpretation is as follows. S regards two events, perceived as simultaneous by S', as separated by a time Δt given by $\Delta t = (v/c^2)\Delta x$, where Δx is the distance between the events measured by S and v is the velocity of S' relative to S.]

Discussion: We pay attention to the physical interpretation of a mathematical calculation and vice versa. This problem focuses on the key notion of simultaneity in the theory of special relativity. There is a note of caution for this problem. The picture of the spacetime diagram (according to the observer S) is to be seen in two ways: (i) the picture as it is, just like a picture one is accustomed to see in school geometry, (ii) the coordinate system of S with coordinates assigned to each event. In the lecture we take good care in denoting points in (i) by letters P, Q, P', O, etc., and events in (ii) by $(x(P), t(P)), (x(Q), t(Q)), (x(P'), t(P')), (x(O), t(O))$, etc. One can read the same in the shoes of the other observer S', in which case events in (ii) will be denoted by $(x'(P), t'(P)), (x'(Q), t'(Q)), (x'(P'), t'(P')), (x'(O),$

$t'(O)$), etc. In the lecture we also explain how $x(P), t(P)$ are related to $x'(P), t'(P)$ and vice versa (by the Lorentz transformation).

8.5 Conclusion

The triumph of Maxwell's theory on electromagnetism resolved many problems and yet introduced new difficulties that were resolved by Einstein's theory of special relativity. The triumph of Einstein's theory of special relativity resolved many problems and yet introduced new difficulties that were resolved by Einstein's theory of general relativity. But then the theory of general relativity introduces a more difficult problem on incompatibility with quantum mechanics, which is not revealed until one comes up with a situation where both the mass involved is very large and the size involved is very small, for instance, a black hole [5, 14]. Physics will march on to solve further problems, and so will mathematics, hand-in-hand with physics, in a harmonious way.

Bibliography

[1] Barnett, L., 1949, *The Universe and Dr. Einstein*, London: Victor Gollancz.

[2] Boyer, C., 1968, *A History in Mathematics*, New York: Wiley (revised by U.C. Merzbach, 2nd edition, 1989).

[3] Einstein, A., Infeld, L., 1938, *The Evolution of Physics: The Growth of Ideas From Early Concepts to Relativity and Quanta*, Cambridge: Cambridge University Press (2nd edition, 1961).

[4] Feynman, R.P., 1995, *Six Easy Pieces: Essentials of Physics Explained by its most Brilliant Teacher*, Reading: Helix Books (selected chapters from *The Feynman Lectures on Physics, Volume 1-3*, R.P. Feynman, Reading: Addison-Wesley, 1963–1965).

[5] Greene, B., 1999, *The Elegant Universe: Superstrings, Hidden Dimensions, and the Quest of the Ultimate Theory*, New York: W.W. Norton.

[6] Hawking, S. (ed.), 2002, *On the Shoulders of Giants: The Great Works of Physics and Astronomy*, London-Philadelphia: Running Press.

[7] Hewitt, P.G., 2006, *Conceptual Physics*, 10th edition, San Francisco: Benjamin Cummings.

[8] Koestler, A., 1959, *The Sleepwalkers: A History of Man's Changing Vision of the Universe*, London: Hutchinson.

[9] Lines, M.E., 1994, *On the Shoulders of Giants*, Bristol-Philadelphia: Institute of Physics Publishing.

[10] Longair, M.S., 1984, *Theoretical Concepts in Physics: An Alternative View of Theoretical Reasoning in Physics*, Cambridge: Cambridge University Press (2nd edition, 2003).

[11] Mills, R., 1994, *Space, Time and Quanta: An Introduction to Contemporary Physics*, New York: W.H. Freeman.

[12] Olenik, R.P., T.M. Apostol, and D.L. Goldstein, 1985, *The Mechanical Universe: Introduction to Mechanics and Heat*, Cambridge: Cambridge University Press.

[13] ——, 1986, *Beyond the Mechanical Universe: ¿From Electricity to Modern Physics*, Cambridge: Cambridge University Press.

[14] Penrose, R., 2004, *The Road to Reality: A Complete Guide to the Laws of the Universe*, London: Vintage Books.

[15] Pólya, G., 1963, *Mathematical Methods in Science*, Washington, D.C.: Mathematical Association of America (reprinted in 1977).

[16] Siu, M.K., 1993, *1,2,3,... and Beyond* (in Chinese), Hong Kong: Joint Publishing (HK).

[17] Smith, D.E., (ed.), *A Source Book in Mathematics*, McGraw-Hill, New York, 1929.

About the Author

Siu Man Keung obtained a BSc as a double major in mathematics and physics from the University of Hong Kong and went on to earn a Ph.D. in mathematics from Columbia University. Like the Oxford cleric in Chaucer's *The Canterbury Tales*, "and gladly would he learn, and gladly teach." Having been doing that for more than three decades he is still enjoying doing the same after retirement in 2005. He has published some research papers in mathematics and computer science, some more papers of a general nature in the history of mathematics and mathematics education, and several books in popularizing mathematics. In particular he is most interested in integrating history of mathematics with the learning and teaching of mathematics, and has been participating with zest in an international community of History and Pedagogy of Mathematics since the mid 1980s.

9

Exposure to Mathematics in the Making: Interweaving Math News Snapshots in the Teaching of High-School Mathematics

Batya Amit, Nitsa Movshovitz-Hadar, Avi Berman
Technion—Israel Institute of Technology, Israel

9.1 Introduction: The Ever Growing Nature of Mathematics

Beyond its glorious past, mathematics has a vivid present and a promising future. New results are published on a regular basis in the professional journals; new problems are created and added to a plethora of yet unsolved problems, which challenge mathematicians and occupy their minds.

Movshovitz-Hadar [13] suggested a classification of mathematical news into five categories which we bring here with examples, many of which can be made accessible to high-school students:

(i) **A recently presented problem of particular interest and possibly its solution.** *E.g.* Herzberg and Murty's paper concerning the mathematical problems related to Sudoku puzzles [10] and Murty's later discovery of a Sudoku puzzle with exactly two solutions [14].

(ii) **Long-term open problems recently solved.** "Recently solved" is defined as past 30 years and "long-term" is defined as at least 100 years. *E.g.* The proof of Kepler's conjecture; The mapping of the E_8 group; The Four Color Problem; Fermat's Last Theorem. For a more comprehensive (yet partial) list see Movshovitz-Hadar [13].

(iii) **A recently revisited problem.** This category includes a new proof to a known theorem, or new findings in an already solved problem, or a new solution to a previously solved problem, or a generalization of a well established fact, or even a salvaged error. *E.g.*, In 1996, Robertson, Sanders, Seymour and Thomas published a proof of the Four-Color problem, freeing it from the doubts about its computer-assisted proof, provided twenty years earlier by Appel and Haken [19]. Another example is the constant race for new prime numbers, elaborated in Section 9.6, below.

(iv) **A mathematical concept recently introduced or broadened,** including concepts that evolved into new areas in mathematics. *E.g.*, The concept of dimension, broadened in 1918 by the German mathematician Felix Hausdorff and later on by Mandelbrot [12] to non-integer dimension and fractals; The notion of Vertex Algebra invented by Richard Borcherds upon which he based his proof for the Conway-Norton "moonshine conjecture," for which he received the Fields Medal of 1998 [5, 24].

(v) **A new application to an already known piece of mathematics.** Mathematics develops as a result of human curiosity, quite often independent of the physical real world. It is fascinating to find-out that a piece of pure mathematics, developed by some intellectually-intrigued mathematicians becomes utterly useful for some real application. Perhaps the ultimate example in this category is the employment of prime factorization to modern Public-Key-Cryptography [18]. Another example is the Linear Algebra application to Google search engine [4].

9.2 The Problem: A Gap Between School Mathematics and Mathematics

At the moment, school mathematics curricula in many countries do not reflect the ever growing, accumulative nature of the field and its problem-based character. Consequently, students graduate high-school having the (wrong) image of mathematics as a "dead-end" discipline, in which all answers are known, and little is left for their creative investigation. This claim is supported by Schoenfeld [20, 21] who studied (among other issues) the perspectives and typical student beliefs about the nature of mathematics, as well as by Leder et al. [11] who found that students do not develop positive attitudes towards mathematics and consequently do not choose to study mathematics beyond compulsory years.

One well-accepted solution is integrating the history of mathematics in the curriculum. Furinghetti and Radford [9] found that biographies of several mathematicians served as a source of motivation for students, regarding motivation as a cardinal-positive attitude. Fried [8] takes it one step further, suggesting that the history of mathematics assumes an essential role in mathematics education both as a subject and as a mediator between historians' and working mathematicians' ways of knowing.

Similar to the role of the history of mathematics—contemporary mathematics may play an important role in providing the taste of mathematics in the making, pointing at paths to mathematical creation and opening the door for future activities in the domain. The next section elaborates on this proposed solution.

9.3 A Proposed Solution and its Rationale

As a way of bridging between school mathematics and the true nature of contemporary mathematics, we propose interweaving 10–15 minutes snapshots of mathematical news in the teaching of high-school mathematics, on a weekly basis.

This proposal stands on the shoulders of giants, among them Henri Poincaré who in 1908 opened his address to the 4th ICM in Rome, by saying: "The true method of forecasting the future of mathematics lies in the study of its history and its present state." [15].

The study of present-state mathematics may take various modes. The one advocated in this paper, i.e., interweaving mathematical-news-snapshots in the ordinary teaching, assumes that a snapshot is a short intermezzo, taking a small portion of the weekly class-time , and preferably, but not necessarily, linked to the particular topic in the curriculum, that occupies the class during that week. It does *not* change the flow of the curriculum. An alternative mode might be whole-lesson-modules, each focusing on a selected topic. Clearly, such modules differ from a collection of short snapshots, interspersed in the curriculum. The alternative mode was considered in planning the study reported in this paper. It was decided to focus on the snapshot mode due to time and schedule constrains. This is similar to the dilemma faced when one wants to introduce history of mathematics in mathematics classes [7]. The significance of, and challenges presented by, this study resemble in many other ways those of educational studies in the history of mathematics, as today's news is tomorrow's history.

One may doubt the possibility of introducing high-school students to contemporary mathematical news, on the grounds of students' lack of sufficient background to cope with advanced topics. We take Hilbert's side in this dilemma:

> "The edifice of science is not raised like a dwelling, in which the foundations are first firmly laid and only then one proceeds to construct and to enlarge the rooms. Science prefers to secure as soon as possible comfortable spaces to wander around and only subsequently, when signs appear here and there that the loose foundations are not able to sustain the expansion of the rooms, it sets about supporting and fortifying them. This is not a weakness, but rather the right and healthy path of development." [6, Quoting Hilbert 1905, p. 119]

In considering the gap between the mathematical knowledge of high-school students and contemporary topics, it seems reasonable to adopt not only Hilbert's idea, but also Courant's "Romantic approach." Courant (in [17] suggested that the essence of mathematics is *not* in the punctuality of technical details, thus opposing the classic approach to the study of mathematics which pays full attention to every detail and insists on polishing and refining every mathematical piece of work. Courant's "Romantic approach" rejects any guilt ignoring these elements and encourages concentration on what is stated beyond the rigor proof and the mathematical completeness.

A similar approach is introduced by William Thurston who prefers introducing mathematical topics in a general manner, allowing emphasis on the image *about* mathematics and not merely entirely on mathematics itself [1].

Siu [25] suggested that the history of mathematics affords perspective and presents a fuller picture of what mathematics is, remarking that the past, present and future of mathematics, being inter-related, make mathematics a cumulative science. Radford [16] argued for making students sensitive to the changing nature of mathematics and reconnecting Knowing and Being. Swetz [26] advocated a continual expanding of the exposure to the *scope* of mathematics, recommending: "Teach more *about* mathematics first, and then teach mathematics." Their views about the history of mathematics are no less relevant to contemporary mathematics and hence to our study.

Indeed, the news-snapshots are not meant to be the complete theory, but rather "a wander around"—intriguing enough to impress students, detailed enough to motivate them to do mathematics, and mind opening enough to yield the desired image of mathematics as a vivid creative domain. They are prepared to introduce mathematical news within the context of human creation, linking it to the people who took part in the invention. Every snapshot is a narrative that exposes the history (past) the novelty (present) and the open ended nature (future) of some new mathematical result by a contemporary mathematician. Most mathematical-news-snapshots also spell out some applications beyond mathematics itself.

9.4 The Challenge for the Teacher

The views presented above and adopted for integrating mathematical-news-snapshots into the teaching of high-school mathematics, present teachers with some serious challenges.

First of all, most teachers are used to emphasize punctuality. They usually insist on performance of mathematical algorithms rigorously, and rarely stop to tell students about mathematics, let alone about contemporary results. Quite often they are also pressed by the need to "cover" the curriculum and assure the success of the students [3]. Adding to these the difficulties students face in the ordinary curriculum and teachers' effort to help them cope, it seems unlikely to expect teachers to interweave any extra-curricular materials, even short interventions like the proposed snapshots.

Additionally, teachers' familiarity with contemporary mathematics may be limited thus inhibiting them from preparing, let alone introducing, mathematical-news-snapshots to their classes. Obviously, professional mathematics journal resources and web sites do not readily lend themselves to implementation in high-school classroom.

Apart from the demanding knowledge *of* mathematics and *about* mathematics, pedagogical-content-knowledge (PCK; [22, 23]) is obviously required. Yet another factor is teachers' own beliefs about the relevance of contemporary mathematics to educating their students mathematically. Furinghetti & Pehkonen [9] see beliefs as belonging to subjective-personal-knowledge that differs from objective-knowledge.

Finally, the question of evaluating the outcome of interweaving mathematical-news-snapshots in the implemented curriculum on a weekly basis, presents a real challenge.

Indeed, new tasks along with new responsibilities are created for the teachers if they take the risk of interweaving mathematical-news-snapshots as we are proposing.

9.5 The Study

As mentioned above, the road to exposing high-school students to mathematical news is strewn with challenges. In order to examine the feasibility of the proposed solution we have been involved in an experimental study in three age-group classes at one high-school in Israel. The first author acts as a teacher-researcher in two of the experimental classes and as an observer-researcher in the third class. The three parallel classes, one in each age-group, that study mathematics at the same level and whose populations are of comparable strength, serve as control groups for various measurements.

A series of 10 PowerPoint Presentations for a weekly-based snapshot-interweaving-program was prepared especially for the study. They are in the following topics:

- The search for prime numbers;

- Mathematics of Sudoku;
- Fermat's Last Theorem;
- The proof of Kepler's Conjecture;
- The mapping of the E8-Group;
- The four color problem;
- The linear algebra behind Google search engine;
- RSA cryptography;
- Benford's Law applied to tax-frauds and to election frauds.
- The Fundamental Theorem of Algebra applied to astrophysics;

During the development period and throughout the implementation we regularly considered the following issues:

- How can one tell that a particular piece of news is worth introducing to high-school students?
- Can we set up a list of *criteria* for selecting news for high-school age-level?
- What about *accessibility* and other pedagogical issues such as *connectivity* to the current curricular topics dealt with in class?
- What means might be used to make high-school students get *interested* in a piece of news?
- How do we make room for students' questions within the restricted time devoted to the snapshot?
- Reflecting upon the gap-bridging goal—how would a teacher *evaluate* such an intervention and how could we evaluate the teaching-learning situation (a goal-oriented evaluation)?

This paper is based on part of an on-going study. More details about the design of the study, the particular research questions, the intervention procedures, the research instruments, the findings and their comprehensive analysis, will be available in due course. In this paper, we present the first snapshot to which the students were exposed to. This will be followed by a brief analysis of *observed* reactions to the snapshot.

9.6 A Snapshot: "The Search for Prime Numbers"

The PowerPoint presentation of this snapshot consists of 22 slides. Table 9.1 presents their verbal contents and their categorization.

Table 9.1 A Snapshot on the Search for Prime Numbers

Slide No.	A Condensed Version of the Text on the Slide	Slide Category
1	**Prime Numbers—The Search and Discovery**	Title
2	**News from Aug, 2008: A Prize of $100,000 on a mathematical discovery.** Edson Smith, 41 years old, the head of the computer team in the Mathematics Department, UCLA, won (together with his colleagues) a prize on discovering the first prime-number with more than 10 million-digits. Actually, the number has almost 13 million-digits! (The slide contains a photo of Edson Smith)	News
3	**The Number's Length in Kilometers...** Printing all 13 million-digits (12 points size each)—can you estimate the number's length? **Answer:** 48 KM The company Perfectly Scientific is involved in research and development, consulting and marketing of products in the Mersenne-Prime-numbers domain. The company produces and sells posters of these numbers. If you wish to purchase a poster—go to the internet: `http://www.perfsci.com`	News

9.6. A Snapshot: "The Search for Prime Numbers"

Slide No.	A Condensed Version of the Text on the Slide	Slide Category
4	**Some of the Digits from the Beginning of The Number and until its End—millions of digits were omitted from the center!** 3164702693302559231434537239493375160541061884752646 4414030417673281124749306936869204318512161183785672 6816539985465097356123432645179673853590577238179357 9008764261039437823764945917429345884971175871469169 7298476115906087325093946208557574075457709862055801 1779529884042198287643319330465064455234988142139565 7854474740235463537585373248018381203876008684165254 0079038128588825668708585545623157752793930592081176 6585308670132129155221804381548625787943020694528015 999221718191557761... (millions of digits omitted) ...06934159709 8036883089983720514634411159760282269091566821920139 8 1830822014104610660911290342036586081253335507924074 42 6181487091805592043237230196201683535946231098006743 4 9846253807872478025327585113335024607788843390340197 0092766395816769890801073610141013699685292570327255 3544622464685928707526568105993689915218073801443404 9450082664259324131398269150840699911592797919083981 3022330482408311909319599980145624563479412021959009 2 8079670729447921616491887478265780022181166697152511	News
5	**Prime Numbers, Composite Numbers and The Unit.** Every natural number > 1 can be expressed as a product of two natural numbers. Some natural numbers can be expressed in one way only, others in more than one. Can you give examples? **Ans.** $2 = 2 \times 1, 3 = 3 \times 1, 5, 7, 11 \ldots$. But $10 = 2 \times 5 = 10 \times 1, 36 = 4 \times 9 = 12 \times 3 = 18 \times 2 = 36 \times 1, \ldots$. Do you know the name for the first type? **Ans.** Prime numbers Numbers of the second type are called Composite Numbers. The number 1 is neither prime nor composite.	**Basic Math Background (Connected to Previous Knowledge)**
6	**The Magic of Prime Numbers** Prime numbers always fascinated mathematicians for two facts proved hundreds of years BC: (i) There is an *infinite* number of primes (ii) It is hard to predict the next prime or just a larger prime, because there is *no* useful generating formula for primes. This was an *intellectual* challenge with *no* application until the 20th century. Then surprisingly an amazing application was found in *cryptographic theory*.	**Historical/ Cultural Background**
7	**There Are Infinitely Many Prime Numbers** Suppose 17 is the largest prime-number, then no prime-number greater than 17 exists. Consider the number: $N = 2 \times 3 \times 5 \times 7 \times 11 \times 13 \times 17 + 1$ *Is N* divisible by 2? **Ans.** No, there is a remainder of 1 Is it evenly divisible by any other known prime? **Ans.** No (same reason) What can we conclude? **Ans.** N possesses no smaller prime factor therefore, N is prime, But N is larger than 17 and this contradicts the assumption that 17 is the largest prime! What can we conclude? **Ans.** That our assumption was wrong, 17 is NOT the largest possible prime. Similarly, one can prove for any number P that it can't be the largest prime. This is how Euclid proved the infinitude of primes. Indeed, the search for primes is an endless challenge.	**Advanced Mathematics**

Slide No.	A Condensed Version of the Text on the Slide	Slide Category
8	**A Priest Interested in Prime Numbers** The French mathematician Marin Mersenne lived in the 17th century. He became famous for a conjecture he phrased in 1644: Numbers of the following form: $2^n - 1$ are primes for $n = 2, 3, 5, 7, 13, 17, 19, 31, 67, 127, 257$. For all other positive integers smaller than 257 they are composite numbers! (The slide includes a photo of Marin Mersenne (1588–1648))	**About Mathematicians (Past or Present)**
9	**Mersenne's Conjecture Had a Long Life** Mersenne's conjecture was proved only in the middle of the 20th century—300 years after it was first phrased! Apparently the conjecture had 5 errors: 3 prime numbers were missing and 2 numbers in the original list were proved to be composite. The correct statement is: Numbers of the following form: $2^n - 1$ are primes for all $n = 2, 3, 5, 7, 13, 17, 19, 31,$ **61, 89, 107**, 127. The 2 numbers omitted are: 67 and 257. The 3 missing primes appear in **bold**.	**Historical/ Cultural Background**
10	**Is Every Prime Number A Mersenne Number?** A *Mersenne Number* is a number of the form $2^n - 1$, where n is a positive integer. In order for a Mersenne number to be prime, n must be prime. Let us look at a few prime numbers: Is 2 a prime number? (Yes) a Mersenne number? (No) 3-? A prime and a Mersenne ($2^2 - 1$) 5- ? A prime but not a Mersenne 7-? A prime and a Mersenne ($2^3 - 1$) *Not* every Prime-Number is a Mersenne number	**Advanced (External to the Curriculum)**
11	**Breakthroughs** Since 1876 it has been known that: $2^{67} - 1$ is a Mersenne number with 21 digits—not a prime itself but a product of 2 smaller numbers. In 1903, after 3 years of weekly Sunday afternoon efforts, Frank Nelsen Cole, a professor of mathematics from Columbia University, NY factored the number into its 2 factors: $2^{67} - 1 = 761,838,257,287 \times 193,707,721$	**Historical/ Cultural Background**
12	**Frank Nelsen Cole (1861–1926)** (Picture Included) Born in Rehoboth, Massachusetts, a mathematician since the age of 17, interested mainly in number theory and group theory. As a teacher he introduced the most recent topics and regarded improving teaching an important issue. After his retirement, Cole established a foundation to encourage young mathematicians in number theory and algebra. The foundation originated from gifts collected by his colleagues, honoring Cole as a mathematician, a teacher, a friend and a kind human being.	**About Mathematicians (Past or Present)**
13	**How Many Mersenne Primes Exist?** Euclid proved that there are infinitely many prime numbers Question: Are there infinitely many Mersenne-Primes?? This is a question that mathematicians have not yet solved! To-date: Only 46 Mersenne prime numbers are known. What are the gaps between the numbers? This is another question that occupies mathematicians researching in this domain, still with no answer, as many other questions referring to the primes themselves.	**Contemporary Mathematics (Previous Findings Related to the News)**

9.6. A Snapshot: "The Search for Prime Numbers"

Slide No.	A Condensed Version of the Text on the Slide	Slide Category
14	**The Story of the Discovery of the 37th Mersenne Prime Number** The number $2^{3,021,377} - 1$ that has 909,526 digits was discovered in January, 1997 by Ronald Clarkson, (his photo appeared) a young 19-year-old student. He used a Pentium-200 for 46 days alternately, to prove it is a prime-number. If he had worked continuously the computerized test would have taken one *full week*.	**Contemporary Mathematics (Previous Findings Related to the News)**
15	**The Most Recent Discovery: $2^{43,112,609} - 1$** This is the 46th Mersenne prime pumber—the largest yet known. Do you wish to participate in the search? If so go to: `www.mersenne.org`. You may win: Prizes, Glory and mainly enjoyment.	**News**
16	**What Drives People to Look for (and find) Prime Numbers?** *Curiosity*: This was the main drive of Euclid (photo included, 365–275 BCE) who proved that there are infinitely many prime numbers, and of Euler and Fermat who followed by studying their distribution. *Collecting rare and beautiful things*: The by-products of these studies are new "jewels," namely rare prime numbers, intriguing problems, and theorems that are surprisingly beautiful. The *glory and personal satisfaction*: similar to climbing Mt. Everest! *New applications!*—Prime search is used to test hardware in super-computers. *Prizes* (money!) and …*Enjoyment*!	**Mathematical Culture**
17	**How do we Search for Prime Numbers?** **Formulas for Prime numbers:** The formula $y = x^2 - x + 41$ generates prime numbers *only* for any value of *x* between 1 and 40. Let's try…The problem—Legendre proved a long time ago that there is NO polynomial formula that generates *only* primes, and certainly not *all* the primes. A Mersenne number with a prime exponent does not ensure a prime number as a result, but gives hope…	**Basic Math Background (Connected to Previous Knowledge)**
18	**Who is Searching for Prime Numbers Today?** The international project GIMPS (Great International Mersenne Prime Search), in which mathematicians and other curious people participate, is a volunteer project. Thousands of people from all over the world are involved and connected with the most powerful computers that exist.	**Collaborations in Mathematics**
19	**The Prize** "…*from the University of California, Los Angeles, discovered a huge Prime number…Mathematicians finally discovered a prime number with 13 million digits. The discovery is a mile-stone that has been searched for a long time and makes them worthy of the prize of $100,000…The team worked for a full month employing a network of 75 computers using Windows XP system.*"	**Prize for Mathematics**
20	**Who Grants the Prizes and What For?** Prizes are granted by the EFF—Electronic Frontier Foundation. Although the Prime Numbers are important for mathematics and for encrypting, the prizes are granted for cooperation. The main theme is that many other problems can be solved using similar methods!	**Prize for Mathematics**

Slide No.	A Condensed Version of the Text on the Slide			Slide Category
21	**Every Prime-Number and The Granted Prize!**			**Prize for Mathematics**
	The Prize (in thousands of $)	The number of digits	Date granted	
	50	1,000,000	Apr. 2000	
	100	10,000,000	Sep. 2008	
	150	100,000,000	Not yet— waiting!	
	250	1,000,000,000	Not yet—waiting!	
22	**Does All This Have an Application?** Prime-Numbers have found interesting application after hundreds of years of merely being an intellectual challenge. *Cryptography*: Briefly, it is easy to see that composing a number as a product of two primes, even huge ones, is not difficult. However decomposing a huge number to find its prime factors is extremely difficult and time consuming, up to the point that it is practically impossible. This is the basis of modern cryptography. www.claymath.org/posters/primes/ *Testing computer hardware*: Factoring some numbers of more than 100 digits involves testing divisibility by more numbers than all the particles in the universe. In the 20th century, processes of searching for Mersenne-Numbers were applied as tests of the capabilities and limitations of computer hardware.			**Applications**

9.7 Analysis of Observed Reactions to the Snapshot

In Grade 11, the classes in which a systematic observation took place by the 1st author, there were 9 boys and 16 girls. They study mathematics towards the matriculation exam, at the 4-units level (out of 5).

One of the data collection mechanisms was an observation of students' reactions by the researcher during a colleague-teacher's activation of the PowerPoint presentation. Verbal-responses were also audio-recorded. Altogether 134 verbal reactions were recorded during the 15 minutes presentation. This indicates a significant involvement of the students in this part of the lesson. They were classified a posteriori by 4 independent professional teachers into 5 main categories: (i) Expressions indicating mathematical knowledge; (ii) Making connections to previously learned topics, everyday life, personal experience; (iii) Expressions of interest and motivation; (iv) Expressions of emotions and attitudes: Admiration; Love; Empathy; Contempt; Humor; (v) Expressions indicating views, perceptions or beliefs about: Formula; Numbers; Open-endedness; Mathematics in general or mathematicians' work; The distribution of students' reactions to various categories of slides in the presentation is presented in Table 9.2.

The data in Table 9.2 indicate that students' reactions were evident both in the cognitive domain (58 in categories i, ii) and in the affective domain (76 in categories iii, iv, v). The most dominant outcomes 49/134 referred to mathematical knowledge (Category i), both procedural and conceptual. The students felt free to share out loud their emotions—(Cat. iv) and views (Cat. v) (39/134, 25/134 respectively), 12 of the 134 reactions expressed their interest and motivation (Cat. iii) and only 9 revealed connections (Cat. ii) (9/134).

While one should wait for a more detailed analysis of the results[1], we already have a strong feeling that the "snapshot-approach" is helpful. To support it we conclude this section with 7 quotes of students:

- An indication of acquiring new mathematical knowledge (Cat. i):
  "What's a Mersenne-number?—I forgot! Can you explain it once again?—OK, I got it!;"

- Expressions indicating views, perceptions or beliefs (Cat. v):
  "What is the benefit in discovering a formula? The process still continues—there is always something else to search for!"
  "Thousands of people working together—This is some Cooperation!"

---
[1] Between submission of this paper and its publication another paper with more data was presented at ESU6 (Vienna 2010) and will be published in ESU6 proceedings 2011. (See [2].)

Table 9.2 Distribution of Students Responses by Categories and Slides

Slides by Categories \ Students Response Category	(i) Knowledge	(ii) Connections	(iii) Motivation	(iv) Emotions	(v) Perceptions	Total
Slide 1 Title	6	1	1			8
Slide 5, 17 Basic Mathematical Background	7	4		4	2	17
Slide 7, 10 Advanced Mathematics	7			2	1	10
Slide 13, 14 Contemporary Mathematics	8	1	6	1	4	20
Slide 8, 12 About Mathematics	8	1		2	1	12
Slide 2, 3, 4, 15 News	9			12		21
Slide 16 Mathematical Culture			5	7	1	13
Slide 6, 9, 11 Historical/Cultural Background	3			2	9	14
Slide 22 Applications	1			1		2
Slide 19, 20, 21 Prizes		2		6	7	15
Slide 18 Collaboration in Mathematics				2		2
Total 22 Slides	49	9	12	39	25	134

- Expressions of emotions and attitudes (Cat. iv):
  "*All their life they learn, work and do mathematics for just one new number… all their life!*";
  "*How come all these clever people do all these clever things and we just do these stupid routines.*"

9.8 Closing Remarks

This paper presents a research and development study the goal of which is to bridge the gap between the ever growing nature of mathematics and the stagnated nature of school curriculum. It is addressing ESU aims:

"…lead to a better understanding of mathematics itself and to a deeper awareness of the fact that mathematics is not only a system of well-organized finalized and polished mental products, but also a human activity, in which the processes that lead to these products, are equally important with the products themselves."
(ESU aims and focus statement, `class.pedf.cuni.cz/stehlikova/esu5/01.htm`)

The proposed solution is interweaving mathematical-news-snapshots in the teaching of mathematics on a weekly basis. Considerations of the advantages and the limitations of this solution are included. Since the second round of the intervention study is presently going on, it is too early to come up with an overall analysis of the 10-week series of experimentation of this solution. Nevertheless, one detailed example of a mathematical-news-snapshot and partial analysis of the observations of students' reactions towards it, are included. This preliminary analysis indicates positive students' reactions, even an enthusiasm about the new window the snapshots open to the nature of mathematics.

Beyond the influence on students, and based upon the experimenting teachers' attitudes, we believe that the suggested approach of interweaving mathematical-news-snapshots in the ordinary teaching, can also help boost high-school teachers' self esteem and status, as well as fight speedy burnout, so common among teachers after relatively small number of years in the profession.

Readers of this paper are invited to join this on-going research and development project conducted at Technion, Israel Institute of Technology.

Bibliography

[1] Albers, D. J., G. L. Alexanderson, and C. Reid, 1990, *More Mathematical People*. HBJ, Publishers.

[2] E. Barbin, M. Krongellner, C. Tzanakis (eds.), 2011 (in press): *History and Epistemology in Mathematics Education: Proceedings of the Sixth European Summer University*, Vienna 2010.

[3] Barton, B., 2009, *The Language of Mathematics*, Springer.

[4] Berman, A. and N. Shaked-Monderer, 2009, "Nonnegative Matrices and Digraphs," in: Meyers, R. A., Editor-in-chief, *Encyclopaedia of Complexity and Systems Science*.

[5] Borcherds, R. E., 1992, "Monstrous Moonshine and Monstrous Lie Superalgebras." *Inventiones Mathematicae* pp. 109, 405–444.

[6] Corry, L., 2004, "The Axiomatic Method in Action 1900–1905." In: *David Hilbert and the Axiomatization of Physics* (1898–1918). Kluwer Academic Publishers.

[7] Fried, M., 2001, "Can Mathematics Education and History of Mathematics Coexist?" *Science and Education* pp. 10, 391–408.

[8] ——, 2007, "Didactics and History of Mathematics: Knowledge and Self-Knowledge" *Educational Studies in Mathematics* pp. 66 (2), 203–223.

[9] Furinghetti, F. and E.Pehkonen, 2002, "Rethinking Characterization of Beliefs," in: Leder, G. C., Pehkonen, E. and Torner, G. (Eds.): *Beliefs: A Hidden Variable in Mathematics Education?* pp. 39–59. Kluwer Academic Publishers. The Netherlands.

[10] Herzberg, A.M., and R.M. Murty, 2007, "Sudoku Squares and Chromatic Polynomials," *Notices of the American Mathematical Society*, pp. 54, 708–717.

[11] Leder, G. C., E. Pehkonen, and G. Torner, 2002, "Setting the scene," in: *Beliefs: A Hidden Variable in Mathematics Education?* pp. 1–10. Kluwer Academic Publishers.

[12] Mandelbrot, B. B., 1977, Fractals: *Form, Chance and Dimension*. San Francisco, CA: W. H. Freeman and Company, 1977, xviii+ pp. 265 (A shorter French version published 1975, Paris: Flammarion).

[13] Movshovitz-Hadar, N., 2008, "Today's news are Tomorrow's history—Interweaving mathematical news in teaching high-school math." In: Barbin, E., Stehlikova, N., Tzanakis C., 2008 (eds.) History and Epistemology in Mathematics Education: *Proceedings of the fifth European Summer University*, Ch. 3.12, pp.535–546, Vydavatelsky Press, Prague 2008.

[14] Murty, M. R., 2008, "The Mathematics of Sudoku Squares," in *Queen's Mathematical Communicator'* spring 2008, p.1–2. Issued by the Department of Mathematics and Statistics, Queens' University Kingston, Ontario, K7L3N6.

9.8. Closing Remarks

[15] Poincaré, H., 1908, Mathematical Discovery (In Science and Method). In: Cipra, B. 1998, *What's Happening in the Mathematical Sciences 1998–1999*, American Mathematical Society, pp 115–126. Also retrievable from: `www.ias.ac.in/resonance/Feb2000/pdf/Feb2000Reflections.pdf`

[16] Radford, L., 2008, "Beyond Anecdote and Curiosity," in panel discussion: "Mathematics of Yesterday and Teaching of Today," Published in: Barbin, E., Stehlikova', N., Tzanakis, C., 2008 (eds.) *History and Epistemology in Mathematics Education*: Proceedings of the 5th European Summer University (ESU 5), Ch. 2.2 pp. 163–167, Vydavatelsky Press, Prague 2008.

[17] Reid, C., 1985, *Hilbert-Courant*. Springer-Verlag, New-York, NY.

[18] Rivest, R. L., A. Shamir, and L.Adleman, 1978, "A Method for Obtaining Digital Signatures and Public-Key Cryptosystems." Retrievable from: http://people.csail.mit.edu/rivest/Rsapaper.pdf, p. 15.

[19] Robertson, N., D. P. Sanders, P. Seymour, R. Thomas, 1996, "A New Proof of the Four-Colour Theorem." *Electronic Research Announcements of the American Mathematical Society*, pp. 2, 17–25. Retrievable from: `www.ams.org/era/1996-02-01/S1079-6762-96-00003-0/home.html`.

[20] Schoenfeld, A. H., 1988, "When Good Teaching Leads to Bad Results: The Disaster of 'Well-Taught' mathematics courses." *Educational Psychologist*, pp. 23 (2), 145–166.

[21] ——, 1992, 'Learning to think mathematically: Problem Solving, Metacognition and Sense Making in Mathematics.' In D. A. Grouws (Ed.): *Handbook of Research on Mathematics Teaching and Learning*, Macmillan, NY, pp. 334–370.

[22] Shulman, L. S., 1987a, "150 Different Ways" of knowing: representations of knowledge in teaching. In: Calderhead, J., 1987, *Exploring Teacher Thinking*. Sussex, Holt, Rinehart & Winston.

[23] ——, 1987b, "Knowledge and Teaching: Foundations of the New Reform." *Harvard Educational Review*, pp. 57 (1), 1–22.

[24] Singh, S., 1998, "Interview with Richard Borcherds," *The Guardian*, 28 August, 1998. (Retrievable from: `www.simonsingh.net/Fields_Medallist.html`, December 25, 2008).

[25] Siu, M., 2000, "The ABCD of using history of mathematics in the (undergraduate) classroom." (Retrievable from: `hkumath.hku.hk/~mks/ABCD.pdf`, July 30, 2009).

[26] Swetz, F., 2008, "Comments," in panel discussion: "Mathematics of Yesterday and Teaching of Today." In Barbin, E., Stehlikova', N., Tzanakis, C., 2008 (eds.) *History and Epistemology in Mathematics Education*: Proceedings of the 5th European Summer University (ESU 5), Ch. 2.2, pp. 167–169, Vydavatelsky Press, Prague 2008.

About the Authors

Batya Amit is Ph.D. candidate in mathematics education at Technion—Israel Institute of Technology, due to graduate in 2011. Her thesis focuses on interweaving contemporary mathematics in the high-school teaching practice. Her main research interest is the impact of introducing Mathematical News Snapshots on students and also on mathematics teachers. She finds it a challenge to develop MNSs. Batya Amit earned her B.Sc. in Mathematics, Statistics and Computer Science from The Hebrew University, Jerusalem, Israel (1977) and M.Sc. in Computer Science from The Department of Applied Mathematics, The Weizmann Institute of Science, Rehovot, Israel (1980). Since 1993 she has been a mathematics teacher in high-school and for several years participated in "MatiMachshev"—a team project developing and experimenting with new mathematical materials at the Weizmann Institute. Batya has presented various facets of her Ph.D. study in five international conferences. She won three prizes on innovations and excellence in mathematics practice and three scholarships on excellence in research.

Nitsa Movshovitz-Hadar was Professor of Mathematics Education at Technion—Israel Institute of Technology and has been Emeritus since 2004. She entered the faculty of the Technion in 1975, having had 12 years of high school mathematics teaching experience, and has shared her research and development interests with prospective mathematics teachers. She has headed major curriculum development projects from 1977 to 2011, has supervised 25 graduate students, and from 1986 to 2006 served as academic director of Kesher-Cham—Israel R&D centre for improving and reviving mathematics education. In 2003 she was Laureate in Residence at La Villa Media, Grenoble, France, while from 1998 to 2002, she was Director of the Israel National Museum of Science. Since 2010 she has been on the advisory council of MOMATH—the museum of mathematics in New York. She has published one book (with J. Webb) and many papers in professional journals (one of which received in 1995 the MAA Lester Ford award with I. Kleiner). Since 2009 she has been giving public lectures in mathematics.

Avi Berman is a Professor of Mathematics at the Technion, where he holds the Israel Pollak Academic Chair. His research interests include Nonnegative Matrices, Spectral Graph Theory, Mathematics Education, and Giftedness and Creativity. He was the Head of the Department of Education in Technology and Science and the Pre-University Academic Center at the Technion and held the Chair of the Israeli Society for Research on and Promotion of Giftedness.

10

History, Figures and Narratives in Mathematics Teaching

Adriano Demattè and Fulvia Furinghetti
University of Genova, Italy

10.1 Introduction

This article relates to a project that has been developed in various classrooms of secondary school. The main assumptions inspiring this project are the following:

- students' motivations to learn mathematics are enhanced if they develop an image of mathematics that encompasses the 'social sense of mathematics' (that is they see mathematics as a human process embedded in culture)
- the history of mathematics may provide them with a context suitable to develop such an image.

The concern about introducing a human and cultural dimension in the image of mathematics is explicitly stated by many teachers and researchers who plan to use the history in classroom, see [2, 12, 15, 22, 27] for example.

An important component of the social sense of mathematics is the 'historical sense of mathematics', which includes the perception that mathematics was built over time and through various civilizations and an appreciation of its past as a component of the present. We are aware that the historical sense is rather elusive for most people, including those who have a mathematical culture, and, even more, is particularly difficult to define in the case of young students, whose education (not only mathematical) is still in progress. Nevertheless, in previous studies with students of lower secondary school, see [6], we have found some traces of the historical sense of mathematics in their image of mathematics. We claim that the development of the historical sense may be enhanced by linking students' experience with the mathematicians' experience, in particular, by bringing to the fore the socio-economical environment and the debate among different streams of thought in which the mathematical development happened. 'Historical-like' questions such as "What were the reasons that induced mathematicians to investigate isosceles triangles?" found in [4] may stimulate students in this direction. But other ways have been devised, more explicitly based on history. Percival [22] asked her students aged 12 and 13 to make their own constructions of objects (such as Babylonian tablets) and to use ancient calculating devices, albeit in modern reconstruction. Her goal was to show the "humanist, 'human-made' side of arithmetic" (p. 21). Smestad [27] used old and modern paintings when teaching perspective. Thus his students (prospective teachers) had the chance to see "that this particular part of mathematics has developed over time as people met artistic and mathematical problems that they needed to solve. They also see that there are important connections between mathematics and art" (p. 552).

10.2 Using Pictures as a Special Case of Using Original Sources

In our project we use various types of original sources, as advocated by [18]. In the experiment here reported our original sources, however, are special, since they are not verbal, but only pictorial. Actually, in many cases pictures are an important aspect of mathematical documents of the past, as shown in [20] and in the studies on Greek diagrams by Netz [21] and Saito [26]. Sometimes the historical pictures are only decorative elements without any reference to the written text; in other cases they have an explanatory function, and, consequently, are important elements for mathematical communication.

The pictures we used represent people who are doing actions that have some relation with mathematics, e.g. they are measuring with different kinds of instruments, or writing numbers for performing computations and so on. Figures 10.1, 10.2, and 10.4 are examples of such pictures. The pictures were shown to the students and they were required to write what is happening in the represented scenes. It was recommended that they focus on the mathematics present in the picture. Through this task the students were set in a hypothetical historical situation. As a consequence they did not act with certain and consolidated knowledge, but are in the process of constructing new knowledge in a concrete context. The method used for carrying out this activity is inspired by some suggestions coming from [11], in particular, those concerning history and the idea of 're-actualization' here applied in the case of interpreting a non verbal text.

The task proposed to our students involves a hermeneutic activity, which is typical of history. As Jahnke [17, p. 173] put it, "History of mathematics is essentially a hermeneutic effort: Theories and their creators are interpreted. Interpretation comprises a circular process of forming hypotheses and checking them against the material given." A main part of the chapter on the use of original sources in the ICMI Study book [10] is dedicated to the process of interpretation and hermeneutics, see [18]. This is due to the fact that a source has to be interpreted following a process that may be described by a twofold circle. In the primary circle a scientist (or a group of scientists) is acting and in a secondary circle the modern reader tries to understand what is going on. There is an intertwining between the interpretations by the modern reader of a certain concept or theory and the interpretation of the original authors. Jahnke et al. [18, p. 299] claim that "As a crucial point in hermeneutics, the student's self will unavoidably enter the scene, not as a disturbing factor, but as a decisive prerequisite to insight." Students who think themselves into a person doing some form of mathematics (no matter whether the person is the author of a written text or a figure in a picture) must themselves do mathematics; they move in a mental game in the primary circle reflecting what the person under study might have had in mind and mobilize their imagination to generate hypotheses about that. Of course, certain aspects of the historical persons and their ideas will be easily accessible, while other aspects will remain alien.

The texts produced by our students may be considered a kind of 'mathematical essay', created at a naïve level. Some of these texts had the nature of narratives in which characters act in a given context, use tools to achieve a given goal and have mental states. To produce a plot of a story is a way of reducing the complexity of a description. Even if the story teller stops on some details, the need for going on in the story inspires us to follow the thread of the story. A story stresses the presence of a goal and is structured in accordance with this goal. Also in mathematical reasoning goals are present: for example, the thesis of a theorem or the statement that is the object of an explanation give direction to the reasoning [7]. According to Bruner [5] the difference between scientific and narrative thinking is that the former leads to descriptions which are outside the original context, while the latter keep the richness of the context. Bruner deems that the difficulty in doing mathematics is originated by the absence of narratives, by the lack of stories. For Bruner the most natural way of organizing our experience and knowledge is the narrative mode. On one hand the events of the story receive their meaning from the whole story, and on the one hand the whole story is made by its parts. This is what constitutes the "hermeneutic circle."

Bruner [5] claims that a story may be subject to interpretation. Also the reverse is true: Smorti [28] ascribes to stories an inferential and interpretative function and thus the story itself may be the form of an interpretation. We note that a description inserted in a context requires a story involving the essential elements. The interpretation has to be coherent and to possess the characters of verisimilitude. Too many details make the interpretation less significant, as if it were not an organized perception. The identification of a description that is coherent and aimed at the essentials appears as a step in the passage to decontextualization that for Bruner is the feature characterizing scientific knowledge. Of course, the choice of the details on which the story is constructed is subjective, but must help to provide the story with a sense.

In literature the association of mathematics and narrative is considered from different perspectives and with different approaches. Alexander [1] discusses the development of mathematics in the early nineteenth century through romantic narratives. Thomas [30] compares theorems and proofs with narratives, both fictional and historical. In mathematics education, Brown [3] discusses the role of language and narratives in the learning of algebra, by taking as a main reference the work of Paul Ricoeur on hermeneutics. Solomon and O'Neill [29] examined mathematics writing—in particular, the letters of William Rowan Hamilton (1844)—in historical perspective and concluded that mathematicians operate within a number of non-narrative genres through which mathematical meanings are constituted. They use this argument against those who argue that narrative approaches to mathematics are both desirable and possible to solve the problem of under-achievement in mathematics. Our position is that narratives may be a good means for an *initial approach* to mathematical knowledge, since the students' spontaneous development of specific narratives may give a meaning to what they learn. This happens because students are provided with contexts that make the use of the mathematics emerging from the figures an act of productive thinking. They do not answer questions posed by someone, as is usual in school, but questions asked by themselves.

10.3 The Experiment

The activity presented in this paper was implemented in Italian classes of upper secondary school (with students aged 14–18). A copy of an ancient picture such as those in Figures 10.1, 10.2, 10.4 was given to each student together with a form explaining the task, see Table 10.1:

Table 10.1. Form given to students participating to the experiment

---

Research project: WHAT INFORMATION A PICTURE RELATED TO THE HISTORY OF MATHEMATICS MAY CONVEY?
Students' ideas

- In our society we communicate by using visual images: television, computer, posters and so on.

- Mathematics was born and developed in a social context. Many people had used it.

I ask you to interpret the attached picture.
Look at the people in the picture and describe what you think they are doing (examine the context and the details; try to use your mathematical knowledge).
Thank you.

The first three rows of the form acquainted students with the fact that their performance was part of a research project: this information was given with the aim of motivating students to fill a task rather unusual for them. They knew that no mark was given. Students worked in small groups. The activity was carried out in different situations, which we describe when reporting students' excerpts of their writings.

Though the request in the form addressed verbal descriptions, there was not an explicit reference to the use of narratives. Narratives, indeed, were not present in all students' answers. As a general impression we note that the students showed some interest in the activity. In the following we comment on some of their excerpts that may illustrate how they interpreted the pictures by building stories.

Figure 10.1. Egyptian mural at Sheikh Abd el-Qurn (Egypt), around 1400 BCE[1].

Chiara and Francesca about Figure 10.1

Chiara and Francesca were students in the first year of upper secondary school (Lyceum with socio-psycho-pedagogical orientation) aged 14. The task was given during an extracurricular time. They had never encountered topics in the history of mathematics. They wrote:

> In the first part of the picture the Egyptians are measuring how many hectares (with a different unit of measure, of course) of [field of] wheat they have. We have deduced this fact from the rope with equidistant knots displayed along the picture. There are also some children who are helping to measure with a rope. Afterwards they [the Egyptians] calculate the amount of the harvested wheat and the ratio between the harvest and the plots of ground. [In the second part of the picture] at the left the cart carrying two kings' servants is represented. The slaves have the task of recording the data and reporting to the king.

The narrative structure of this excerpt is evident. A story emerges that unifies the two stripes of the picture and sets the events in a temporal sequence. Chiara and Francesca do what a story teller does: the story teller does not have to tell facts that are real, certain and confirmed by proofs, but rather is allowed to resort to imagination by reorganizing and selecting those events that are functional to the goals of the story. The two girls have responded to the task in a synthetic, but exhausting way: they gave a role to the persons represented, have outlined a plausible context and identified some details, for examples the knots in the rope, that few students have noticed. We stress that in the Egyptian painting there were no explicit elements indicating that the course of the events was finalized with some kind of calculations. The students introduced some conjectures that make the story appear believable and interesting. They also have exploited some pieces of their mathematics knowledge. In their story it is just the need to collect the data for performing some calculations that defines the course of events: the goal, indeed, is to give an account of the productiveness of the wheat fields.

---

[1] The figure is taken from p. 7 of the chapter "Lengths, Areas, and Volumes" in [19]. It reproduces the mural in the tomb of Menna (probably Eighteenth dynasty), who was the "scribe of the Fields of the Lord of the Two Lands." The mural "shows surveyors or "rope-stretchers" measuring a field of grain with a knotted rope .... Although they may also have surveyed sites for pyramids and temples, it is believed that the primary purpose of these "rope-stretchers" or *harpedonaptai* was to survey the farmland along the Nile River after the yearly floods in order to assess taxes fairly." [19, p. 7].

Anna and Lucia about Figure 10.1

Also Anna and Lucia were students in the first year of upper secondary school (Lyceum with socio-psycho-pedagogical orientation) aged 14. The task was given during an extracurricular time. They had never encountered topics in the history of mathematics. They wrote:

> In the first part of the picture there are slaves who are measuring the area of a wheat field. There are also children who are helping to measure with a rope. In the second [part] the horse is taking the wheat to the granary, where other slaves are arranging it.

The two scenes are connected in a unitary story: the guiding thread relies on the conjecture that the horse is taking the wheat to the granaries, but the narrative structure of this story is slighter than that told by the previous students. The reference to mathematics is given only by the information precisely shown in the picture—people who are measuring—but it is not used to create a guiding thread in the story imagined by the students.

Both pairs of students make a conjecture deriving from the perception that people who measure are children. This fact recalls Peirce's theory of perception as inference that involves also abduction, as discussed in [31]. When the same picture was shown to two university mathematics students, who have a wider mathematical knowledge, the small figures were interpreted as adults in the background represented in a naïve form of perspective[2].

In the two stories inspired by the Egyptian picture, mathematics plays a different role. In the first the concept of ratio is used as a means to interpret the picture, by giving to the picture a meaning that goes beyond the information directly provided by it and by constructing a story that goes toward a final goal. In the second narrative the reference to mathematics concerns only the first scene. The students produce a conjecture that gives to the story a partial goal: the slaves are measuring with the aim of finding the area of the field.

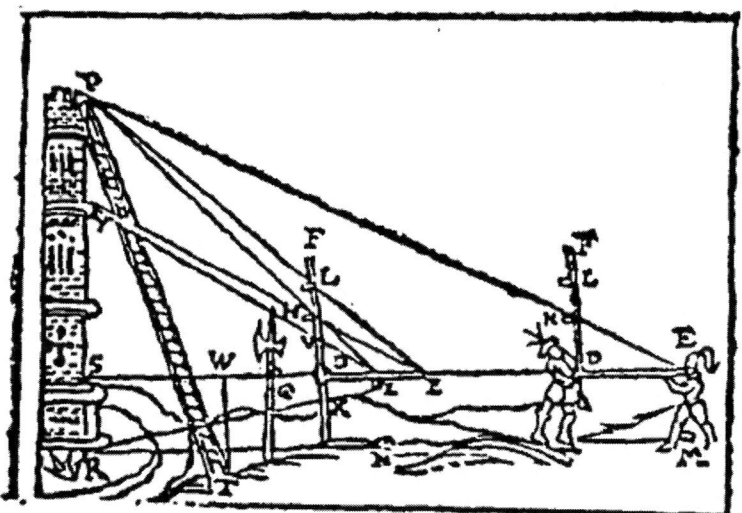

Figure 10.2. Measurements in a field manual (1590)[3].

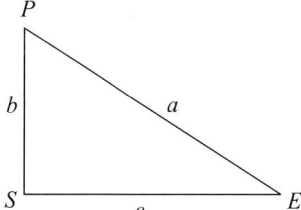

Figure 10.3. The triangle used by Nicola and Vera.

Nicola and Vera about Figure 10.2

Nicola and Vera are students in the fourth year of upper secondary school (Lyceum with orientation to social sciences) aged about 18. In the preceding school year they have used historical sources, in particular Caramuel's drawings (see

---

[2] A widely accepted hypothesis on the differences in dimensions of people is that it is a means to represent the difference in their social status.

[3] Source: *Baculum Familliare, Catholicon sive Generale. A Booke of the making and use of a Staffe, newly invented by the Author, called the Familiar Staffe*, by John Blagrave, publ. Hugh Jackson, London 1590—facsimile publ. Da Capo Press, New York 1970. The figure used in the experiment is at p. 103 of the chapter "Trigonometry" in [19] together with good reason given by Blagrave why combatants measured from afar: "If a Wall or tower were to be scaled ... How ... to get height thereof, thereby to make your scaling ladders accordingly, ... Where you dare not come neere the base of the tower for daunger of shot or let by reason of some deepe mote or ditch." As Katz and Michalowicz [19, p. 99] put it, "During the medieval period, waging battle involved studying the motion of cannonballs, muzzle angle in gunnery, distance to enemies, height of their fortifications, distance between two separated outposts, and the depth of ditches." Blagrave used similar triangles and his "familiar staffe" instrument rather than trigonometry. Possible uses in mathematics teaching of Blagrave's work are presented in [25].

Figure 10.4) for approaching trigonometry. They wrote:

> In this picture, people are checking if the tower is perpendicular with respect to the ground. They are using Pythagoras's theorem of the right triangle ($a = \sqrt{b^2 + c^2}$). Another hypothesis: people are trying to measure the height (side b) by using the formula ($b = \sqrt{a^2 - c^2}$). [The student added to the picture some letters, as we show in Figure 10.3]."

Also Nicola and Vera show the disposition to conjecture in order to reconstruct a goal of the represented scene. They explicitly point out that some statements are conjectures, because they are aware that the goals of the activities performed by the people in the pictures may be different from that they have supposed. In the case of the Egyptian picture the students appeared less aware of the fact that they were using conjectures. This different disposition may be due to the different ages, but also to the type of representation: Figure 10.2 shows a geometrical construction rather complex to be interpreted in all details.

Figure 10.4. Picture in Caramuel's treatise on theoretical and practical mathematics[4].

Michela about Figure 10.4

In Caramuel's treatise the figure ("Lamina IX") consisted of two parts: a naturalistic part (see Figure 10.4) and a geometrical part (see Figure 10.5) explaining the use of the instrument. In our experiment only the naturalistic part was shown to students. According to all students, Figure 10.4 represents a war story. The picture is full of details and it was difficult to give a role to all of them in a story related to this scene. The dotted triangle added by the painter addresses the interpretation.

Michela was a student of the fourth year of upper secondary school (Lyceum with socio-psycho-pedagogical orientation) aged 18. She had some experience in the history of mathematics, since she participated in the preparation of an exhibition in her school by preparing posters on historical figures and constructing instruments such as Blagrave's baculum and the quadrant. She wrote:

> We are in a war camp close to a town. The soldiers are setting the guns and through trigonometric rules are calculating the right position to strike the targets. The triangle, which has two vertexes in the men who are

---

[4]Caramuel, Johannes, 1670, *Mathesis biceps vetus et nova*. Campania. The instrument described in the figure was called "Tripleuron" by Caramuel (see the Greek word in Figure 10.5). In the section (pp. 356–358) of the chapter on geometry, which is devoted to explain methods for surveying lands, the author explains how to use it for solving triangles. Figure 10.4, which is taken from *Il giardino di Archimede*, was employed in other classes with purposes different from those illustrated in this paper, *e.g.* for introducing trigonometry, or in final tests for assessing students.

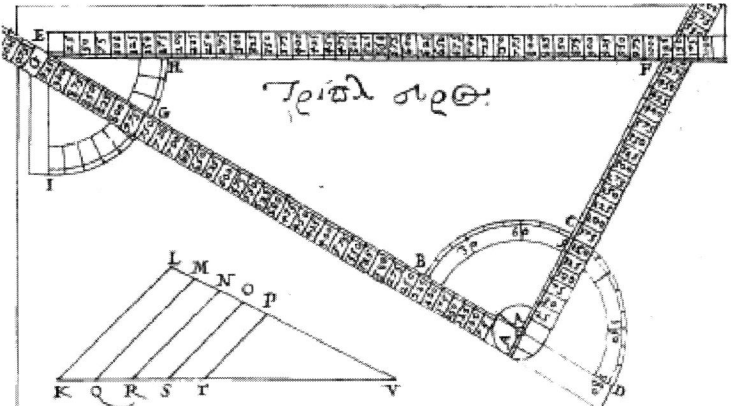

Figure 10.5. Geometrical explanation of Caramuel's instrument in "Lamina IX."

taking sight and the third vertex in the tower, is not right angled, but the gun between the two men divides it into two triangles, which, maybe, are right angled. .... Moreover it should be noted the presence of a compass close to the 180° protractor: this stresses the importance of the direction and the position.

I wonder at the presence of a rowing boat just in the middle of a military attack. The absence of vegetation sufficiently thick to hide the attack may mean the author's simplification or the impossibility for the town to be defended. ... the presence of a quite big ship hints that the town may be an important centre of commercial exchanges.

Michela's words reveal the effort to use mathematics to explain the scene represented in the picture, but she does not exploit in depth the mathematical information provided by the painter, such as the use of a protractor and the possible applications of trigonometry by the people in the scene, as other students did about Pythagoras' theorem in commenting on Figure 10.2. The student's words ("the gun may divide...") suggest that triangles that are not right angled are a disturbing element for the mathematical interpretation. Nevertheless the story is reconstructed through a narrative rich with references to the picture's details. In the final part Michela focuses on two elements that she inserts with difficulty into the story: the rowboat and the absence of vegetation. The two attempts at interpretation are different; the representation of the boat is mainly referred to as an artistic choice made by the painter, while the second tries to insert the absence of vegetation into the story. The coexistence in the student's writings of considerations concerning both the action represented and the painter's choices shows a double level of reflection. One reflection focuses on the intrinsic coherence of the figure considered as a real object; the other reflection concerns the informative and explicatory intentions of the artist. This latter type of reflection is little present in the other students' writings.

10.4 Didactical Implications and Conclusions

In our experiment we have explored and tested an unusual way of using history in mathematics teaching. Eagle [9, p. 388] discusses the "semiotic potential" in the case of figurative arts and pictorial signs. The excerpts previously reported show how this semiotic potential addressed in a direct way students' attention to the real situations represented. The power of pictures, indeed, is acknowledged by one of the authors we used (Caramuel). In the text accompanying Figure 10.4 he wrote that he will use just a few words because Figure 10.5 was explaining mathematically the use of the instrument[5].

Our activity is a special case of approaching original sources, with the advantage that in the case of historical pictures the problem of dealing with unknown languages in which sources may be written is bypassed. Following Jahnke [16, p. 154 ff.] we may say that the use of historical pictures is relating the 'synchronous' and the 'diachronous' mathematical culture to each other. Synchronous culture refers to the classroom as well as to the role of mathematics in public life, in economy, technology, science, and, generally speaking, in all form of culture. The diachronous culture is related to the development of these elements through history.

---

[5]"picturamque impraesentiarúm paucis verbis exponam" (p. 356).

Jahnke et al [18] stresses that the aim of school activities with original sources is not at all an imitation of the professional historian in regard to rigor and sophistication. Rather, the students are led to asking questions, which, in general, they had never asked before. In our experiment, indeed, the cognitive actions carried out by our students involved particular operations, "which can be identified as *argumentative operations* because of their linkage of social arguing and individual reasoning" [23, p. 368]. Such actions, according to Pontecorvo and Girardet, resemble those carried out by experts when interpreting historical events and documents.

As a matter of fact the writings of the students reveal their inclination to produce conjectures, when the context is favorable to this activity. The conjectures are used to introduce pieces of information not explicitly given in the pictures. In this way the meaning of the story is constructed by filling the gap provoked by the lack of information in the pictures and, consequently, setting continuity in it. The conjectures are sometimes accompanied by students' expressions such as "It makes me think" (Michela), "Another hypothesis" (Nicola and Vera), etc.

In summarizing the main features characterizing our project, we may say that historical figures are a way of using the history of mathematics that, if correctly implemented, may have some educational benefits, because:

- it provides students with an efficient way to approach the history of mathematics, since it encompasses the hermeneutic approach to the original documents through a mode rather accessible and motivating for students

- stresses the socio-cultural aspects of mathematics

- links the narrative and the logical thinking

- fosters the skill of producing conjectures.

At the ground of our project there is the goal to give to students the idea of what in [24, p. 510] is called "the ineluctably sociocultural embeddedness of the ways in which mathematics is carried out." This idea has an educational value not only for its influence on students' image of mathematics, but also for the potentiality of linking mathematics to the other school disciplines and, consequently, of promoting interdisciplinary projects that involve teachers of all school subjects.

Acknowledgment. Our project was firstly inspired by the *Historical Modules for the Teaching and Learning of Mathematics* edited by Katz and Michalowicz [19]. As noted in the comment to the historical modules in [8], the easy access to original sources and the historical materials suitable for the classroom provided by these modules allows teachers to be autonomous in planning their teaching as regards the use of history in practice, and enhance the collaboration with researchers in the field.

Bibliography

[1] Alexander, A. R., 2006, "Tragic mathematics. Romantic narratives and the refounding of mathematics in the early nineteenth century," *Isis* 97, 714–726.

[2] Bidwell, J. K., 1993, "Humanize your classroom with the history of mathematics," *Mathematics Teacher* 86, 461–464.

[3] Brown, T., 2001, *Mathematics Education and Language—Interpreting Hermeneutics and Post-Structuralism*, Dordrecht / Boston / London: Kluwer Academic Publishers.

[4] Brown, S. I. and M. I. Walter, 1983, *The Art of Problem Posing*, Hillsdale NJ: Erlbaum.

[5] Bruner, J., 1996, *The Culture of Education*, Cambridge MA: Harvard University Press.

[6] Dematté, A., 1994, "Storia, pseudostoria, concezioni," *L'Insegnamento della Matematica e delle Scienze Integrate* 17B, 269–281.

[7] ———, 2006, "Historical images, narrative, and goals in mathematical thinking," in *Proceedings of CIEAEM 58—Changes in Society: A challenge for Mathematics Education*, J. Coufalová (ed.), Plzeň: University of West Bohemia, Faculty of Education, pp. 283–288.

10.4. Didactical Implications and Conclusions

[8] ——, 2009, "Moduli di storia per insegnare e apprendere la ma-te-ma-ti-ca," *L'Insegnamento della Matematica e delle Scienze Integrate* 32A, 565–580. This article outlines the main features of (Katz & Michalowicz 2004).

[9] Eagle, H., 1977, "The semiotics of art: A dynamic view," *Semiotica* 19, 367–396.

[10] Fauvel, J. and J. Van Maanen (eds.), 2000, *History in Mathematics Education: The ICMI Study*, Dordrecht / Boston / London: Kluwer.

[11] Gadamer, H. G., 1989, *Truth and Method*. 2nd rev. edition. Trans. J. Weinsheimer & D. G. Marshall. New York NY: Crossroad.

[12] Glaubitz, M. R., 2008, "The use of original sources in the classroom. Theoretical perspectives and empirical evidence," in *Proceedings of the 5th European Summer University ESU 5*, E. Barbin, N. Stehlíková, & C. Tzanakis (eds.), Plzeň: Vydavatelský servis, pp. 373–381.

[13] Hamilton, W., 1844, "Letter to Graves on quaternions; or on a new system of imaginaries in algebra," *Philosophical Magazine* 25, 489-95.

[14] *Il giardino di Archimede—Un museo per la matematica. La matematica su CD-rom. Una collezione di volumi antichi e rari di matematica e scienze affini*, Firenze.

[15] Isoda, M. 2006, "Why we use historical tools and computer software in mathematics education: mathematics activity as a human endeavor project for secondary school," in *Proceedings HPM 2004 & ESU 4 Revised edition*, F. Furinghetti, S. Kaijser, & C. Tzanakis (eds.), Iraklion: University of Crete, pp. 229–236.

[16] Jahnke, H. N., 1994, "The historical dimension of mathematical understanding: objectifying the subjective," in *Proceedings of the eighteenth international conference for the Psychology of Mathematics Education*, J. P. da Ponte, & J. F. Matos (eds.), Lisbon: University of Lisbon, vol. I, pp. 139–156.

[17] ——, 1996, "Set and measure as an example of complementarity," in *History of Mathematics and Education: Ideas and Experiences*, H. N. Jahnke, N. Knoche, & M. Otte (eds.), Göttingen: Vandenhoeck & Ruprecht, pp. 173–193.

[18] Jahnke, H. N. with A. Arcavi, E. Barbin, O. Bekken, F. Furinghetti, A. El Idrissi, C. M. S. da Silva, and C. Weeks, 2000, "The use of original sources in the mathematics classroom," in *History in Mathematics Education: The ICMI Study*, J. Fauvel & J. Van Maanen (eds.), Dordrecht / Boston / London: Kluwer, pp. 291–328.

[19] Katz, V. and K. D. Michalowicz, 2004, *Historical Modules for the Teaching and Learning of Mathematics*. Washington, DC: Mathematical Association of America.

[20] Mazzolini, R. G. (ed.), 1993, *Non-Verbal Communication in Science prior to 1900*, Firenze: Leo S. Olschki.

[21] Netz, R., 1999, *The Shaping of Deduction in Greek Mathematics*, Cambridge: Cambridge University Press.

[22] Percival, I., 2001, "An artefactual approach to ancient arithmetic," *For the Learning of Mathematics* 21(3), 16–21.

[23] Pontecorvo, C. and H. Girardet, 1993, "Arguing and reasoning in understanding historical topics," *Cognition and Instruction* 11, 365–395.

[24] Radford, L., 2006, "The cultural-epistemological conditions of the emergence of algebraic symbolism," in *Proceedings HPM 2004 & ESU 4 Revised edition*, F. Furinghetti, S. Kaijser, & C. Tzanakis (eds.), Iraklion: University of Crete, pp. 509–524.

[25] Ransom, P., 2006, "John Blagrave, gentleman of Reading," in *Proceedings HPM 2004 & ESU 4 Revised edition*, F. Furinghetti, S. Kaijser, & C. Tzanakis (eds.), Iraklion: University of Crete, pp. 177–184.

[26] Saito, K., 2006, "A preliminary study in the critical assessment of diagrams in Greek mathematical works," *SCIAMVS* 7, 81–144.

[27] Smestad, B., 2011, "History of mathematics for primary school teacher education or: Can you do something even if you can't do much?," in *Recent Developments on Introducing a Historical Dimension in Mathematics Education*, V. Katz & C. Tzanakis (eds.), Washington, DC: Mathematical Assocation of America, pp.203–212.

[28] Smorti, A., 1994, *Il Pensiero Narrativo. Costruzione di Storie e Sviluppo della Conoscenza Sociale*, Firenze: Giunti.

[29] Solomon, J. and L. O'Neill, 1998, "Mathematics and narrative," *Language and education* 12(3), 210–221.

[30] Thomas, R. S. D., 2002, "Mathematics and narrative," *The Mathematical Intelligencer* 24(3), 43–46.

[31] Tuzet, G., 2003, "L'abduzione percettiva in Peirce," *Aquinas* 46, 307–327.

About the Authors

Adriano Demattè is a secondary school teacher in Trento—North Italy. He collaborates with GREMG—Group of Research on Education in Mathematics Genoa—coordinated by Fulvia Furinghetti. His research deals with the integration of the history of mathematics in teaching and the mathematics view of students. In his last book (in Italian) possible connections between these domains are analysed through the presentation of historical images taken from ancient mathematics books. He edited a book that collects original sources for the classroom accompanied by proposals of activities for students and suggestions for teachers on how to use them in the classroom; he also authored manuals for junior high school students, articles for teacher training, and contributions in international conferences proceedings. His current work focuses on the role of narration as a unifying theme with respect to students' construction of sense in mathematical learning. In this concern word problems taken from history are used.

Fulvia Furinghetti is full professor of Mathematics Education in the Department of Mathematics (University of Genoa, Italy). Her research concerns mathematics education and history of mathematics education. In the first years of her career she carried out research in projective-differential geometry. She has developed projects on the use of history of mathematics in teaching, the history of mathematics education, beliefs, the public image of mathematics, proof and problem solving, algebra, technology in mathematics education, teacher professional development. She has organized the Symposia celebrating the Centenary of *L'Enseignement Mathématique* (Geneva, 2000) and of ICMI (Rome, 2008) and edited the proceedings of both Symposia. She is the author of the website on the first hundred years of ICMI (co-author Livia Giacardi) http://www.icmihistory.unito.it/. In 2000–2004 she chaired HPM, the International Study Group on the relations between History and Pedagogy of Mathematics affiliated to ICMI (International Commission on Mathematical Instruction).

11

Pedagogy, History, and Mathematics Measure as a Theme

Luis Casas and Ricardo Luengo
Universidad de Extremadura, Spain

11.1 Introduction

Interdisciplinary research in history is a resource that can provide knowledge in breadth and in depth not only of the subjects that have traditionally been studied—historical facts—but of others related to different areas of knowledge.

In this sense, the History of Education can contribute valuable understanding of the phenomena, institutions, and academic disciplines of schools. Research in this field has included analyses of the development and evolution of particular concepts throughout history, and studies of the pedagogical approaches that have been taken in their teaching and learning. Other work has reviewed, and in many cases retrieved, the materials that have been used in schools in teaching a particular area of science.

Indeed, in our opinion, these approaches [15] represent the main thrust of studies in the field of the History of Education, including much interesting work in the area of mathematics teaching and learning. But our own interest is principally didactic, so that our focus has been on how the results of these inquiries can be transferred to the classroom as an educational resource for teachers [12, 13]. Similar work carried out by other researchers has been centred on the context of undergraduate education [21, 22, 23, 19].

One of these researchers, [16, p. 5], makes a number of suggestions for the use of the History of Mathematics in the classroom:

1) Mention anecdotes of mathematics from the past.

2) Provide historical introductions to concepts that are new for the pupils.

3) Encourage the pupils to understand the historical problems to which the concepts they are studying were a response.

4) Give some specific lessons on the "History of Mathematics."

5) Design exercises for the classroom or for home that use mathematics texts from the past.

6) Organize drama activities that relate to mathematics.

7) Encourage the creation of poster exhibits or other projects with a historical theme.

8) Prepare projects on mathematics activities that relate to the past of their own local environment.

9) Make use of specific examples from the past to illustrate techniques or methods.

(10) Explore misconceptions / errors / alternative approaches from the past that might be of help to the pupils of today in resolving their difficulties in mathematics.

(11) Design the pedagogical approach to a topic in terms of its historical development.

(12) Design the organization and structure of the curriculum's topics in accordance with their historical basis.

Another researcher, [9], stresses that the incorporation of elements of History or other subjects into the teaching of Mathematics enables pupils to see the profound relationship between this field of science and everyday reality. This mitigates the at times far too formalistic vision of Mathematics, and helps the teacher present it as an activity that specifically responds to human needs.

Our intention in developing the Educational Innovation to be described in this communication was to put some of these suggestions into practice. With respect to the proposal of designing exercises using mathematical texts from the past, the difficulty that we encountered was that the content of these texts (demonstrations, problems,...) were not usually adapted to primary education. We therefore believed it would be necessary to select just those questions that were appropriate for this level. With respect to the suggestion of encouraging the creation of poster exhibits and similar projects with a historical and mathematical theme, we believe that our project is ideally suited to this end. Given the rural context in which we work, in which traditional practices have been preserved, we believed it would indeed be feasible to carry out a project dealing with mathematical activities in the past. With this, we could then also address another of the above suggestions: the use of specific examples from the past to illustrate techniques or methods. The theme chosen, *Weights and Measures*, is capable of providing a wide range of activities and educational resources with a mathematical content, and hence of bringing Mathematics closer to pupils as an everyday activity.

The main point of interest in our project, however, is that expressed by [16, p. 4]: It is not hard to find good reasons for using history in teaching mathematics, but the question is how to incorporate the activities into the classroom, and above all how to adapt them to primary education since the possibilities open to a primary school teacher are very different from those in secondary or university education.

From our teaching experience in primary schools in small rural towns, we know the great interest that is aroused in the pupils when resources from their family and social environment are integrated into the teaching. Ethnographic inquiry can be a further source of enrichment of school work, since it naturally leads to such integration. In this sense, our approach may be described as one of "Ethnomathematics" [14].

11.2 The Innovation Project

The Educational Innovation project "Recovery of the Units and Instruments of Weights and Measures that were Traditional in Extremadura" was conducted in collaboration with the teachers of two primary schools in Extremadura. These two schools had different characteristics. One was a small school in a rural town, and the other was a large urban school. These different characteristics complemented each other in the project. In the former, it was possible to consult people from the rural context who still conserved various of the old instruments and in many cases had used the old measures. And in the latter, it was possible to consult sources conserved in specialized library and document collections. The project covered two school years, and involved a group of 10 teachers and 96 pupils of different educational levels ranging in age from 10 to 12 years.

The scheme of phases and activities of the project is shown in Figure 11.1.

The first part of the project consisted of a group of activities outside the classroom (bibliographic research, survey of informants, recovery of tools, and mounting the exhibit) which involved teachers and parents. These activities provided information relating to History, and recovered traditional measuring instruments which allowed activities to be designed for the pupils and the participating teachers. The information obtained was used in the second part of the project which consisted of various classroom activities. These included drafting informational texts for teachers and pupils with equivalence tables between ancient and modern units, and the design of activities and problems to put to the pupils.

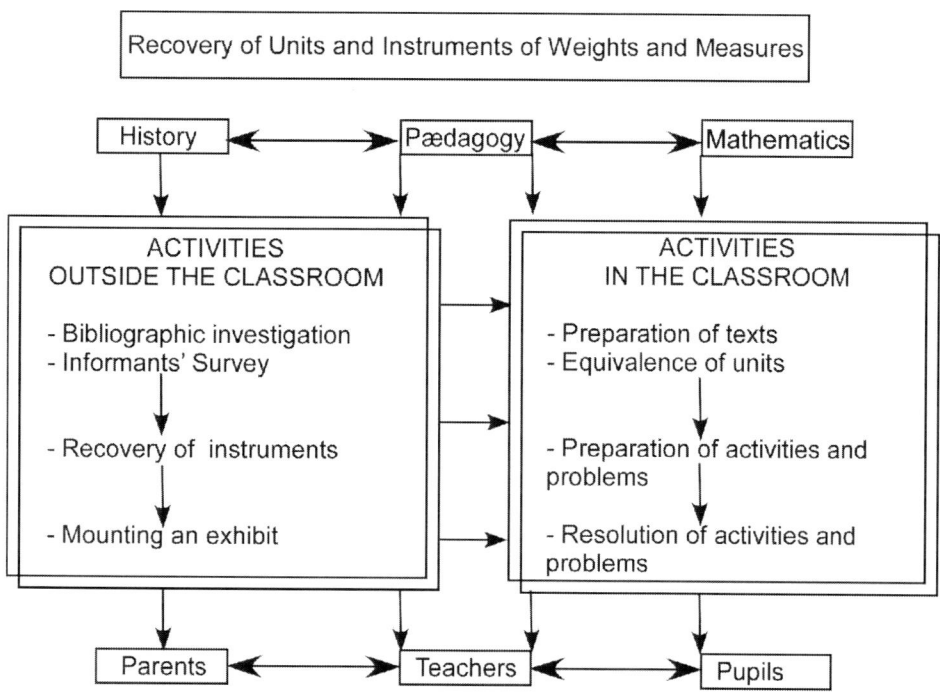

Figure 11.1. Outline of the project.

11.2.1 Activities outside the classroom

Two sources of information were used: consulting informants and a bibliographic review.

For the consultation of informants, a survey questionnaire was drafted that was given to older inhabitants of the two locations involved in the project. The resulting data revealed forgotten aspects of folk knowledge, and an extensive body of information on the instruments and units of measurement that were still known and used in the region's villages and towns. A second effect was that the survey led to the pupils' parents, and even more so their grandparents, taking serious interest in the school work that the children were going to do.

The bibliographic review involved three types of source:

- Textbooks from the XVIII to XX centuries of different levels of education. These were reviewed for information on the teaching methods used for content relating to weights and measures (units and instruments) of the periods before and after the introduction of the metric system. Some of these textbooks were found to be particularly useful [18, 17, 8, 10].

- Texts from the same era, such as [26], written for professionals such as builders, surveyors, and traders, and which described the construction and use of measuring instruments. Many of these were anonymous even to there being no mention of the place of publication, and some were handwritten. They were all conserved in the Library of the Real Sociedad Económica de Amigos del País, in Badajoz[1].

- Texts, some from the XIII century, consisting of the municipal ordinances and bylaws of different locations that referred to regulations of weights and measures. In particular, we used the Ordinances and *Fueros* (Bylaws) of the municipalities of Usagre[2], Plasencia[3], and Badajoz[4] (Figure 11.2).

These sources, following lengthy and interesting study, yielded a wealth of information reflecting the use and transformations of units and measuring instruments in Extremadura throughout its history.

---

[1] This is a Society established in many cities in Spain, that since the XVIII century has been dedicated to the development of culture, science, and industry. Included in the list of References of the present work are some anonymous books found in its Library.

[2] We used the edition of Osuna, A.J. [20], published by the Usagre Town Council. There is an earlier edition, of Ureña and Bonilla, published in Madrid in 1907, which we include in the list of References, and which is available on the Web.

[3] We used the edition of the Plasencia Town Council (1987), cited in the list of References.

[4] We used the edition of the Unión de Bibliófilos Extremeños (1993) cited in the list of References.

Figure 11.2. Ordinances of the city of Badajoz

This information was then written up in different types of documents.

First, documents were prepared at different levels of complexity for the use of teachers and pupils. These presented the historical information that had been collected. By way of example, the following is part of one of the texts drafted by the participating teachers:

> The weights and measures standards were kept in monuments representative of religious or civil power. Such was the case of the Jews in their Temple, the Greeks in the Acropolis, and the Romans in the Capitol.
>
> Acts of measurement, particularly of weighing, also often appear in the iconography of temples. The actions of ancient Egyptians were weighed by Amon, and of Christians by the archangel Michael who is portrayed in numerous churches weighing souls and thus determining their destiny for eternity. Measurement and the possession of measuring instruments, particularly of sets of the standards, soon became a symbol of power. Indeed, since possession of the standards meant being able to modify them to one's own advantage, in all countries there were numerous disputes over weights and measures (those used by the church, by the lord, by the municipality,...) throughout their histories.
>
> The need for strict control over weights and measures led to the appearance of laws and their corresponding sanctions designed to regulate their use. There are many instances of these norms in the Ordinances and Fueros of the major towns of Extremadura, all of them extremely meticulous. These documents also show us what measures were in routine use in a given area at a given time. For example, consider what the Fuero of Usagre (XIII century) has to say in bylaw number 378: "All tilemakers in fabricating roof-tiles, will make them in accordance with the *marco*[5] of the council. And give a thousand tiles per maravedí[6] ..."
>
> Similarly, in the Fuero of Plasencia (1290), one reads: "The taverner who uses some measure other than those of the council, or has a measure which is reduced in size, shall pay four maravedís. If the taverner does not fill the measure brimful while holding it firmly upright, he shall pay two maravedís. What spills over is the purchaser's. The measurer who puts his thumb in the measure shall pay two maravedís."

Second, a glossary was prepared for the pupils to have available. This included vocabulary of old terms related to units of measurement or to agricultural work.

---

[5] The *marco* or frame was the standard mould which the measurements of the roof tiles had to match.

[6] Old Spanish currency.

[7] The top figure shows a tool used by shoemakers to take the measurement of the feet. The second, a set of weights used for high-priced products such as saffron. The third, a set of measures for liquids such as milk or wine. And the last, a measure used for grain.

11.2. The Innovation Project

	Tonel	Tonelada	Pipa	Moyo	Cántara or Arroba	Cuartilla	Azumbre	Botella	Cuartillo	Cuarterón or Copa	LITRES
Tonel	1	1 + 1/5	2 + 2/5	4 + 1/8	66	264	528	1408	2112	8488	1064.7
Tonelada		1	2	3 + 7/16	55	220	440	1173 + 1/3	1760	7040	887.31
Pipa			1	1 + 23/22	27 + ½	110	220	586 + 2/3	880	3520	443.65
Moyo				1	16	64	128	341 + 1/3	512	2048	258
Cántara or Arroba					1	4	8	21 + 1/3	32	128	16.13
Cuartilla						1	2	5 + 1/3	8	32	4.03
Azumbre							1	2 + 2/3	4	16	2.02
Botella								1	1 + ½	6	0.76
Cuartillo									1	4	0.5
Cuarterón or Copa										1	0.13

Figure 11.3. Equivalence of units of capacity for liquids other than oil.

Figure 11.4. Items from the exhibition of weights and measures[7].

And third, equivalence tables between the old and modern units were prepared based on the documentation obtained from the historical texts and from consultations with older people. One of them is presented in Figure 11.3.

These tables and the write-up of the information on the use of different measures were both greatly appreciated by the pupils' parents, and especially by their grandparents, since they often represented what was essentially a rescue of a culture that was on the verge of being lost.

In parallel with the information processing phase described above, extensive ethnographic work was carried out collecting weighing and measuring instruments that, now in disuse, were put away in the attics of the pupils' houses. These were instruments that had been used in various trades, mostly in relation to agriculture, and were remembered by older people who had themselves used them or had seen them used. Many of the instruments recovered were restored—a task in which many teachers, parents, and individuals of the participating localities participated. Figure 11.4 shows some of these pieces.

Several exhibits have been mounted throughout Spain with these restored instruments.

11.2.2 Classroom activities

Activity proposals were designed using the materials and documents described above, including both traditional and recreational mathematics problems. The content was organized to have references to various aspects of the mathematics curriculum and to the curricula of other areas. The activities were proposed in a form that allowed them to be resolved in two ways: using the informational text that the pupils had available, or, in many cases, resorting to the information that could be obtained by consulting parents or grandparents of the locality.

By way of example, the following are three of the activities put to the pupils:

1. Oral Tradition:
   In the town of Barcarrota in Extremadura, the procedure to change from fanegas[8] to hectares and vice versa is as follows:

   a) To go from hectares to fanegas, divide the hectares by two and then add the results to the number of hectares, and the result is fanegas.

   b) To go from fanegas to hectares, the procedure consists of taking the number of fanegas and subtracting a third of it.

   According to this procedure, how many fanegas are 40 hectares? And how many hectares are 90 fanegas?

2. From the book "Practical Guide for Surveyors and Farmers" by Francisco Páez Verdejo [26, page 155]:
   Verdejo poses the following problem: "You want to plant a vineyard in 5 fanegas of land that you own. You have computed the cost of each plant and the labour for its planting, and the result is half a real[9] per vine. How much will the planting cost you, leaving two-and-a-half varas[10] between every two plants?"

3. From the book "History of Extremadura" by Cardalliaguet [11, page 356]:
   Rodrigo Alonso, a *recuero*[11] who was assaulted by servants of the mayor of Portezuelo as he was passing the fortress southwards, was transporting 210 varas of *sayal*[12] and 100 varas of Breton canvas.

   a) Find out about the terms *recuero* and *sayal*.

   b) How many metres of each type of goods was the *recuero* transporting?

Another type of activity that the pupils found to be of great interest was the construction of rudimentary measuring instruments like those that were used historically, and others that could be used to imitate the methods used for different types of measurement. For this purpose, low-cost or recycled materials were used preferentially. For example, in the photograph in Figure 11.5, the pupils are measuring a plot of ground using a surveyor's chain that they themselves had made.

---

[8] Old unit of farm land area, equivalent to 6439 square metres.
[9] Old Spanish currency.
[10] Old unit of length, equivalent to 83.59 cm.
[11] A muleteer.
[12] A coarse cloth.

Figure 11.5. Measuring a tract of land using a surveyor's chain.

Also as classroom activities, an interactive CD was prepared that included the information gathered during the project, and the proposed activities. The internal structure of the CD was designed to allow access to its content by level of difficulty. The user would thus not be limited to browsing through it at random, but in order to access information of greater difficulty would first have to resolve the activities proposed at the previous, simpler levels.

11.3 Evaluation of the Project and Conclusions

To evaluate the pupils' knowledge, a test was applied at the beginning and at the end of the experiment.

To evaluate the change in pupils' attitudes, and the project's impact on them as well as on their teachers and family members, classroom observation logs, questionnaires, and interviews were used. The classroom observation logs were maintained by all the teachers during the performance of the activities, and their results were analyzed during the project's regular coordination meetings. The questionnaires were sent to all the participants, and a very high response rate was obtained. For the interviews, voluntary informants were selected from among the pupils, teachers, and parents.

Other data were also obtained for evaluation, although more informally, from visitors to the weights and measures exhibits, based on their comments collected in forms made available for that purpose.

The results of the test evaluating the pupils' knowledge indicated that there had been improvements in their estimation of lengths, areas, volumes, and weights, in the level of information about traditional units and instruments, and in their ability to resolve problem situations involving weights and measures.

It was concluded from the analysis of the interviews and questionnaires given to the pupils that their attitude to Mathematics had improved. Before the project, many of them considered Mathematics not only to be a difficult subject, but also one that was abstract and of little use. In their responses following the project, however, they noted the usefulness of mathematics for everyday life, and showed that they had discovered many of its relationships with other areas of knowledge. Many of them also noted that their parents, and especially their grandparents, had been very keen to help them in the tasks related with the project.

In the interviews and questionnaires given to the teachers, and in the project's coordination meetings, the teachers emphasized that they had observed an improvement in the understanding of the concept of proportionality, and above all a greater use of mental calculation for the conversion of units when using bases different from the decimal, such as with the old units. They also noted that the older pupils used procedures for the conversion of units that were different from those proposed in the textbooks, and made fewer mistakes.

From the results of the analyses of the observation logs, there also stood out that the pupils' attitude of self-confidence in approaching the proposed activities had improved, as had their perseverance in resolving them. The teachers' opinions that were collected also showed a marked interest in experimenting with a new working dynamic and that they had discovered aspects of popular culture that merited further exploration and educational use, not only in the field of Mathematics, but also in History, Literature, and Art. Similarly, the interactive CD was in their view an interesting working tool.

Regarding the project's repercussion in the communities in which it was implemented, the parents and grandparents surveyed and interviewed stressed that it had been very interesting for them to be able to show the children their knowledge about the traditional measures and instruments that had been particularly useful in farm work, but which they had never considered to be important as school knowledge. They had also found particularly appealing the collection of measuring instruments that they themselves had used, and the recollection and reminiscing about something that had formed part of their traditional culture. In this sense, the staging of the Exhibition of Traditional Weights and Measures aroused great interest not only among the pupils, but also among older people, and involved the school being brought closer to popular culture, as was confirmed by the comments collected in the visitors' forms.

By way of conclusion, we would note that the evaluation of the project by all its participants—pupils, teachers, and parents—was very positive, and was recognized as a novel approach to enriching Mathematics activities by means of contributions of History. In addition to the treatment of Weights and Measures as an element of the History of Mathematics and its introduction into the classroom, we would highlight in this project its use of new working methods, such as the recovery of a people's cultural tradition, or the use of new technologies. We furthermore believe that this experience can be readily translated, with a very similar working strategy, to other regions and even other countries, since many of them have an ethnographic and cultural wealth that can be recovered and retransmitted to their societies.

Bibliography

[1] Anonymous, 1767, *Ordenanzas de la Muy Noble y Muy Leal ciudad de Badajoz*, Real Sociedad Económica de Amigos del País.

[2] Anonymous, 1850, *Prontuario práctico del sistema métrico legal de pesos y medidas*, Library of the Real Sociedad Económica de Amigos del País, in Badajoz.

[3] Anonymous, 1850, *Tabla de monedas, pesos y medidas de Castilla*, manuscript conserved in the Library of the Real Sociedad Económica de Amigos del País, in Badajoz.

[4] Anonymous, 1850, *Manual para maestros de obras*, manuscript conserved in the Library of the Real Sociedad Económica de Amigos del País, in Badajoz.

[5] Anonymous, 1852, *Cuentas ajustadas por libras desde 1 a 33 cuartos*, Library of the Real Sociedad Económica de Amigos del País, in Badajoz.

[6] Anonymous, 1875, *Tablas de reducción al sistema decimal de las pesas y hedidas antiguas de la provincia de Badajoz*, Library of the Real Sociedad Económica de Amigos del País, in Badajoz.

[7] Ayuntamiento de Plasencia, 1987, *Edición Facsímil del Fuero de Plasencia (1297)*, Plasencia: Excmo. Ayuntamiento.

[8] Benot, E., 1893, *El consultor del sistema métrico*, without place of publication.

[9] Bidwell, J., 1993, "Humanize your classroom with the history of mathematics," *The Mathematics Teacher. An Official Journal of the National Council of Teachers of Mathematics*, 86(6), 461–64.

[10] Calleja, S., 1900, *Método completo de primera enseñanza cíclica o progresiva. Tomo IV*, Madrid: Saturnino Calleja.

[11] Cardalliaguet, M., 1988, *Historia de Extremadura*, Badajoz: Universitas.

[12] Casas, L. and C. Sánchez, 1996, "Recuperación de instrumentos y unidades de medida tradicionales como motivación al estudio de un tema curricular: La medida," in Consejería de Educación (ed.): *Premios Joaquín Sama 1995 a la Innovación Educativa*, 87–303, Mérida: Junta de Extremadura.

[13] Casas, L., R. Luengo, and C. Sánchez, 2000, "Cultura, Historia y Matemáticas: el tema de la Medida," *Cátedra Nova*, n° 11, pp. 277–304.

[14] D'Ambrosio U., 1993, *Etnomatemática*, São Paulo: Editora Atica S.A.

[15] Escolano, A., 2006, *Historia ilustrada de la escuela en España*, Madrid: Fundación Germán Sánchez.

[16] Fauvel, J., 1991, "Using history in mathematics education," *For the Learning of Mathematics*, 11 (2), 3–6.

[17] García, J. J., 1821, *Elementos de aritmética, algebra y geometría. Tomos I y II*, Madrid: Ibarra.

[18] Malo de Medina, F., 1787, *Guía del niño instruido y padre educado*, Madrid: Imprenta Real.

[19] Meavilla, V., 2007, *Selección de Problemas lineales y cuadráticos rescatados de los elementos de Álgebra de Leonhard Euler*, Badajoz: Servicio de Publicaciones de la FESPM.

[20] Osuna, A.J., 1994, *Fuero de Usagre (Versión libre)*, Usagre (Badajoz): Excmo. Ayuntamiento.

[21] Sierra, M., 1993, " La introducción y enseñanza de las ideas algebraicas en España durante los siglos XVII y XVIII," in E. Filloy, T. Rojano & L. Puig (eds.): *Historia de las ideas algebraicas*, 153–158, México: Centro de Investigación de Estudios Avanzados.

[22] Sierra, M., L. Rico, and B. Gómez, 1997, "El número y la forma: libros e impresos para le enseñanza del cálculo y la geometría," in A. Escolano (dir.): *Historia ilustrada del libro escolar en España*, 373–398, Fundación Germán Sánchez Ruipérez. Madrid.

[23] Sierra, M., 1999, "Uso de la historia de la Matemáticas en el aula," in T. Ortega (ed.): *Temas controvertidos en educación matemática*, 13–26, Universidad de Valladolid. Valladolid.

[24] UBEX Unión de Bibliófilos Extremeños (ed.), 1993, *Ordenanzas de la M. N. y M. L. Ciudad de Badajoz. Edición facsímil del original de 1767*, Badajoz: UBEX.

[25] Ureña, R. and A. Bonilla, 1907, *Fuero de Usagre*, Madrid: Hijos de Reus. Available at www.archive.org/texts/flipbook/flippy.php?id=fuerodeusagresig00ureuoft

[26] Verdejo y Páez, F., 1848, *Guía práctica de Agrimensores y labradores*, Madrid: Imprenta José Repullés.

About the Authors

Luis M. Casas García received his Ph.D. in Educational Psychology from the Faculty of Education of the University of Extremadura (Spain), where he is now a professor in the area of Theory and History of Education. He also has extensive experience as a teacher of Primary Education and has directed several projects for educative innovation and software development. His research has focused on mathematics education in areas such as the cognitive aspects of learning, the use of new technologies or the use of history to teach mathematics. He is the author of numerous articles and presentations at national and international conferences related to the research areas mentioned above. He is also a co-author of several books and software for teaching. He is a member of several Education Societies, including the Spanish Society for Research in Mathematics Education, the Spanish Society of the History of Education, and the Society of Mathematical Education in Extremadura, of which he is the director.

Ricardo Luengo González has a Ph.D. in Mathematics and holds the Chair of Teaching of Mathematics at the University of Extremadura (Spain). He has extensive experience in management over 20 years as Director of the Teacher Training Center and Vice Chancellor of the University. He was Chairman of the Spanish Federation of Teachers of Mathematics and is now Director of its National Service of Publications and President of the Society for Mathematical Education in Extremadura. He coordinates a group of research and is a member of the Spanish Society for Research in Mathematics Education. Its lines of research focus on the NT and Mathematics Education. Co-author of several books and magazine articles, he belongs to several editorial committees and research projects. He has extensive teaching experience and has participated in over one hundred conferences and attended more than forty courses.

12

Students' Beliefs About the Evolution and Development of Mathematics

Uffe Thomas Jankvist
University of Southern Denmark, Denmark

12.1 Introduction

Beliefs about the history of mathematics is a topic which is touched upon from time to time in the literature on history in mathematics education, *e.g.*, in Furinghetti [8] and Philippou and Christou [23]. However, when scanning these samples, one soon finds that these concern the beliefs of in-service or pre-service teachers. Studies on students' beliefs about the history of mathematics seem to be rather poorly represented in the literature, if not altogether absent[1]. One reason for this that I can think of is that, in general, studies of beliefs in mathematics education are conducted with the purpose of improving mathematical thinking, learning, and instruction.[2] Beliefs, both cognitive and affective ones[3], are investigated in order to identify the 'ingredients' which do or do not make students capable of solving mathematical tasks or teachers capable of teaching differently and/or more effectively. Certain beliefs are identified as advantageous in the learning of certain mathematical contents, the solving of related tasks, etc., and educational studies are then conducted on how to change already existing beliefs into these more favorable ones. In this sense beliefs are regarded as means—or *tools*—to achieve understanding in the individuals' constructive learning process. Only rarely is providing students or teachers with certain beliefs, *e.g.*, by changing existing ones, about mathematics or mathematics as a discipline considered as a *goal* in itself. And when this is done, the term 'beliefs' is usually not used. Instead mathematical appreciation, mathematical awareness, or providing students with a more profound *image* of what mathematics is, are the words or phrases more commonly used (*e.g.*, [7, 20, 5]).

It seems to me that the beliefs discussion in mathematics education lacks a goal-oriented dimension, a dimension that addresses students' mathematical world views and proposes and evaluates some desirable beliefs in order to turn students into more critical citizens by providing them with intelligent and concerned citizenship and with some *Allgemeinbildung* in general [20]. That is to say, it is necessary to provide students with coherent images of mathematics as a discipline, of the influence of mathematics in society and culture, of the impact of society and culture on mathematics, and of the historical evolution and development of mathematics as a product of time and space, to mention a few of the more 'pressing' ones. Occasionally researchers will touch upon these issues in the form of personal opinions,

---

[1] A few exceptions are the Danish study of Christensen and Rasmussen [1] and the Italian one by Demattè and Furinghetti [4].
[2] A few examples are [25, 26, 27, 18].
[3] I shall not here enter the discussion of defining 'beliefs'. I do, however, implicitly base my understanding of beliefs on the definition given by [22].

e.g., in curriculum development. However, a dimension about 'beliefs about desirable beliefs'—meta-beliefs we may call them—can only be addressed properly if the meta-beliefs are articulated as such, i.e. as goals in themselves.

In this paper I shall first present some extracts from the 2007 regulations for the Danish upper secondary mathematics program and the Danish report on competencies and learning of mathematics, the so-called KOM-report, which may serve as such a goal-oriented dimension for students' beliefs. Especially I shall focus on students' beliefs concerning the history of mathematics. Secondly, I shall report on a piece of empirical research in which a number of students were asked about their beliefs concerning the evolution and development of mathematics[4]. Thirdly, these students' beliefs are analyzed and evaluated against the goal-oriented descriptions. The paper concludes with some final remarks and reflections on the empirical data presented and the larger research study of which they are a part.

12.2 The Danish Context

Since 1987 history of mathematics has been part of the formal regulations for the Danish upper secondary mathematics program (see *e.g.*, [6, pp. 5–7], and with the newest reform and the present regulations this part has become more dominant. Students are now expected to be able to "demonstrate knowledge about the evolution of mathematics and its interaction with historical, scientific, and cultural evolution," knowledge acquired through teaching modules on history of mathematics [28, my translation from Danish]. The official regulations for the Danish upper secondary mathematics program of 2007 are to some extent based on the Danish report *Competencies and Learning of Mathematics*, the so-called *KOM-report*, [21, title translated from Danish][5] where it says the following about history:

> In the teaching of mathematics at the upper secondary level the students must acquire knowledge about the historical evolution within selected areas of the mathematics that is part of the level in question. The central forces in the historical evolution must be discussed including the influence from different areas of application. Through this the students must develop a knowledge and an understanding of mathematics as being created by human beings and, in fact, having undergone an historical evolution—and not just being something that has always been or has suddenly arisen out of thin air. [21, p. 268, my translation from Danish]

In the report, the focus of integrating history of mathematics is discussed in terms of a certain kind of overview and judgment that the students should acquire as part of their mathematics education.

> The form of overview and judgment should not be confused with knowledge of 'the history of mathematics' viewed as an independent subject. The focus is on the actual fact that mathematics has developed in culturally and socially determined environments, and subject to the motivations and mechanisms which are responsible for this development. On the other hand it is obvious that if overview and judgment regarding this development is to have solidness, it must rest on concrete examples from the history of mathematics. [21, p. 68, my translation from Danish]

The 2007 regulations describe the "identity" of mathematics in the following way:

> Mathematics builds upon abstraction and logical thinking and embraces a long line of methods for modeling and problem treatment. Mathematics is indispensable in many professions, in natural science and technology, in medicine and ecology, in economics and social sciences, and as a platform for political decision making. At the same time mathematics is vital in the everyday. The expanded use of mathematics is the result of the abstract nature of the subject and reflects the knowledge that various very different phenomena behave uniformly. When hypotheses and theories are formulated in the language of mathematics new insight is often gained hereby. Mathematics has accompanied the evolution of cultures since the earliest civilizations and human beings' first considerations about number and form. Mathematics as a scientific discipline has evolved in a continual interrelationship between application and construction of theory. [28, my translation from Danish]

---

[4] The full questionnaire consisted of 20 questions covering historical and developmental, epistemological and philosophical, sociological, and more personal affective matters of mathematics. Questions 1 to 7 are a variety of these. For a more thorough discussion of the questions, see [16].

[5] The word 'demonstrate' in Danish has a dual meaning; it may be used both as the word 'prove' and as the word 'display'. Thus, students may only need to display knowledge.

Thus, when the students are to "demonstrate knowledge about the evolution of mathematics" etc., as stated in the academic goals of the regulations, one must assume that it is within the frame of this "identity" that they are expected to do so. Another way of phrasing this is to say that one purpose of the teaching of mathematics at the Danish upper secondary level is to mold the students' beliefs about mathematics according to the above description of identity. The purpose of including elements of the history of mathematics has to do with showing the students that mathematics is dependent on time and space, culture and society, that mathematics is not 'God given', that humans play an essential role in the development of it, etc., etc.

12.3 Students' Beliefs About the 'Identity' of Mathematics

In the beginning of 2007, I conducted a questionnaire and interview research study of second year upper secondary students' (age 17–18) beliefs about the 'identity' of mathematics. A number of these questions had to do with the evolution and development of mathematics. In the following I shall present the students' answers to seven of these questions. All in all 26 students answered the questionnaire. The students' questionnaire answers have been indexed in the following manner: one; few; some; many; the majority; the vast majority; a partition which roughly corresponds to the percentage intervals: 0–5%; 6–15%; 16–35%; 36–50%; 51–85%; 86–100%. Based on the questionnaire answers 12 students were chosen as representatives for the class in general, and these 12 students were interviewed about their answers. All quotes from the questionnaires and the interviews have been translated from Danish.

1. *How do you think that the mathematics in your textbooks came into being?*
   The majority of the students believe that the mathematics is due to people in history who have been wondering or been curious about something, and therefore attempted to explain what they observed. Many of the students who think so believe the people responsible for the mathematics are some special, wise persons and great "minds of ideas," a few mention Pythagoras. One student suggests that the ones responsible are "some very patient, half autistic people who have been wondering about the connections, rules, etc. between things, *e.g.*, the angles of triangles, the lengths of the sides, etc." Some believe the mathematics to be an accumulation of experiences, observations, and experiments, possibly anchored in nature. A few of these emphasize the cumulative nature of mathematics. Others believe that mathematics was created because of a need, for instance in connection to trade or "in order to make things more manageable." The interviews provided no additional information.

2. *When do you think it came into being?*
   The majority believe that the mathematics in their textbooks came into being "sometime long ago." The suggestions to exactly when are, however, many and different: "from even before da Vinci's time!"; "when the numbers were invented"; "when we began using Arabic numerals"; "way before it says in the books." Some point to antiquity and provide as argument that "the construction of, for instance, the pyramids must have required at least some mathematics." One of the more interesting answers goes: "Long, long ago it all began and since then it has continued. But I am confident that the development goes more and more slowly, because you eventually know quite a bit."

   Out of this majority of students, some share the perception that mathematics always has existed, or at least has existed as long as human beings have been around. One says: "Mathematics in general has existed since the dawn of time, but highly developed [mathematics] has only emerged within the last 200–100 years." Only one student believes the mathematics in the textbooks to be of a more recent date, and he is not afraid to fix this to "40 years ago."

   In the followup interviews, events in the history of mathematics were occasionally fixed with some kind of accuracy, for instance, the beginning of mathematics to 4000–5000 years ago; Pythagoras to the first couple of centuries; and Fermat's last theorem to "the Middle Ages or something." But only few students were able to do this. Whether this is due to lack of knowledge about history of mathematics or lack of knowledge about history in general, or maybe both, is difficult to say. Finally, one of the students seemed very strong in her belief that it was impossible to practice mathematics without the Arabic numerals. When asked why not, she replied: "the mathematics you do today, you wouldn't have been able to do that ..."

3. *Why do you think it came into being?*
   The majority of students believe that there was a need to have mathematics at one's disposal. A few even talk about

a necessity: "For example with constructions it has been important to be able to predict/calculate if, for instance, the walls can support the roof etc. Better to find errors on the drawing board than when the final construction collapses." Many students mention the development of society and related aspects as the main causes. Some again mention that people have been wondering, been curious about something, and followed their ideas and impulses. One student ascribes the cause to "The will of God—or Big Bang, if you like." In the follow-up interviews one student said: "Because people had too much time on their hands, for example, so they were given jobs as mathematicians."

4. *Are the negative numbers discovered or invented? Why?*

In the answering of this question the class was divided in two approximately equal parts, one in favor of discovery and the other in favor of invention. The arguments provided were quite different though. A few students believed the negatives to have been discovered in connection with or immediately after the positive numbers. Others believed that they always had been there, but that it might have taken some time to "learn to express them" or that people were "able to see it, but might have had difficulties explaining it." Among the arguments for invention we find: "They are invented, I think, because you would get something wrong if they weren't there"; "On the face of it, invented because you can't have something which isn't there"; "They are invented because you needed values smaller than 0"; "Think they are invented since it appears strange that a number all of a sudden should fall from the sky or something." From time to time the same arguments were used for both discovery and invention: "Discovered. If we imagine a man who has bought a cow, but doesn't have enough money, so that he owes money away, i.e., a negative number"; "Discovered. If you were in debt to someone, maybe." One student plays it safer: "I'd think they were invented because almost all mathematics is invented, but at the same time also discovered." The interviews provided no additional information.

5. *Do you believe that mathematics in general is something you discover or invent?*

The majority of the students believe that mathematics in general is something you discover. Only a few believe that it is something you invent. More students, though, believe that it might be a combination of the two. Many of those who believed that negative numbers were something discovered stick to this point of view for mathematics in general. A few of the answers are: "Discover. I don't think you can invent mathematics—it is something 'abstract' you find with already existing things."; "Discover. Because mathematics is already invented. What happens today is only that you discover new elements in it." A lot of those who believed that negative numbers were discovered and a few of those who believed them to be invented now seem to think both: "Many things might begin as an invention, but afterwards they are explored and people discover new elements in the 'invention' in question"; "Both, [I] think that you discover a problem and then solve it by inventing a solution or applying already known rules of calculation"; "You discover formulas after having discovered relationships." Some of those who believed negative numbers to be invented now believe mathematics in general to be discovered: "Mathematics is all over—in our society, our surroundings and in the things we do. Therefore I do not believe mathematics to be something you invent, but on the contrary something you discover along the way. Of course, it might be difficult to say precisely, because where is the line drawn between discovery and invention?" One of the answers touch upon the question of what mathematics 'really' is: "Good question... very philosophical. I think there are many different standpoints to this. I personally believe that it is something you discover. Numbers and all the discoveries already made are all connected. So for me it is more a world you enter into than one you make."

In the followup interviews the student responsible for the last remark explained further: "Well, I see it as if mathematics is just there, like all natural science as, for instance, outer space. Outer space is there and now we are just discovering it and learning what it is. That's what I think: It's the same thing with mathematics." When the remaining interviewees in favor of discovery were asked if the 'exploration' of mathematics corresponds to the exploration of the universe they all confirmed this belief. That is to say that they believed mathematics to always have existed, or as one student phrased it: "Mathematics has always been there, in the form of chemistry or something like that at the creation of Earth. And then we haven't found out about it until later." Or another one: "I think it has always been there, but I just think that the human beings are exploring mathematics more and more and are discovering new things." One of the students who believed mathematics in general to be discovered thought that negative numbers were invented: "It is something you have made up because you had to. Well, it isn't something written down somewhere from all eternity or from God, and which has just been there. It is something you invent because there is a need. If you invent a chair, right, then it is because you have a need to sit down." As an answer this even links to the previous question of why mathematics came into existence.

6. *Do you think mathematics has a greater or lesser influence in society today than 100 years ago?*
   The vast majority of the students believe greater. This answer is in general based on the increased amount of technology in our everyday society. Answers as "definitely, more computer=more mathematics" and "everything develops and everything has to be high-technology" are often provided. A few of those who believe that mathematics has a greater influence today also point to economic affairs as the reason, or that "the use of mathematics has become more advanced in our time." Some think that mathematics has the same influence today as it had 100 years ago, and only very few believe that the influence today is less. One of the more exceptional answers of the latter kind is: "No, I don't believe that, because even though we use mathematics a lot more in space etc. we have modern machines to do it."

The followup interviews to a large degree confirm the beliefs described above. To the deepening question of why a student found the influence today to be greater, she answered:

"Because today you can, for instance, get an education at… or study mathematics at the university and things like that, and that you couldn't do a hundred years ago. […]
So it is something relatively new that you can study mathematics at the university?
No not new, but I do believe at a higher level. That is, you didn't know as many things back then as you do today.
And you couldn't get an education as a mathematician in the same way, you think?
No."

The student who argued lesser influence due to the use of modern machines is also given the opportunity to expand on her view in the interviews. She finds, among other things, that mathematics appears less at present because we rely on technical aids to a great extent, and because the use of mathematics mostly is about "pushing some buttons."

7. *Do you think that mathematics can become obsolete? If yes, in what way?*
   To this question the vast majority answers a clear no or that it appears unlikely, for instance: "a proof is a proof" or "the basic things we build our mathematical development on are so used and tested that it won't become obsolete." Some provide a no with modifications: "Don't think that it can become obsolete, but that theorems/theories can be disproved and thereby provide a foundation for 'new mathematics'." Or more striking: "No, but there are probably some things which will not be used so much in the future: Such as vectors." Only a few answer yes or maybe. The followup interviews provided no additional information.

12.4 Evaluating Students' Beliefs Against the 'Goals'

How does the above presentation of students' beliefs about the evolution and development of mathematics correspond with the goal-oriented description of overview and judgment in the KOM-report and the 'identity' of mathematics in the 2007 regulations? For example, are students able to "demonstrate [display] knowledge about the evolution of mathematics and its interaction with historical, scientific, and cultural evolution?" Overall the students' answers to some of the questions appear rather diffuse, but let us take them from the beginning. As an answer to question 1 the majority seem to believe that mathematics has developed and evolved as a result of peoples' personal curiosity and wonder. Only few mention extrinsic reasons such as trade. In question 3, however, there is an agreement that mathematics has come into existence because of a need or even as a necessity. Thus, there is a slight incongruence between the answers of the majority to questions 1 and 3. Of course, one may interpret it as if the inner motivation and curiosity of people to get involved with mathematics have been turned on by outer circumstances, which for certain incidences in history would be correct. The fact that mathematics itself besides being driven by outer driving forces also is driven by inner driving forces (forces which not only concern the personal motivation of a single mathematician, but the intriguing problems within mathematics itself) is not an aspect which the students seem aware of. And concrete examples from the history of mathematics, in the form of the KOM-report's talk of "solidness" (cf. page 124), are not something that the students seem able to provide either. One student mentions the building of the pyramids as an example of the need of mathematics in older times, but if this answer is founded in actual knowledge about mathematics in ancient Egypt, or if it is merely evidence of the student being able to think for himself is not to say. In any case, he does not provide any concrete examples of Egyptian mathematics.

In the answers to question 2 there seem to be an agreement that mathematics is 'old'. One student implies that da Vinci is old and that mathematics is older than him. However, only very few are capable of providing years on the origin of mathematics as well as on specific mathematical results. That some students believe mathematics only could come into existence by aid of the Arabic numerals does not strengthen the interpretation that the students possess knowledge about the evolution of mathematics in interplay with historical and cultural events either.

Despite some discrepancies with the answers of question 4, the majority in question 5 give expression to the fact that they believe mathematics in general to be discovered. In a Danish educational context this may appear surprising since, as Hansen [10, p. 71, my translation from Danish] puts it: "it is clear that the strong position of constructivism in school circles fertilizes the ground for a more radical constructivist perception of the entire nature of mathematics. Because of the pedagogical constructivism in schools, children and young people are likely to have difficulties believing in special existence of mathematical quantities, figures, and concepts." Of course there are students who are inclined toward a view of mathematics in general as something invented, but they are few in number. The majority give expression to a more Platonic stance. With the words of one of the students, it is "a world you enter into"—a world of ideas—where you explore the already existing mathematical objects in a similar way as we are exploring the Milky Way and the rest of the universe our planet is part of. Such a view may result in a playing down of the creative side of mathematics as a human activity, as opposed to something being created by human beings and not just suddenly having popped out of thin air (cf. page 2). On the other hand, the students seem to have a quite good understanding of the fact that mathematics today has a much greater influence in society than it did 100 years ago (question 6). Again it is computers and other technology that are given credit for this. The fact that students only pay scant attention to economic affairs and political decision-making, *e.g.*, based on mathematical models, may be seen as a consequence of the invisibility of mathematics in society [20]. One student touched upon this when she said that mathematics appears less present due to use of technology. Another example is the student who in question 2 believed that the development of mathematics was happening at a slower and slower pace and who in the interviews explained herself:

> "Yes, but they just discovered more a long time ago, didn't they? It isn't very often you hear about someone who has discovered something new within mathematics, is it? Maybe it's just me who isn't enough of a mathematics geek to be told about it. But it just seems to me that nothing is really happening. Stuff is happening more often within natural science: now they have found a method to see the fetus at a very early stage by means of a new type of scanning or something."

Besides relating to the invisibility of mathematics and the fact that it does not often find its way into the public media, the student's remark also touches upon one of the differences between mathematics and the natural sciences: just because mathematics now is able to prove Fermat's last theorem or the Poincaré conjecture, this is not something that will change our everyday life or society either tomorrow or in 50 years (most likely), something which would be far more likely for discoveries in physics, chemistry, or biology—and to a larger extent for technology basing itself on these disciplines.

The students also seem to have an understanding of mathematics as a science that is not likely to become obsolete (question 7). This has to do with the ability of mathematics to most often include previous results as special cases in more general and abstract new constructs. Of course, one might mean many different things with 'obsolete', and the inclusion in theory building is only one aspect. Another way to see it would be to think of concrete applications of mathematics. In this respect, the history of mathematics provides examples of very old pure mathematics that suddenly finds its way into an application (*e.g.*, the use of old number theoretic results in RSA cryptography), but probably the history contains even more examples of pieces of mathematics that until now have not, or maybe never will, find their way into applications. But the point is that one can never tell what will and what won't. In any case, one could have hoped that the students would have been able to provide some concrete examples of one or the other supporting their views—the only example was the student who thought that vectors were unlikely to be used for anything in the future. This supports the above mentioned lack of 'solidness' of the students' beliefs.

12.5 Final Remarks and Reflections

According to Lester, Jr. [19, p. 352], Kath Hart at a PME conference once asked: "Do I know what I believe? Do I believe what I know?" Lester's version of this question is: "Do students know what they believe?" Furinghetti and Pehkonen [9] argue that one should take into consideration both the beliefs that students hold consciously as well as unconsciously. But how to do this? Lester, Jr. [19, pp. 352–353] sows doubt about some of the more usual methods for doing this: "I am simply not sure that core beliefs can be accessed via interviews [...] or written self-reports [...] because interview and self-report data are notoriously unreliable. Furthermore, I do not think most students really think much about what they believe about mathematics and as a result are not very aware of their beliefs." Thus, the results above must perhaps be viewed in this light. However, other researchers (*e.g.*, [24]) argue that questionnaires, interviews, etc. are perfectly well suited to access students' beliefs about mathematics as long as the usual precautions, for example the interviewee trying to please the interviewer, are taken into account.

In the research reported in this paper, the students knew nothing about my personal viewpoints on the evolution and development of mathematics; they were not familiar with the descriptions in the KOM-report, or the 'identity'-description in the regulations for that matter. So it seems reasonable to say that none of these views could have affected the students' answers. Of course, they knew that the interviewer was a mathematician which might have led them to alter some of their views. Also, it is true that many students do not have a clear and conscious idea regarding their beliefs about mathematics, as Lester says. When asking the interviewed students to deepen or expand their questionnaire answers some of them would have trouble remembering what they answered, some would be puzzled about their own answers, and some would take on different viewpoints in the interviews than what they had expressed in the questionnaire. Especially the question of invention versus discovery was one that seemed to puzzle the students; often they would have difficulty in making up their minds. From an educational perspective, this is, however, the power of this precise question: that there is no correct answer to it. It is a matter of conviction, whether you are a Platonist, a formalist, a constructivist, a realist, an empiricist, or something else. Thus, students will have to *reflect* about the question on their own in order to take a standpoint.

Reflection, in particular, and the ability to perform reflection are considered to be major factors in changing beliefs [3, 2]. Thus, if the students who took part in the research presented above were to have their beliefs 'molded' or 'shaped' in such a fashion that they would fit the previously presented goal-oriented descriptions, then one way of doing this would be to set a scene which enabled them to perform reflections. In fact, the students' questionnaire and interviews reported above are an initial part of a larger research study, one purpose of which was to provide the students with classroom situations in which they were expected to work actively with and reflect upon issues related to, amongst other, the previously discussed aspects of the evolution and development of mathematics (questions 1 through 7). More precisely, these situations consisted of two larger teaching modules which the upper secondary class was to engage in over a longer period of time[6]. During and after the period of implementation, the changes in students' beliefs were evaluated through more questionnaires and interviews but also by means of videos of classroom situations taking as the point of departure the 'initial' student beliefs as presented in this paper[7]. A comparison of the questionnaire and interview results presented in this paper, i.e., those from before implementing the modules, with the later research findings, those from during and after the implementations, may be found in [16].

As a very final remark, I shall point to my own belief that reflections ought not only be considered as a means for changing existing beliefs, or creating new ones. A students' image of mathematics should include an awareness of mathematics as a discipline that consists of and gives rise to questions to which there are no correct answers (*e.g.*, that of invention versus discovery), and for this reason the ability to reflect is equally important. That is to say that not only is the act of providing students with an image of, or a set of beliefs about, mathematics as a discipline a goal in itself, the act of making the students capable of reflecting about their images is a goal as well.

---

[6]Descriptions of and results from this research study may be found in Jankvist, [11, 13, 14, 15, 17]. The complete study is presented in [17]. See also [12] for a more general introduction to history in mathematics education and the idea of using 'history as a goal'.

[7]*E.g.*, beliefs on question 5 were evaluated by posing more specific questions relating to the cases of the two modules (see [16]).

Bibliography

[1] Christensen, J. and K. L. Rasmussen, 1980, *Matematikopfattelser hos 2.G'ere—en analyse*, No. 24A in Tekster fra IMFUFA. Roskilde: IMFUFA.

[2] Cooney, T. J., 1999, "Conceptualizing teachers' ways of knowing," *Educational Studies in Mathematics* 38, 163-187.

[3] Cooney, T. J., B. E. Shealy, and B. Arvold, 1998, "Conceptualizing belief structures of preservice secondary mathematics teachers," *Journal for Research in Mathematics Education* 29, 306-333.

[4] Demattè, A. and F. Furinghetti, 1999, "An exploratory study on students' beliefs about mathematics as a socio-cultural process." In G. Philippou (ed.): *Eighth European Workshop, Research on Mathematical Beliefs—MAVI-8 Proceedings*, Nicosia: University of Cyprus, pp. 38-47.

[5] Ernest, P., 1998, "Why Teach Mathematics?—The Justification Problem in Mathematics Education." In J. H. Jensen, M. Niss, and T. Wedege (eds.), *Justification and Enrollment Problems in Education Involving Mathematics or Physics*, Roskilde, Denmark: Rosklide University Press, pp. 33-55.

[6] Fauvel, J. and J. van Maanen (eds.), 2000, *History in Mathematics Education—the ICMI Study*, Dordrecht: Kluwer Academic Publishers.

[7] Furinghetti, F., 1993, "Images of Mathematics Outside the Community of Mathematicians: Evidence and Explanations," *For the Learning of Mathematics* 13(2), 33-38.

[8] ——, 2007, "Teacher education through the history of mathematics," *Educational Studies in Mathematics* 66, 131-143.

[9] Furinghetti, F. and E. Pehkonen: 2002, "Rethinking Characterizations of Beliefs." In G. C. Leder, E. Pehkonen, and G. Törner (eds.), *Beliefs: A Hidden Variable in Mathematics Education?*, Dordrecht: Kluwer Academic Publishers, pp. 39-57. Chapter 3.

[10] Hansen, H., 2001, "Opfindelse eller opdagelse?." In M. Niss (ed.), *Matematikken og Verden*, København, Forfatterne og Forlaget A/S, pp. 65-96. Kapitel 3.

[11] Jankvist, U. T., 2008, "A Teaching Module on the History of Public-Key Cryptography and RSA," *BSHM Bulletin* 23(3), pp. 157–168.

[12] ——, 2009a, "A categorization of the 'whys' and 'hows' of using history in mathematics education." *Educational Studies in Mathematics* 71(3), 235-261.

[13] ——, 2009b, "Evaluating a Teaching Module on the Early History of Error Correcting Codes." In M. Kourkoulos and C. Tzanakis (eds.), *Proceedings 5th International Colloquium on the Didactics of Mathematics*, Rethymnon: The University of Crete, pp. 447–460.

[14] ——, 2009c, "History of Modern Applied Mathematics in Mathematics Education," *For the Learning of Mathematics* 29(1), 8-13.

[15] ——, 2009d, "On Empirical Research in the Field of Using History in Mathematics Education," *ReLIME* 12(1), 67-101.

[16] ——, 2009e, *"Using History as a 'Goal' in Mathematics Education,"* Ph.D. thesis, Roskilde University, No. 464 in Tekster fra IMFUFA, Roskilde. `milne.ruc.dk/ImfufaTekster/pdf/464.pdf`

[17] ——, 2010, "An empirical study of using history as a 'goal'," *Educational Studies in Mathematics* 74(1), 53-74.

[18] Leder, G. C. and H. J. Fortaxa, 2002, 'Measuring Mathematical Beliefs and their Impact on the Learning of Mathematics: A New Approach', In G. C. Leder, E. Pehkonen, and G. Törner (eds.), *Beliefs: A Hidden Variable in Mathematics Education?*, Dordrecht: Kluwer Academic Publishers, pp. 95-113. Chapter 6.

[19] Lester, Jr., F. K., 2002, "Implications of Research on Students' Beliefs for Classroom Practice." In G. C. Leder, E. Pehkonen, and G. Törner (eds.), *Beliefs: A Hidden Variable in Mathematics Education?* Dordrecht: Kluwer Academic Publishers, pp. 345-353. Chapter 20.

[20] Niss, M., 1994, "Mathematics in Society." In R. Biehler, R. W. Scholz, R. Sträßer, and B. Winkelmann (eds.), *Didactics of Mathematics as a Scientific Discipline*, Dordrecht: Kluwer Academic Publishers, pp. 367-378.

[21] Niss, M. and T. H. Jensen (eds.), 2002, *Kompetencer og matematiklæring—Ideer og inspiration til udvikling af matematikundervisning i Danmark*, Undervisningsministeriet, Uddannelses-styrelsens temahæfteserie nr. 18.

[22] Philipp, R. A., 2007, "Mathematics Teachers' Beliefs and Affect," In F. K. Lester,Jr. (ed.), *Second Handbook of Research on Mathematics Teaching and Learning*, Charlotte, NC: Information Age Publishing, pp. 257-315. Chapter 7.

[23] Philippou, G. N. and C. Christou, 1998, "The effects of a preparatory mathematics program in changing prospective teachers' attitudes towards mathematics," *Educational Studies in Mathematics* 35, 189-206.

[24] Presmeg, N., 2002, 'Beliefs about the Nature of Mathematics in the Bridging of Everyday and School Mathematical Practices'. In G. C. Leder, E. Pehkonen, and G. Törner (eds.), *Beliefs: A Hidden Variable in Mathematics Education?* Dordrecht: Kluwer Academic Publishers, pp. 293-312. Chapter 17.

[25] Schoenfeld, A. H., 1985, *Mathematical Problem Solving*. Orlando, Florida: Academic Press, Inc.

[26] ——, 1992, 'Learning from Instruction'. In D. A. Grouws (ed.), *Second Handbook of Research on Mathematics Teaching and Learning*. New York: Macmillian Publishing Company, pp. 334-370. Chapter 15.

[27] Törner, G., 2002, 'Mathematical Beliefs—A Search for a Common Ground: Some Theoretical Considerations on Structuring Beliefs, Some Research Questions, and Some Phenomenoligical Observations'. In G. C. Leder, E. Pehkonen, and G. Törner (eds.), *Beliefs: A Hidden Variable in Mathematics Education?* Dordrecht: Kluwer Academic Publishers, pp. 73-94. Chapter 5.

[28] Undervisningsministeriet, 2007, 'Vejledning: Matematik A, Matematik B, Matematik C'. Bilag 35, 36, 37.

About the Author

Uffe Thomas Jankvist holds a master's degree in mathematics and computer science from Roskilde University, Denmark (2005), and a Ph.D. in the didactics of mathematics, also from Roskilde (2009). The topic for his master's thesis was the use of mathematics in the Mars Exploration Rover mission, in particular error-correcting codes and data compression as well as the historical origins of these mathematical disciplines. The topic of his Ph.D. was the use of history of mathematics in mathematics education, which he considered both theoretically and empirically, the latter by video monitoring the implementing of two historical teaching modules in a Danish high school. In 2010 he received a prestigious government grant from the Danish Agency for Science, Technology and Innovation, which funds a two-year postdoctoral project at the University of Southern Denmark, where he will design, implement, and assess high school teaching activities between the history of mathematics, the application of mathematics, and the philosophy of mathematics using original sources.

13

Changes in Student Understanding of Function Resulting from Studying Its History

Beverly M. Reed
Kent State University, United States of America

13.1 Introduction

Professional mathematical societies are deeply concerned about the mathematical education of our teachers and are continuing to search for effective means to deepen students' understanding of fundamental mathematical concepts [6]. National reports call for better preparation of our mathematics teachers [3, 9].

The purpose of this study was to discern if students learn mathematics by studying its history. In particular, it investigated the changes in pre-service secondary school teachers' thinking about functions resulting from their studying the history of the concept. This study addressed the following questions.

- Does studying the history of the concept of function deepen a student's *understanding* of the concept in any way and if so, in what way? In particular, does studying the history facilitate his or her move from an *action* level understanding to a *process* level understanding as described by APOS theory (**A**ction, **P**rocess, **O**bject, and **S**chema)?

- Does a student's studying the history of the concept of function facilitate his or her move from a process level understanding to an *object* level understanding?

13.2 Theoretical Basis for the Study: APOS Theory

This Section explains APOS Theory, a constructivist approach to the learning and understanding of mathematics at the post-secondary level. The developers of APOS theory used the idea of theoretical cognitive structures from Piaget and relate them to observable behaviors in college-level students [2]. They created a model for conducting research in mathematics education. APOS is the cognitive aspect of the model. It guides the theoretical analysis of a student's understanding of a mathematical concept.

The acronym APOS stands for **A**ction, **P**rocess, **O**bject, and **S**chema–mental constructions made by students in their attempts to understand mathematics. *Actions* lead to *processes*, which must come before seeing a concept as an *object*.

A student who has an *action* understanding of functions sees an algebraic expression as a command to calculate. Such a student can carry out a transformation only by reacting to *external cues* (textbook directions, teacher suggestion, etc.) that give exact details on what to do. Though it is considered the lowest level of abstraction, it is a necessary beginning to the understanding of functions.

A *process* is an internal construction that performs the same transformation as the action, but it is internal and hence under the control of the individual. She no longer needs the external stimuli, no longer needs to actually evaluate an expression to think of its result. With this understanding, a student can link one or more processes to construct a composition, for example, or reverse the process to obtain inverses of functions.

Once students have practice working with processes, groundwork is laid for them to begin thinking about *sets* of inputs in relation to sets of corresponding outputs. APOS theorists say students are then ready to begin to reason more formally about functions–they *encapsulate* the process to form an *object*. An *object* understanding of a concept sees it as "something to which actions and processes may be applied" [13, p. 19].

The *schema* construct is the highest level of abstraction and can be thought of as a coherent collection of processes and objects that are organized in a structured manner. The current study was not concerned with the *schema* construct.

13.3 Procedure for the Study

Students enrolled in the History of Mathematics course participated in a five-week unit on the history of the concept of function. The unit consisted of nine worksheets: three created by the researcher, one adapted from Usiskin et al., [14] and five from Anderson et al. [1]. Students worked collaboratively on the function project, sometimes in class, but much of it outside of class. The groups were informal and consisted of three or four students; students self-selected their groups. Each individual student handed in a completed project. During class, the instructor provided background and related information. Since only one section of the course was taught per semester, the researcher was the instructor of the course.

All students completed a questionnaire before and after the instructional program. The researcher chose seven students to interview based on responses on the initial questionnaire, thus obtaining participants operating at a variety of APOS levels. In the initial interview, the researcher asked each student to explain her thinking as she completed the questionnaire. Each interview was audio taped and lasted between 60 and 90 minutes. One participant dropped out of the study because of illness.

Students worked collaboratively to complete the readings and worksheets. After each worksheet, each student wrote a one page reflection indicating

- her understanding of the concept of function;
- if and how the worksheet led to new insights concerning the concept of function.

The researcher conducted another individual interview with each participant at the conclusion of the unit. The researcher asked each participant to explain her thinking in detail as she worked through the second questionnaire and asked if she would change any of her answers on the initial questionnaire. These interviews, the completed questionnaires, the student worksheets, and the individual reflections comprised the data for this study.

13.4 The Research Instruments

The initial questionnaire consisted of three parts. Part 1 consisted of three questions probing students' general understanding of functions: *What is a function? Describe the ways a function can be represented. Give several examples of functions.* Part 2 consisted of 19 situations, some obviously mathematical and others not obviously so. Students were to describe each using one or more functions. The situations were adapted from Dubinsky's and Harel's work [7] and consisted of the following.

1. Two finite sequences: one with integer values and one with Boolean, true/false values: $\{3n + n^2 : n \text{ in } [1\ldots 100]\}$ and $\{2^n > n^2 + 3n : n \text{ in } [1\ldots 100]\}$.

2. Two character strings: "Kent State's Men's basketball team makes it to the Sweet Sixteen" and "ABCDEFG."

3. Three graphs: a single valued continuous curve (with respect to both axes), a discrete graph, and a continuous curve.

4. Three sets of ordered pairs: $\{(x, 3x + 2) : x$ in the set of all integers$\}$; $\{(x, y) : x, y$ in the set of all rational numbers$\}$; $\{(1, 2x) : x$ in $[1 \ldots 100]\}$.

5. One two-column table: The first column contained student names and the other contained the amount of dues money they owed to their club.

6. Five equations: $y^4 = x^2$; $y^4 = x^3$; $2x^3 y - \sqrt{x} \log y = 2$;

$$\left. \begin{array}{l} x = t^3 + t \\ y = 1 - 3t + 2t^4 \end{array} \right\} \, t, \text{ a real number}$$

7. Three statements: "A swimmer starts from shore and swims to the other side of Walden Pond;" "A square in the plane centered at the origin is rotated clockwise by $90°$;" "A record of all MAC women's basketball teams giving, for the 2003–2004 season, each team's field goal shooting percentage."

The sequence and character string questions are indicators of a process conception of function because the input, the ordinal position, is not visibly given. Recall that a student with an action conception of function needs perceptual input in order to recognize a function. Graphs have potential to indicate the ability to use a process conception of function, particularly if the graph required use of values on the vertical axis as the domain. Given ordered pairs or a table, a student can construct a function only by identifying the domain as the first element and taking the second element as the result of the function process. The ordered pair and table representations do not suggest this construction, so it must come from the student himself. While discussing the situations with equations, a student who insisted upon solving it for one variable in terms of the other, probably displayed an action conception of function. If he described the process without actually doing it, he exhibited at least the beginning of a process conception. The statements were the most open-ended questions and included simply to test the students' creativity and observe what type of functions they would construct given very little structure.

Part 3 of the questionnaire consisted of 5 graphing tasks, three of which asked students to analyze graphs of speed vs. time. One required interpretation of a position vs. time graph and another provided a sketch of a bottle and asked students to create a graph of the height of water vs. the volume of water in the bottle.

The post-questionnaire was similar in content and structure to the initial questionnaire, but shorter in length. Students again gave their definition of function, indicated functions in situations, and interpreted graphs. Below is a sample question from the post-questionnaire.

A caterpillar is crawling around on a piece of [coordinatatized] graph paper, as represented below. If we were to determine the creature's location on the paper with respect to time, would this location be a function of time? Why or why not?

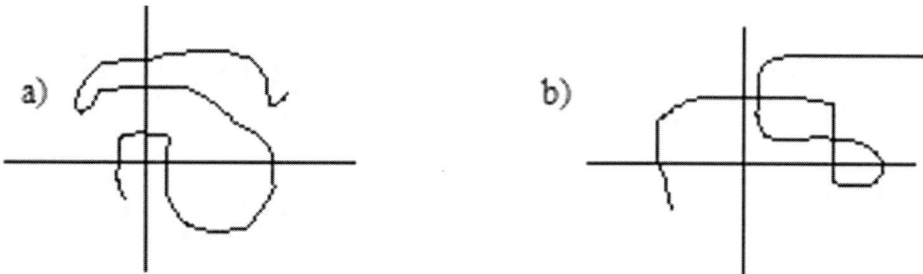

13.5 The Worksheets and Readings

The unit on functions consisted of nine worksheets and readings: three created by the researcher, one adapted from Usiskin et al., [14] and five taken from Anderson et al. [1].

Worksheet #1 concerned the history of graphs of functions, focusing on the work of Oresme. It offered a combination of historical readings, comparisons of Oresme's techniques to modern ones, and Oresme's proofs with details to be supplied by the student. The following is an excerpt from the Oresme worksheet.

Activity 3 Reading from Oresme (c.1350) in [5]:

> And so every uniform quality is imagined by a rectangle and every quality uniformly difform terminated at no degree is imaginable by a right triangle. Further, every quality uniformly difform terminated in both extremes at some degree is to be imagined by a quadrangle having right angles on its base and the other two angles unequal. Now every other linear quality is said to be "difformly difform" and is imaginable by means of figures otherwise disposed according to manifold variation. (p. 193)

1. Make a sketch of a configuration for each of the following:

    a) Uniform quality

    b) Uniformly difform quality (two possibilities—one whose initial value of the output is 0, and one whose initial (and final values are greater than 0.)

    c) Difformly difform quality

2. Given the same two variables as in the previous activity (marital satisfaction vs. number of years of marriage), make a new table of data for each of the configurations you sketched in problem #1 above.

3. Suppose that you were tutoring a middle school student in math. Create another real-world scenario for each configuration and explain how you would clarify the difference between each type of rate of change.

Worksheet #2 consisted of readings from Kleiner [12], Youschekevitch [16], and Katz [11] concerning Fermat's role in the development of the function concept. Students translated Fermat's symbols to modern notation, created an example using Fermat's coordinate system, and supplied reasons for his analysis of the two-point case of a theorem of Apollonius. Here is an excerpt:

> In the following brief reading, assume B and D are known, but arbitrary quantities. Consider the constant ratio $\frac{D}{B}$ as the point Z moves to the right along the line NZM.
>
> Let NZM be a straight line given in position, with the point N fixed. Let NZ be equal to the unknown quantity A, and ZI, the lines drawn to form the angle NZI, the other unknown quantity E. If D times A equals B times E, the point I will describe a straight line given in position. [11, p. 435]

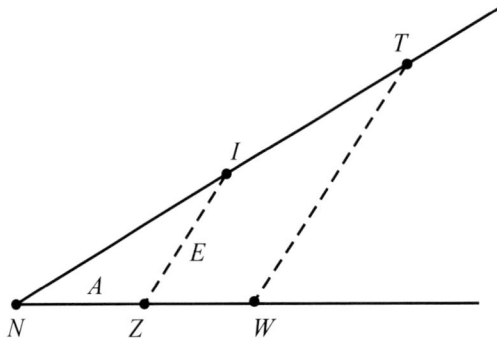

He assumes also that $\angle INZ$ is fixed.

1. If Fermat used Viète's convention of vowels representing unknowns and consonants representing known quantities, translate the last statement above using modern notation.

Note that Fermat uses a single axis, a constant ratio D/B, and an equation, which he wants to show represents a straight line. He generates his curve by the motion of the "endpoint I of the variable line segment ZI as Z moves along the given axis." [11, p. 435]. His method is unlike our method of plotting points with respect to two axes.

2. Explain why that for any such points Z and W along the segment NM and the corresponding points I and T on NI, $\triangle NIZ \sim \triangle NTW$.

3. Create a straight line using Fermat's method with $D = 3$ and $B = 2$. Show at least 3 different positions of Z.

13.5. The Worksheets and Readings

Worksheet #3 consisted of a reading and activity concerning Descartes' analytic geometry from "Functions" [1, pp. 90, 91, 94, 97]. Students worked with Descartes' oblique coordinate system and read about his generalization of Apollonius' problem of loci with respect to three or four lines. They worked through the loci problem with three or four lines in a manner similar to Descartes, except using a rectangular coordinate system. The researcher hoped that by studying early coordinate systems, students would gain flexibility in their understanding of loci and use of a coordinate system.

Worksheet #4 was an adaptation of a problem from Usiskin et al. [14, p. 130]. Students read about the first use of the term "function" and worked with Leibniz's six functions of a point P on a curve C (abscissa, ordinate, tangent, normal, subtangent, subnormal). Students reflected on the difference between the modern notion of function and Leibniz's use of the term. The researcher hoped that seeing the term in a different context might expand student thinking, particularly helping those students who held the "function as formula" conception.

Worksheets #5 and #6 dealt with Fibonacci sequences and were from the module, "Functions" [1, pp. 72–73, 76]. Many students did not recognize sequences as functions on the initial questionnaire and these worksheets provided an opportunity to refresh their memories and/or expand their thinking about these functions.

Worksheet #7 involved a short reading from Chapter 1 of Euler's *Introduction to Analysis of the Infinite*. Students compared Euler's description of function with the modern definition, identified transcendental vs. algebraic functions, and worked with Euler's definitions of rational functions, single-valued functions, and multiple-valued functions. Students sometimes confuse the uniqueness criterion for functions with the notion of one-to-one functions and the hope was that the discussion following work on this reading would clarify such misconceptions. The researcher hoped that students would identify with Euler's early definition of function as formula, then eventually move beyond it as Euler did. This worksheet was from the module, "Functions" [1, pp. 59–62].

Worksheet #8 dealt with Fourier series. Students read from Boyer's chapter, "Arithmetization of Analysis," [4] and then graphed the Fourier representations of two simple functions for various values of n ($n = 1, 2, \ldots 10, 20$). The hope was to give students a better understanding of Fourier representations of functions and give them an opportunity to reflect on the fact that a single function can have several different representations. This worksheet was from the module, "Functions" [1, pp. 99, 103, 104].

Worksheet #9 traced the definition of functions from Euler's time to Bourbaki's. Students compared and contrasted function definitions, conjectured about what may have caused the definition to change, then analyzed the situations from the initial classroom questionnaire in light of these definitions. They read a history of the function concept from Euler's time to the nineteenth century, including the vibrating string problem and Fourier's work on heat conduction. Following is an excerpt:

Euler (1748): *A function of a variable quantity is an analytic expression composed in any way whatsoever of the variable quantity and numbers or constant quantities. Hence every analytic expression, in which all component quantities except the variable are constants, will be a function of that x; thus $a + 3x$; $ax - 3x^2$; $ax + b\sqrt{a^2 - x^2}$; c^x; etc. are functions of x.*

Euler (1755): *If however, some quantities depend on others in such a way that if the latter are changed the former undergoes changes themselves then the former quantities are called functions of the latter quantities. This is a very comprehensive notion and comprises in itself all the modes through which one quantity can be determined by others. If, therefore, x denotes a variable quantity then all the quantities which depend on x in any manner whatsoever or are determined by it are called functions.*

1. Compare Euler's second definition with his first. Recall that you worked with the first definition in an earlier set of worksheets. What idea expressed in the first definition is absent in the second definition?

2. Recall the "function situations" in the function assessment you took before you started this project. Several are listed below. Which of the following would be functions as defined originally by Euler (1748). (We'll call this definition "*early Euler*"). Which of the following might be functions as defined above by Euler ("*late Euler*"). Briefly explain your reasoning.

The situations from the initial questionnaire of the study followed.

13.6 Findings

This section analyzes the thinking of the six participants (DB, CW, MJ, BG, MS, and CS) and notes emergent patterns regarding growth in their understanding of functions. It has four subsections. The first discusses changes in students' definitions of function, the second notes their increased abilities to recognize functions in situations, and the third summarizes changes in APOS levels. The last discusses changes in their understanding graphs of functions.

The interpretation of data is consistent with the literature [7] concerning indicators of the APOS function conceptions. Sample descriptions, evidence, and student responses for each APOS conception follow.

APOS Level	Description of Level	Evidence	Sample student responses to "What is a function?"
Prefunction	Student has little if any conception of function.		
Action	"A repeatable mental or physical manipulation of objects" [8, p. 251]. Would involve the ability to plug numbers into an algebraic expression and and calculate. A static conception; student needs to think about it one step at a time.	Emphasizes the act of substituting numbers in for variables and calculating to obtain a number but not referring to any overall process.	"an equation that can have numbers plugged in to form a graph." "this function has certain values to be inputted for x and then that x is put into $3x + 2$ and an answer is given, that is, $x = 2$ gives $3(2) + 2 = 8$." A statement is "not a function because its just a statement. There is no equation here."
Process	Involves a "dynamic transformation of objects according to some repeatable means, that given the same original object, will always produce the same transformed object." (p. 251)	Subject can think about process without having to perform it. Can combine the process and even reverse it. Mentions input, transformation, and output in general fashion.	"function is an equation in terms of one or more variables such that any given set of variables will yield a specific output." it doesn't necessarily need to be an equation, um ... but it is whenever you have, uh ... [pause] one thing leading to another." "the mapping can be arbitrary, that is, as long as the mapping explains how individuals elements in two different sets are related you have a function."
Object	When a student can perform an action on a process, the authors say the student has *encapsulated* the process as an object.		While discussing the chain rule: "It is interesting that the functions can be broken up into basic functions that you ... have $f'(x)$ and $g'(x)$. That you break it up into as small a function as you can."

Table 13.1. Description of APOS Levels

13.6. Findings

13.6.1 Definitions of functions

An emphasis on equations, numbers, and evaluating expressions is evidence of an action conception of function. On the initial questionnaire, several participants exhibited this tendency, which was strongest in CW and CS. DB and MJ exhibited the tendency to a lesser degree and DB thought that a graph or a formula is the function, rather than just a representation of a function. Interestingly, DB and MJ at times discussed a function as a general process, but reverted to a "function as equation" as they attempted to recognize functions in situations. This tendency suggests that DB and MJ had an emerging, but not strong, process conception initially. MS at times referred to the need for "some operation."

On the second questionnaire and interview, DB made no mention of function as equation. In her words, "now I feel that a function simply stated is something that takes an input and produces an output." She came to understand the uniqueness criterion, something she overlooked on the initial instruments. Similarly, CW's new definition of function is "a relationship between variables where one input produces one output and the relationship can be shown in a number of ways, not just an analytic expression." MJ and CS also let go of their tendency to look for an equation, with MJ noting, "the mapping can be arbitrary" and CS claiming that a "function is a relationship that provides us with a unique output for a given input." Note that there is no mention of equation in these definitions.

These results suggest that studying the history of functions broadened the participants' definition of function. After working on the worksheet entitled "Definitions of Functions" and the readings associated with it, the class had an extensive discussion about the change in the definition of function over the years and what prompted such change. It is reasonable to conclude that this worksheet and discussion facilitated a move away from the "function as equation" notion in the participants. Table 13.2 summarizes these results.

	Before	After
DB	*	
CW	*	
MJ	*	
BG		
MS		
CS	*	

Table 13.2. Participants Holding the Notion of "Function as Formula" Before and After History of Functions Unit

13.6.2 Functions in situations

Not only did the participants' definitions change, but their ability to find functions in situations improved as well. The following discussion focuses on four tasks on the initial questionnaire and the corresponding tasks on the second questionnaire.

Table 13.3 indicates which participants recognized a function in the given scenario. An asterisk (*) means that participant clearly articulated an appropriate function. The phrase "considered the possibility" indicates that the par-

	String as function	Arbitrary pairing as function, i.e., table or list	Recognizes a Boolean (true-false output) function	Discrete set of points on graph as function
DB			considered the possibility	considered the possibility
CW				
MJ	*			
BG	*	*	*	*
MS	considered the possibility	*		considered the possibility
CS				

Table 13.3. Appropriate Answers to Specific Tasks Before History of Functions Unit

ticipant had a vague notion of function in the scenario, but was either unable to articulate it or gave an inappropriate formulation.

Only two participants considered an arbitrary mapping as a function. The "function as equation" notion was evident in DB ("you apply the function of subtraction"), CW ("if you'd have the previous one plus five, you would have a function"), and MJ ("starting to graph it a little bit and see if there was any kind of relationship"). CS left the task completely blank. On the discrete set of points graph, MS did successfully find a function, but insisted on connecting the dots and trying to find a formula so that he could "find a pattern of how to change it" [the inputs]. Note also that other than BG and MS, the participants had little success with these tasks as a whole.

Table 13.4 indicates the participants who successfully recognized a function in these scenarios on the second questionnaire or while revisiting the initial questions during the second interview.

	String as function	Arbitrary pairing as function, i.e., list	Recognizes a Boolean (true-false output) function	Discrete set of points on graph as function
DB	*	*	*	not discussed
CW	*	*	*	*
MJ	not discussed	*	*	*
BG	*	*	*	*
MS	considered the possibility	*		*
CS	considered the possiblity	*		considered the possibilty

Table 13.4. Correct Answers to Specific Tasks after the Unit on History of Functions

Other than BG, each participant exhibited an increased ability to recognize a function in the given scenarios. The data shows remarkable progress in the abilities of DB, CW, MJ, and CS. It is worth noting that BG, who recognized functions in each of the initial scenarios, exhibited evidence of an *object* conception of function in the second interview. Also noteworthy is the fact that MS, who showed the least growth, had put forth the least effort during the functions project, insisting on working alone, and turning in work of mediocre quality. The others, DB and MJ in particular, worked incredibly hard on the project, seeking help when necessary, struggling with the ideas presented. Not surprisingly, they exhibited impressive growth.

Interestingly, the class never discussed Boolean functions, nor was topic covered on any of the worksheets. It is possible that participants discussed this scenario in their small groups. It is also likely that the worksheet entitled "Definitions of Functions" and the subsequent class discussion facilitated an ability to recognize this type of function.

13.6.3 APOS Level changes

Table 13.5 summarizes the initial APOS conception of each participant. Note that only BG appeared to be operating with a process conception.

Table 13.6 lists each participant's conception after the unit on the history of functions. A double asterisk (**) indicates a change in level from the initial conception. Note that four participants notably strengthened their function conceptions. Two participants moved an entire level: CS from an action conception to an emerging process one and

	Action	Emerging Process	Process	Object
DB		*		
CW	*			
MJ		*		
BG				*
MS		*		
CS	*			

Table 13.5. APOS Conception Before History of Functions Unit

13.6. Findings

	Action	Emerging Process	Process	Object
DB			**	
CW	*			
MJ			**	
BG				**
MS		*		
CS		**		

Table 13.6. APOS Conception After History of Functions Unit

BG from a process conception to an object. Admittedly, the evidence from this study is insufficient to claim that the unit on history of functions enabled BG's move to an object level, since the initial questionnaires did not test for this understanding. All one can claim is that after the unit on the history of functions, BG appeared to hold an object conception. DB and MJ appear to have strengthened their process conception. Recall that though CW did not advance a level in her APOS conception, she appeared to be moving toward a process conception. Recall also that MS was the weakest student among the participants and not surprisingly, showed little growth.

13.6.4 Changes in understanding graphical representations

Table 13.7 provides a description of APOS levels for graphs and evidence used to characterize the levels.

A summary of the participants' understanding of graphs before the history of functions unit is in Table 13.8.

Notable is the low level of graphical understanding for two or more questions in three of the participants, DB, CW, and MJ. Compare these levels with those following the study of the history of functions, in particular, after completion of the worksheet concerning Oresme's techniques. Table 13.9 lists only those conceptions which changed during the course of study.

APOS Level	Description of Level	Evidence
Prefunction	Student has little if any conception of graphs of functions	Is unable to see graph as a representation of a relationship between variables; sees a graph as a picture of the event.
Action	A static conception; student needs to think about it one step at a time.	Emphasizes a reference to specific outputs and corresponding inputs on the graph. No reference to global features of a graph.
Process	Involves a "dynamic transformation of objects according to some repeatable means, that given the same original object, will always produce the same transformed object." [8, p. 251]	Can discuss global features of a graph without referring to specific inputs and outputs. Is able to compare intervals or gradients without reference to specific inputs and outputs.
Object	When a student can perform an action on a process, the authors say the student has *encapsulated* the process as an object.	Subject considers area under a curve significant. The act of finding the area is an *action*, which she is performing on the *process* represented by the graph. Considers area under a graphical representation of speed vs. time as distance traveled, for example. Recognizes a graph as one of several representations of a function.

Table 13.7. APOS Characterizations for Understanding Graphical Representations of Functions

Participant	Task 1 position vs. time	Task 2c speed vs. time (distance traveled)	Task 3a speed vs. time (compare position of 2 cars)	Task 3d speed vs. time (relative position of 2 cars)	Task 5 speed vs. time (bends in a racetrack)
DB	(no answer)	Pre-Function	Pre-Function	Pre-Function	Pre-Function
CW	Process	Process	Pre-Function	Pre-Function	Process
MJ	Action	Object	Object	Pre-Function	Pre-Function
BG	Process	Object	Object	Process	Process
MS	Process	Object	Process	Process	Process
CS	Process	Pre-Function	Object	Pre-Function	Pre-Function

Table 13.8. Interpretation of Graphs Before History of Functions Unit

Participant	Task 1	Task 2c	Task 3a	Task 3d	Task 5
DB	Process	Object	Object	Process	Process
CW			Action		
MJ	Process			Process	Process
BG					
MS					
CS				Process	Process

Table 13.9. Interpretation of Graphs After History of Functions Unit

13.7 Discussion

There appears to be little question that a marked improvement in understanding functions occurred during the course of studying its history for four of these six participants.

Growth was most profound in the area of graphical representations of functions. Of all the worksheets, the Oresme worksheet was the one most dependent upon original sources and provided the most in-depth information about the germination of the concept. It appears to have cured the "graph as picture" tendency in DB and CW and enabled the understanding of area under the curve for MJ and DB. According to MJ, it helped him understand the "connection between/motivation for integration and area." DB commented that this particular set of exercises clarified her thinking about graphs. One can reasonably conclude that this use of primary sources revealing the germination of an idea along with activities relating original methods to modern-day ones enabled conceptual growth. This finding validates the work described by Jahnke [10] concerning the benefits of using original sources.

The worksheet about Leibniz's work concerned his use of the word "function" in a geometric context. This exercise may have suggested to students that a function need not be definable by equations alone. A worksheet about Fibonacci was not historical in the sense that it did not delve into the beginnings of an idea. It was just a problem from history. The learning that occurred as a result of this worksheet supports the claim that history is a good source of problems. Both MJ and CW had not considered sequences to be functions until after their work on this assignment. Perhaps it facilitated their new-found ability evidenced on the second questionnaire and in the second interview to see strings as functions as well.

The worksheet on the Definitions of Function showed participants that mathematical definitions can be dynamic, not static. By comparing and contrasting the definitions, students came to terms with the different notions expressed. The subsequent readings may have expanded their thinking about the limitations of the "function as formula" notion.

The above evidence suggests that a wide variety of materials may pull students along in their understanding of a concept: primary source readings about the germination of a concept, problems from history, or simply reading about the changes of a concept over time. The emphasis here is clearly on a *thematic* approach to history–the germination and development a single mathematical concept.

13.8 Conclusion and Summary

The data in this study indicates that students underwent mental development and learned some mathematics as a result of the use of history. History *can* refresh and strengthen a student's understanding of a mathematical concept. A thematic approach, reading original sources, and working on problems of former mathematicians enabled insight into the conceptual base for the participants in this study. Four of the six participants notably strengthened their function conceptions. Two moved an entire APOS level. Five of the six exhibited an increased ability to recognize a function in a given scenario. Growth was most profound in the area of graphical representations. One claimed that the unit helped refresh ideas long forgotten. This study, then, provided evidence that studying the history of mathematics enables a deep reflection of ideas, or as Von Glasersfeld [15] suggested, a "re-presentation" or reconstruction of ideas.

Bibliography

[1] Anderson, D., R. Berg, A. Sebrell, and D. Smith, 2005, "Functions" in *Historical Modules for the Teaching and Learning of Mathematics* (compact disc), V. J. Katz & K. D. Michalowicz, (eds.), Washington, DC: The Mathematical Association of America.

[2] Asiala, M., A. Brown, D. DeVries, E. Dubinsky, D. Mathews, and K. Brown, 1997, "A Framework for Research and Curriculum Development in Undergraduate Mathematics Education." In E. Dubinsky, D. Mathews, & B. Reynolds (eds.), *Readings in Cooperative Learning for Undergraduate Mathematics*, Washington, DC: The Mathematical Association of America, pp. 37–53.

[3] Ball, D.L. (chair), RAND Mathematics Study Panel, 2003, *Mathematical proficiency for all students: Toward a Strategic Research and Development Program in Mathematics Education*, Santa Monica, CA: RAND Corporation

[4] Boyer, C., 1968, *A history of mathematics*, New York: John Wiley & Sons.

[5] Clagett, M., 1968, *Nicole Oresme and the medieval geometry of qualities and motions. A treatise on the uniformity and difformity of intensities known as Tractatus de configurationibus qualitatum.* Madison: University of Wisconsin Press.

[6] Conference Board of the Mathematical Sciences, 2001, *The Mathematical Education of Teachers*, Washington, DC: The Mathematical Association of America.

[7] Dubinsky, E., and G. Harel, 1992, "The nature of the Process Conception of Function." In G. Harel & E. Dubinsky (eds.), *The Concept of Function: Aspects of Epistemology and Pedagogy*, Washington, DC: Mathematical Association of America, pp. 85–106.

[8] Breidenbach, D., E. Dubinsky, J. Hawks, and D. Nichols, 1992, "Development of the process conception of function," *Educational Studies in Mathematics*, 23, 247–285.

[9] Glenn, J., (commissioner), U.S. Department of Education, 2000, "*Before It's too Late: A Report to the Nation from the National Commission on Mathematics and Science Teaching for the 21st Century*," Washington, DC: Author. Retrieved on March 4, 2006, from http://www.ed.gov/inits/Math/glenn/report.pdf

[10] Jahnke, H.M., 2000, "The Use of Original Sources in the Mathematics Classroom." In *History in Mathematics Education, The ICMI Study*, J. Fauvel, & J. van Maanen, (eds.), Dordrecht-Boston-London: Kluwer Academic, pp. 291–328.

[11] Katz, V. J., 1998, *A history of mathematics. An introduction* (2nd ed.), New York: Addison-Wesley.

[12] Kleiner, I., 1989, Evolution of the function concept: A brief survey. *The College Mathematics Journal*, 20, 282–300.

[13] Selden, A., and J. Selden, 1992, "Research perspectives on conceptions of function: Summary and overview." In G. Harel and E. Dubinsky (eds.), *The Concept of Function: Aspects of Epistemology and Pedagogy*, Washington, DC: Mathematical Association of America, pp. 1–16.

[14] Usiskin, Z., A. Peressini, E. A. Marchhisotto, and D. Stanley, 2003, "*Mathematics for High School Teachers, an Advanced Perspective*," Upper Saddle River, NJ: Pearson Education.

[15] Von Glasersfeld, E., 1995, *Radical Constructivism: A way of Knowing and Learning*, Washington, DC: Falmer Press.

[16] Youschkevitch, A. P., 1976, The concept of function up to the middle of the 19th century. In C. Truesdell (Ed.), *Archive for history of exact sciences*, 16(1), 37–85.

About the Author

Beverly Reed is an assistant professor in the Department of Mathematical Sciences at Kent State University. She teaches a variety of courses, many at the freshmen level, but one of her favorites is the History of Mathematics. She is particularly interested in how studying the history of a topic changes student thinking about it, and the work represented here is her dissertation work. She earned a Master's Degree in mathematics from John Carroll University in Cleveland in 1979 and her Ph.D. in 2007 from Kent State University. Other work at Kent State involves supervising first year graduate students, developing online courses, and coordinating and redesigning freshmen level mathematics courses. In her spare time, she enjoys cycling, her young grandtwins, and her volunteer position as a Trailblazer for the Cuyahoga Valley National Park.

14

Integrating the History of Mathematics into Activities Introducing Undergraduates to Concepts of Calculus

Theodorus Paschos and Vassiliki Farmaki
University of Athens, Greece

14.1 Introduction

The history of mathematics may be a useful resource for understanding the processes of formation of mathematical thinking, and for exploring the way in which such understanding can be used in the designing of classroom activities. Such a task demands that mathematics teachers be equipped with a clear theoretical framework for the formation of mathematical knowledge. The theoretical framework has to provide a fruitful articulation of the historical and psychological domains as well as to support a coherent methodology. This articulation between history of mathematics and teaching and learning of mathematics can be varied. Some teaching experiments may use historical texts as essential material for the class, while on the other hand some didactical approaches may integrate historical data in the teaching strategy, and epistemological reflections about it, in such a way that history is not visible in the actual teaching or learning experience.

We used a teaching approach inspired by history. In particular, we used a *genetic approach to teaching and learning*. According to Tzanakis and Arcavi [17, p. 208]:

> It is neither strictly deductive nor strictly historical, but its fundamental thesis is that a subject is studied only after one has been motivated enough to do so, and learned only at the right time in one's mental development. ...Thus, the subject (*e.g.*, a new concept or theory) must be seen to be needed for the solution of problems, so that the properties or methods connected with it appear necessary to the learner who then becomes able to solve them. This character of *necessity of the subject* constitutes the central core of the meaning to be attributed to it by the learner.

From such a point of view, the historical perspective offers interesting possibilities for a deep, global understanding of the subject, according to the following scheme [17, p. 209]: (1) Even the teacher who is not a historian should have acquired a basic knowledge of the historical evolution of the subject. (2) On this basis, the crucial steps of the historical evolution are identified, as those key ideas, questions and problems which opened new research perspectives. (3) These crucial steps are reconstructed, so that they become didactically appropriate for classroom use.

In our case the reconstruction enters history *implicitly*. It means that a teaching sequence is suggested in which use may be made of concepts, methods and notations that appeared later than the subject under consideration, keeping always in mind that the overall didactic aim is to understand mathematics in its modern form.

14.2 The Historical Background of the Teaching Experiment

We focus on historical elements from the mathematical study of motions during the later Middle Ages (14th Century), and mainly on the role of both the geometric representations of motions and Euclidean geometry in the emergence of Calculus concepts. The study of motion in the 14th century was based on the study of movement in antiquity. The unique mathematical tool of study and representation of movement was Euclid's *Elements*.

14.2.1 Genesis of the mathematization of physics

The philosophical problem which gave stimulus to kinematics was the problem of how *qualities* (or other *forms*) increase in *intensity*. In the technical vocabulary of the schoolmen, this was called the problem of the *intension* and *remission* of *forms*, that is, the increasing and decreasing of the intensity of qualities or other forms. *Form* is every quantity or quality *e.g.*, the local motion, qualities of every kind, the light, the temperature, the velocity....

Duns Scotus, during the early years of the 14th century assumed a *quantitative* treatment of variations in intensity of qualities suffered by bodies. It was accepted by the successors of Scotus that the increase or decrease of qualitative intensity takes place by the addition or subtraction of degrees of intensity. With this approach to qualitative changes accepted, the Merton schoolmen applied various numerical rules and methods to qualitative variations and then by analogy to kindred problems of motion in space.

Tomas of Bradwardine in his *Treatise on the Proportions of Velocities in Movements* of 1328, using the theoretical considerations of William Ockam, made the distinction between dynamics and kinematics, saying that the temporal nature of movement demands only *extension* or *space* through which the movement takes place. Bradwardine's younger contemporary **Richard Swineshead** explicitly added time as a kinematic factor:

> .... it should be known that its velocity is measured simply by the line described by the ... moving point in such and such time.... [2, p. 243].

We can say that the interest concerning the quantitative study of the qualitative variations led to the (gradual) mathematization of physics.

14.2.2 The emergence of kinematics at Merton College (Oxford, circa 1320–1350)

The most famous mathematicians at Merton in the first half of 14th century were: (a) Thomas Bradwardine (1295-1349), and (b) the mathematicians–logicians William Heytesbury (1313–1372), Richard Swineshead (flourished ˜1344–1354), and John Dumbleton (flourished ˜1331–1349), known as Calculators. They considered *intension* or *latitude* of velocity as an arithmetic value (degree) in relation to *extension* or *longitude*, namely the time of the movement.

Let us describe the definitions of motions and the Mean Speed Theorem (MST) of the Merton kinematics [2, p. 235].

William Heytesbury said (*Rules for Solving Sophisms—Part VI. Local motion*):

> ....of local motions, that motion is called uniform in which an equal distance is continuously traversed with equal velocity in an equal part of time.... Non-uniform motion can, on the other hand, be varied in an infinite number of ways, with respect to time.....

The definitions of instantaneous velocity and uniformly and non-uniformly accelerated motion were given by Heytesbury as follows:

> ...In non-uniform motion the velocity at any given instant will be measured (*attendur*) by the path which would be described by the moving point if, in a period of time, it were moved uniformly at the same degree of velocity (*unifirmiter illo gradus velocitatis*) with which it is moved in that given instant...

> ...For any motion whatever is uniformly accelerated (*uniformiter intenditur*) if, in each of any equal parts of the time whatsoever, it acquires an equal increment (*latitudo*) of velocity.

14.2. The Historical Background of the Teaching Experiment

> ...But a motion is non-uniformly accelerated when it acquires a greater increment of velocity in one part of the time than in another equal part.

The *Mean Speed Theorem* (M.S.T) of Merton College is one of the most important results of the Merton studies in kinematics. It gives the measure of uniform acceleration in terms of its medial velocity, namely its velocity at the middle instant of the period of acceleration.

Furthermore, he said, in the same work, [2, p. 262]:

> ...Thus the moving body, acquiring or losing this latitude (increment) uniformly during some assigned period of time, will traverse a distance exactly equal to what it would traverse in an equal period of time if it were moved uniformly at its mean degree of velocity. ...For every motion as a whole, completed in a whole period of time, corresponds to its mean degree—namely, to the degree, which it would have at the middle instant of the time.

Swineshead in *De motu* said [2, p. 244]:

> ...Furthermore, any difform motion corresponds to some degree [of velocity]...

The uniform acceleration theorem and the above statement of Swineshead lead to the emergence of the mean value theorem of the Integral Calculus.

In the 14th century, there were many attempts to give a formal proof of the M.S.T. These proofs were basically of two kinds: arithmetical, which arose out of Merton College activity, and geometrical, mainly by N. Oresme at Paris (1350–60).

14.2.3 The application of two-dimensional geometry to kinematics given by Nicole Oresme

Oresme (1323-1382) used the definitions of motion expressed by the Calculators at Merton College. As examples of Oresme's geometrical model of motion representation, let us consider the accompanying rectangle and right triangle (Figure 14.1).

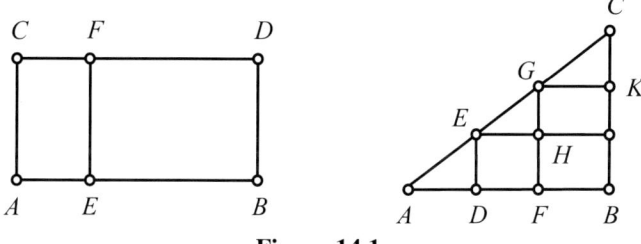

Figure 14.1.

Each figure measures the quantity of some quality (velocity). Line AB in either case represents the *extension* (time) of the quality. But in addition to extension, the *intensity* of the quality from point to point in the base line AB has to be represented; this Oresme represented by erecting lines perpendicular to the base line, the length of the lines varying as the intensity varies. Thus at every point along AB there is some intensity of the quality, and the sum of all these lines is the figure representing the quality globally. Now the rectangle $ABDC$ represents a uniform quality, since the lines AC, EF, BD representing the intensities of the quality at points A, E, and B (E being any point at all on AB) are equal, and thus the intensity of the quality is uniform throughout. In the case of the right triangle ABC, it is equally apparent that the lengths of the perpendicular lines representing intensities uniformly increase in length from zero at point A to BC at B, in accordance with Merton College's definition of uniformly accelerated motion.

Oresme designed the limiting line CD (or AC in the case of the triangle) as the line of summit or the *line of intensity*. This is comparable to a 'curve' in modern analytic geometry. He suggested the fundamental idea of *the total quantity of velocity*, which arises from considering both speed and time through which the motion continues. The total quantity of velocity is measured by the *area* of the figure, is also known as the *total velocity*, and represents the distance traversed.

We can say that this idea of Oresme was a genetic moment of the two-dimensional representation of a function that led to Cartesian representation two centuries later. See the general Figure 14.2:

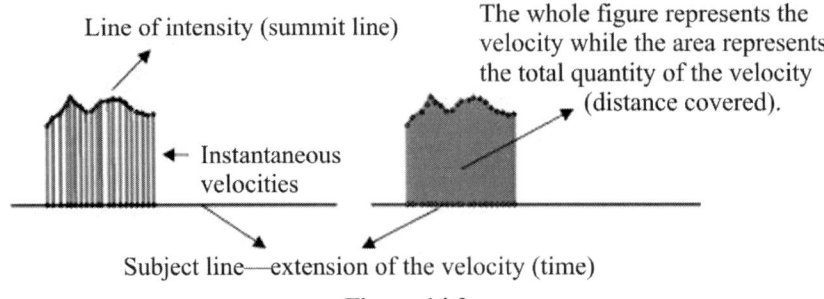

Figure 14.2.

Notice that: (1) The curve or summit line represents a 'function' expressed verbally, instead of algebraically, the verbal expressions of the functions being 'a uniform velocity', 'a uniformly non-uniform velocity', etc. (2) The variables of these 'functions' of Oresme are: (i) the intensity of the velocity, (ii) the extent (time), and (iii) the quantity of the velocity, represented by the area of the figure (distance covered), known as total velocity.

Translating, now, the definitions of instantaneous velocity, uniformly accelerated and non-uniformly accelerated motions given by the Calculators and applying the representation model of Oresme on the Cartesian axes, we obtain:

(1) A discrete approximation of constant changing velocity in which, in equal chosen time intervals we have equal increments of velocity (Figure 14.3). At the instant A of the time axis, the instantaneous velocity is represented by the line AB. The instantaneous velocity of a particle can be measured by the distance covered, if, in a period of time, the particle moves uniformly at the same degree of velocity (i.e., the shaded rectangle $ABCD$).

(2) Uniformly accelerated motion (Figure 14.4): In each of any equal parts of time the particle acquires an equal increment of velocity.

(3) A discrete approximation of non-uniformly accelerated motion (Figure 14.5): The particle acquires a greater increment of velocity in one part of time than in another equal part.

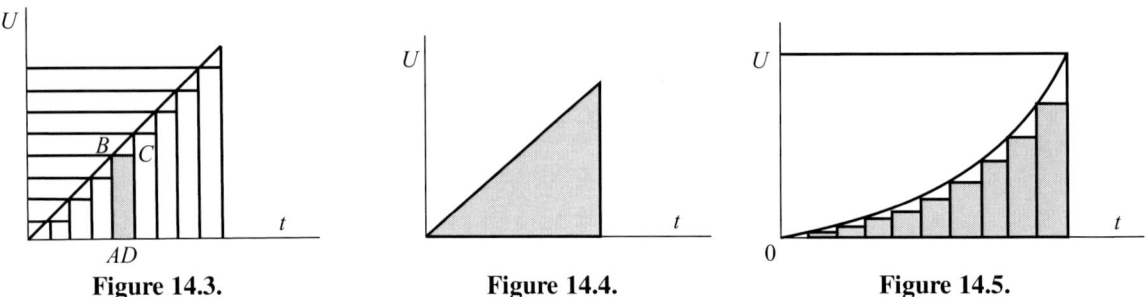

Figure 14.3. **Figure 14.4.** **Figure 14.5.**

Making the transition from the geometric representations, to the algebraic context using modern symbols, we obtain easily the algebraic formulas concerning the uniform (Fig. 14.6) and uniformly accelerated motion (Figure 14.7, 14.8).

Now since the basic kinematic acceleration theorem (M.S.T) equates a uniformly accelerated velocity with a uniform speed equal to its mean *in so far as the same space is traversed in the same time*, the geometric proof of this theorem,

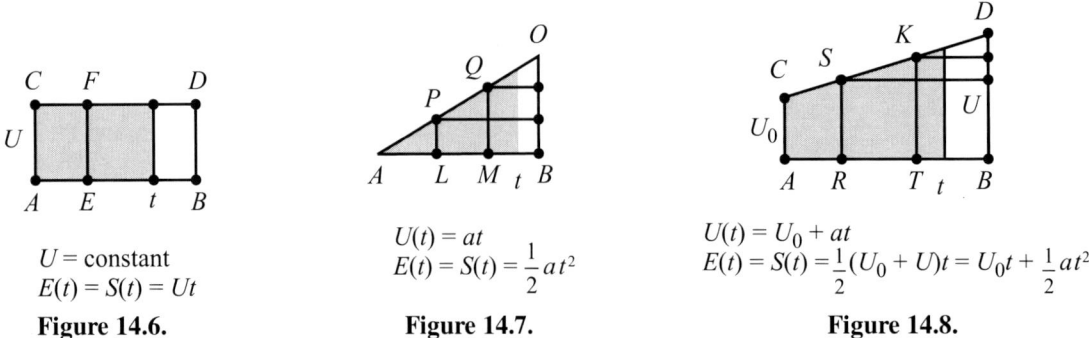

U = constant
$E(t) = S(t) = Ut$

$U(t) = at$
$E(t) = S(t) = \frac{1}{2}at^2$

$U(t) = U_0 + at$
$E(t) = S(t) = \frac{1}{2}(U_0 + U)t = U_0 t + \frac{1}{2}at^2$

Figure 14.6. **Figure 14.7.** **Figure 14.8.**

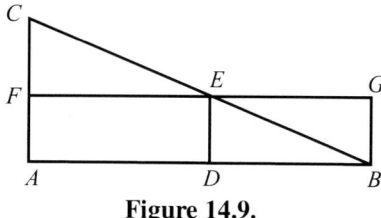

Figure 14.9.

using Oresme's system, must show that a rectangle whose altitude is equal to the mean velocity, is equal in area to a right triangle, whose height represents the whole velocity increment, i.e. a line equal to twice that of the altitude of the rectangle (Figure 14.9).

Oresme's proof is based on propositions of Euclidean geometry (triangle FEC equal to EBG, etc., *Elements* Book I, Proposition XXVI).

14.3 Designing Didactic Activities Inspired by History of Mathematics

The experimental teaching briefly described in this Section forms part of a wider qualitative action research which seeks to introduce students to advanced mathematical concepts using activities based on motion problems. Our research project also included an educational intervention aimed at improving the motivation of fifteen year-old students from an early stage, and at encouraging the development of algebraic, geometric and analytical thought in tandem through the solving of uniform motion problems based on the velocity-time graph. Such problems lead students to: (1) obtain the geometric solution to a problem on the basis of Euclidean geometry alone and using simple geometric transformations; (2) obtain an algebraic solution to the problems as well, thus making the transition from a geometric/graphic to an algebraic approach; and (3) realize the intuitive connection between velocity and distance covered on the same graph by relating the distance covered to the areas of the corresponding rectangles. Such problems should thus help students gain a better grasp of the essence of the Fundamental Theorem of Calculus at a later stage [5]. When applied to a wider range of motion problems as described below, this approach paves the way for the introduction of basic Calculus concepts (such as the integral, derivative and their interrelation).

In this paper, we present the designing of activities which use historical elements from the mathematical study of motions in the later Middle Ages to introduce first-year undergraduates to the definite integral concept and the fundamental theorem of Calculus. The activities are based on motion situations and problems, which are familiar to students' experience, and particularly on velocity-time graph representations of motions The velocity-time graph on which all the varied magnitudes of motion (time, velocity, distance covered) are represented plays a central role in the designing of the activities. Students are led to approach intuitively the mathematical concepts. This process aims at: (1) the stimulation of students' mathematical reflections via the velocity-time representations of motion problems, (2) the understanding of the connection of distance covered with the area of figures and the interrelation of velocity with the distance on the same graph as a first contact with the fundamental theorem of Calculus. The central aim is to give the opportunity to let formal mathematics emerge from the mathematical activity of the students, instead of trying to bridge the gap between informal and formal knowledge, and to understand concepts, not only as tools for solving problems, but also as mathematical objects. At this point, we shall briefly describe the theoretical perspectives that inform the designing of the activities and the analysis of the data collected:

(a) Realistic Mathematics Education (RME) [8], in which the core mathematical activity is 'mathematizing', which stands for organizing from a mathematical perspective. Freudenthal sees this activity of the students as a way to reinvent mathematics. Note that the students are not expected to reinvent everything by themselves. In relation to this, Freudenthal [7] speaks of *guided reinvention*; for him, the emphasis is on the character of the learning process rather than on invention as such. The idea is to allow learners to come to regard the knowledge they acquire as their own private knowledge. In relation to this, Treffers [18] discerns *horizontal* and *vertical* mathematization. Horizontal mathematization refers to the process of describing a context problem in mathematical terms—to be able to solve it with mathematical specific solution methods which are modeled ("*models of*"). Vertical mathematization refers to mathematizing one's own mathematical activity in which the models can be generalized and formalized to develop into entities of their own, which as such can become models for mathematical reasoning ("*models for*"). It is in the process of *progressive mathematization*—which comprises both the horizontal and vertical component—that the students construct (new) mathematics.

(b) A mathematical concept as a *"tool"* and an *"object,"* and their dialectic relation [3]: According to Douady "...*a concept is a tool when the interest is focused on its use for solving a problem. A tool is involved in a specific context, by somebody, at a given time."*

The object perspective in this case means the reified tool and corresponds to the object perspective in the process–object duality, where object denotes the reified process. A suitable collection of reified tools and a high degree of flexibility, supporting changes into a tool perspective, is prerequisite for solving non-standard problems in mathematics. The teacher must organize the transformation of tool status into object and vice-versa. The goal is to put the pupils in a position to capitalize on the knowledge available in situations that have meaning for them.

(c) The three worlds of mathematics [16]. Tall introduced the terms '**embodied, proceptual** and **formal** mathematical worlds'. He says:

"The first ('conceptual-embodied world' or '*embodied* world') grows out of our *perceptions* of the world and consists of our thinking about things that we perceive and sense, not only in the physical world, but in our own mental world of meaning. By reflection and by the use of increasingly sophisticated language, we can focus on aspects of our sensory experience that enable us to envisage conceptions that no longer exist in the world outside... By formulating the embodied world in this way, it includes not only our mental perceptions of real-world objects, but also our internal conceptions that involve visuo-spatial imagery. It therefore applies not only the conceptual development of Euclidean geometry but also other geometries that can be conceptually embodied such as non-Euclidean geometries that can be imagined visuo-spatially on surfaces other than flat Euclidean planes...

The second (*proceptual*) world is the world of symbols that we use for calculation and manipulation in arithmetic, algebra, calculus and so on. These begin with *actions* (such as pointing and counting) that are encapsulated as concepts by using symbols that allow us to switch effortlessly from processes to *do* mathematics to concepts to *think* about. The third (*formal-axiomatic*) world is based on *properties*, expressed in terms of formal definitions that are used as axioms to specify mathematical structures. It turns previous experiences on their heads, working not with familiar objects of experience, but with axioms that are carefully formulated to *define* mathematical structures in terms of specified properties [16]."

(d) Reflective abstraction is drawn from what Piaget [12], called *the general coordinations* of actions, and as such, its source is the subject and it is completely internal. This kind of abstraction leads to a generalization which is constructive and results in "new syntheses in midst of which particular laws acquire new meaning" [15, p. 299]. Piaget distinguishes various kinds of construction in reflective abstraction:

(1) The *interiorization*, as a construction of internal processes, as a way of making sense out of perceived phenomena; as "translating a succession of material actions into a system of interiorized operations" [13, p. 90],

(2) The **coordination** or composition of two or more processes for the construction of a new one,

(3) The **encapsulation** or the conversion of a (dynamic) process into a (static) object, in the sense that, "...actions or operations become thematized objects of thought or assimilation" [14, p. 49]. Piaget considered that "...mathematical entities move from one level to another, an operation on such 'entities' becomes in its turn an object of the theory..." [12, p. 70],

(4) When a subject learns to apply an existing schema to a wider collection of phenomena, then we say that the schema has been **generalized**. Generalization can also happen when a process is encapsulated to an object. The schema remains the same except that it now has a wider applicability. Piaget referred to all of this as a reproductive or generalizing assimilation [12, p. 23] and he called the generalization **extensional** [15, p. 299].

The activities were presented to students in the Mathematics Department at Athens University during two successive summer semesters as an introduction to Integral Calculus. During the first of these semesters, we applied our didactic approach to 43 students. We then used the conclusions arising from this research activity as hypotheses for

14.3. Designing Didactic Activities Inspired by History of Mathematics

the main research that followed the next year, when we applied our teaching approach to 83 students. The course consisted of eight one-hour teaching sessions based on Brousseau's [1] theoretical context for didactic situations in a didactic milieu. During the experimental teaching, students worked in pairs using worksheets. This experimental intervention followed the formal introduction to concepts of Integral Calculus as specified in the official curriculum of the Department of Mathematics.

Sixteen students were interviewed at the end of the last semester. Our aim was to investigate the students' difficulties, the degree of understanding of the concepts, the connections between the initial activities and the subsequent formal mathematical knowledge. This means that we wished to investigate whether the students could justify mathematically their initial intuitive choices in the activities.

14.3.1 The activities (I)

A series of thirteen activities were given to the students. We briefly discuss the didactic aims of a part of them: The aims of the three initial activities were: (1) the representation of a given motion using velocity-time graphs, (2) the transition from a table, or a graph, to the algebraic formula of the velocity function, and (3) the calculation of the distance covered and its interrelation with the area of the figure under the velocity curve. In these activities we used step functions, keeping in mind two things: (a) the definitions of instantaneous velocity and uniformly accelerated motion at Merton College and, (b) the construction by the students, right from the beginning, of models of successive rectangles to partition curvilinear regions in order to calculate their areas. From a modeling perspective, we could say that graphs of step functions come to the fore as *models of* situations, in which velocity and distance vary, while these graphs later develop into *models for* formal mathematical reasoning about calculus.

The 4th activity was important. Not only did the students approximate the linear velocity function (in the case of uniformly accelerated motion) by step functions, but also they proved that the position function and the area function of the region below the velocity curve are equal. It is a 'geometric' proof of the fundamental theorem of Calculus using the velocity—time graph and the introductory hypotheses of the activities. We give an example of the worksheet and an excerpt of the interview given by Peter, a first-year undergraduate [6]:

The worksheet, Peter's interview, and its content analysis

(a) *The worksheet*

> **4th Activity:**
> Consider that a material point begins its motion from rest and moves so that, *in each of any equal parts of time, it acquires an equal increment of velocity*. Moreover, consider that the time intervals are infinitely small.
> (a) Give a graphic representation of the velocity function vs. time, if $t \in [0, 1]$, and $V_{fin.} = 2$ m/s, (t in sec).
> (b) Express the velocity as a function of time (give the formula).
> (c) Calculate the distance covered using the graphic representation.

Peter and his collaborator wrote on the worksheet without any explanation (Figure 14.10).

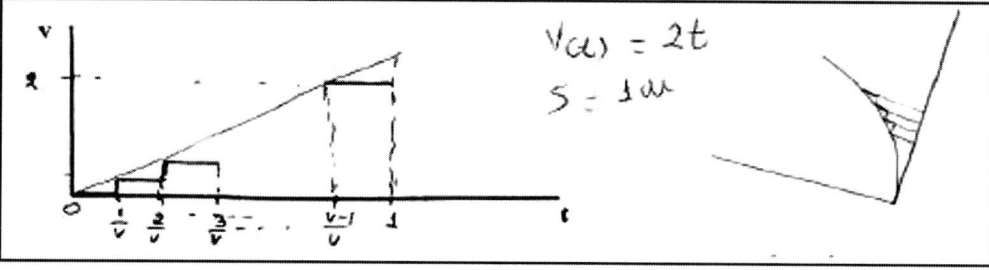

Figure 14.10.

(b) *An excerpt of the interview*

> We asked Peter: 'why did you draw a straight line for the representation of the velocity function vs. time?'
>
> (1) **Peter:** because the hypothesis says that the time intervals are infinitely small we
> (2) consider a denominator v, so that each [time] interval is increased by $1/v$. As v increases,
> (3) $1/v$ tends to zero, that is to say, for very big v this becomes almost infinitely small...
> (4) thus we can draw the velocity on the V-axis, increasing [the velocity] at every instant by
> (5) an equal width, because we know that in each of equal parts of time, it acquires an
> (6) equal increment. Hence the slope, in these small triangles, which are created, is the
> (7) same.

Analyzing Peter's statements we can say that:

He justifies mathematically their choice to draw the velocity as a linear function, exploiting the assumption and the graphic representation of the step function. He is led intuitively to the creation of a sequence of step functions because the width of the "steps" continuously decreases as $(1/v) \to 0$, as he said in (2–3). He considered that this sequence of step functions "approximates" the required graphical representation of the linear velocity function, using a snapshot of the family of step functions. Peter considered explicitly that the vertical sides of the triangles are equal for the selected partition (3–7), mentioning the constant slope of hypotenuses of all right triangles.

...The researcher asked Peter:

> (26) **Researcher:** Here you have drawn this curve (*the researcher points to the right side*
> (27) *of figure 14.10 above*). This should be a graph of velocity vs. time. Why did you
> (28) draw this graph?
> (29) **Peter:** I think that..., I tried to explain to the girl (*to his interlocutor*), something
> (30) about, ...because we had some disagreement about this. (*Peter points to the graph of*
> (31) *the step function on the worksheet, figure...*).
> (32) **R:** Could you give me an explanation?
> (33) **P:** I do not remember exactly her question...She asked me why these increments of
> (34) velocity are equal. I tried to explain that in equal time intervals the velocity acquires
> (35) equal increments.
> (36) **R:** Why did you draw the curve? (*the researcher points to the curve again on the*
> (37) *worksheet*).
> (38) **P:** Here it is not precisely the same. No,...because this [curve] is not a linear
> (39) function.

(c) *Content analysis & observations:*

From lines (26)–(39), we can make two basic observations:

(a) There is an interaction among the students in the classroom. Their "disagreement" activated Peter to give explanations about the choice of the linear function, which obviously, is Peter's choice.

(b) Peter devises the graphical representation of a function, which does not satisfy the assumption. He draws the graph of a nonlinear function, then divides the time axis into equal intervals and observes that the corresponding increments of the velocity are not equal. Then he compares this graph with the linear function's graphical representation in order to show to his interlocutor that only the linear function satisfies the assumption. We consider that Peter makes one more essential step. Not only does he focus continuously on the assumption by which he is led to the linear function of velocity, but also he recognizes that only the linear function fits in the assumption, giving a suitable counterexample. Indeed, Peter does not rely exclusively on intuitive arguments, but goes on to a mathematical justification.

14.3. Designing Didactic Activities Inspired by History of Mathematics

We could describe the conceptual path followed by Peter, as it appears in the episode, in the following way: he is led by the family of step functions, to the linear function of velocity in order to retain the assumption and conversely. Only for the linear function of velocity we have equal increments in equal time intervals. He says: 'Here it is not precisely the same. No, ... because this [curve] is not a linear function' (38–39).

In this activity, almost all the students successfully drew the linear function of the velocity and provided the correct formula for velocity. While many students provided no explanation at all, some referred to their extant knowledge of Physics and to uniform acceleration motion, and others were based on the definition of motion given in the activity. This latter group approached the linear function of velocity intuitively using the family of step functions from the previous activities referring to processes of limiting approximation.

Vangelis and Christos' worksheet; their interview; an analysis of its content

We give another example of the worksheet on the second part of 4th activity (Figure 14.11) and an excerpt of the interview given by Vangelis and Christos. The students worked together on the activity.

(a) *The worksheet:*

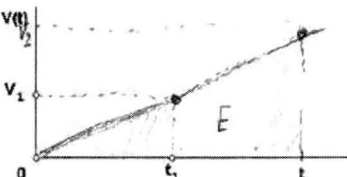

> 2. A material point, starting from rest, is moved so that in any equal time intervals it acquires equal increments of velocity. The total time interval of the motion is $[0, T]$ and V is the final velocity.
>
>   i. Represent graphically the velocity as a function of time, given that at instant t_1 ($t_1 \neq 0$) the velocity is V_1.
>
>   Write the formula for the velocity function $V(t)$. $= \frac{V_1}{t_1} t$
>
>   ii. Write the formula for the function $E(t)$ corresponding to the area of the region which is enclosed between the curve of the velocity function and the time axis, if $t \in [0, T]$ and the value of the velocity at instant t_1 is V_1.
>
> $E(t) = \frac{1}{2} t \cdot \frac{V_1}{t_1} t = \frac{1}{2} t^2 \frac{V_1}{t_1}$
>
>   iii. What may be represented by the function of the area $E(t)$?
>
>   iv. Find the derivative of the area function $E(t)$. $E'(t) = t \frac{V_1}{t_1} = V(t)$
>
>   v. If $S(t)$ is the position function of the moving particle what do you conclude given that $S'(t) = V(t)$. Justify your answer.
>
> $E'(t) = S'(t) \Rightarrow E(t) = S(t) + c \xrightarrow[E(0)=0]{t=0} E(t) = S(t)$

Figure 14.11. 4th activity—second part

(b) *An excerpt from the interview:*

(1) **Researcher:** This activity requires a graphic representation of velocity as a function of time, given that at instant t_1 the velocity is V_1. You wrote that the formula for velocity is $V(t) = \frac{V_1}{t_1}t$, and the formula of the area on the graph is $E(t) = \frac{1}{2}t\frac{V_1}{t_1}t = \frac{1}{2}t^2\frac{V_1}{t_1}$.

(2) **Vangelis:** We calculated the area of the triangle....

(3) **R.:** The next question is: what is represented by the function $E(t)$?

(4) **V.:** The displacement, because it's velocity vs time.

(5) **R.:** What does that mean?

(6) **V.:** If we plot velocity against time, the displacement is represented by the area on the graph.

(7) **R.:** You said in previous activities that Physics tells us that in the case of uniform movement, the distance covered is represented by the area of the rectangles on the graph. Is that the case here with the triangle?

(8) **Christos:** The previous case taught us that if we divide the time axis into very small segments we create rectangles which approximately–not approximately, actually–give the area of the triangle. This means that if we divide the triangle into very small rectangles and calculate the area, this is a limiting approximation of the triangle's area.... Meaning that we use the step functions model from the previous activities.

(9) **R.:** Here, you correctly found the derivative of the area function $E(t)$. ... You wrote that the area function $E(t)$ is equal to the position function $S(t)$?

(10) **C.:** We hypothesized that $S'(t) = V(t)$. We found that $E'(t) = V(t)$. It was then easy to conclude that $S(t) = E(t)$.

(11) **R.:** Do you recognize a theorem from Calculus?

(12) **V & C.:** ...(*they look at each other*).

(13) **C.:** Is this the fundamental theorem of Calculus?

(14) **R.:** Do you remember what the theorem says?

(15) **C & V.:** ...No, no, we don't.

(c) *Content analysis & observations:*

(a) The students identify the position function with the function of the area of the right triangle below the velocity graph [(2), (4)–(6)].

(b) In order to justify this choice, they use the limited approximation to the linear function of velocity provided by the step functions model. This means that they intuitively consider the area of the triangle to be the limit of the sum of the areas of the rectangles, "*if we divide the time axis into very small segments*" [(8)]. Christos reverses this reasoning: he considers the total area of the triangle to represent the displacement of the moving point. In which case, the division "*of the time axis into very small segments*" leads to the creation of rectangles whose sum gives the area of the triangle. We can say that Christos has taken the family of step functions into consideration and their limited approximation as a structural element in calculating the area of the triangle. It is a powerful model which can be generalized to, on the one hand, calculate the area of non-rectilinear regions ('*model of*', [8]) and, on the other, to constitute a modeling-process which is integrated into the definition of the definite integral concept ('model for' mathematical reasoning, [8]). The students' reasoning is based on the velocity graph, and allows them to make the shift from the *embodied* to the *proceptual* mathematical world using symbols and statements from Calculus [16].

(c) The students prove that the position function of the moving object is equal to the function of the triangle's area [worksheet, (10)]. However, they cannot formulate the fundamental theorem of Calculus [(14)–(15)]. Christos's reference to the theorem is rather random [(13)]. They cannot connect the activity with formal mathematics in the case of the Fundamental Theorem of Calculus. We cannot say that the students have made the shift to the formal/axiomatic world [16].

Most of the students found the correct formula for the velocity function and proved that the area function $E(t)$ is equal to the position function $S(t)$ without difficulty.

14.3.2 The activities (II)

Let us return to the activities:

In the next activity the students proved easily the Mean Speed Theorem of Merton College, using propositions of Euclidean geometry in the same manner employed by Oresme.

The 11th activity concerning the calculation of the area of the parabolic region was divided into two phases. In the first phase we gave the students enough time to work on the problem. Some students divided the time interval in equal parts taking upper and lower sums of rectangular areas. It was a process that had been learned during the previous year in high school. Others found it hard to continue. In the second phase (activity 11th, B) the given activity concerning the calculation of the parabolic region area was guided.

Stelios and Dimitra's worksheet; their interviews; an analysis of their content

Let us provide another example of a worksheet (Figure 14.12) from the first part of the 11th activity along with two excerpts from interviews conducted separately with Stelios and Dimitra. The students worked together on the activity in the classroom during the experimental teaching.

(a) *The worksheet (Activity 11th)*

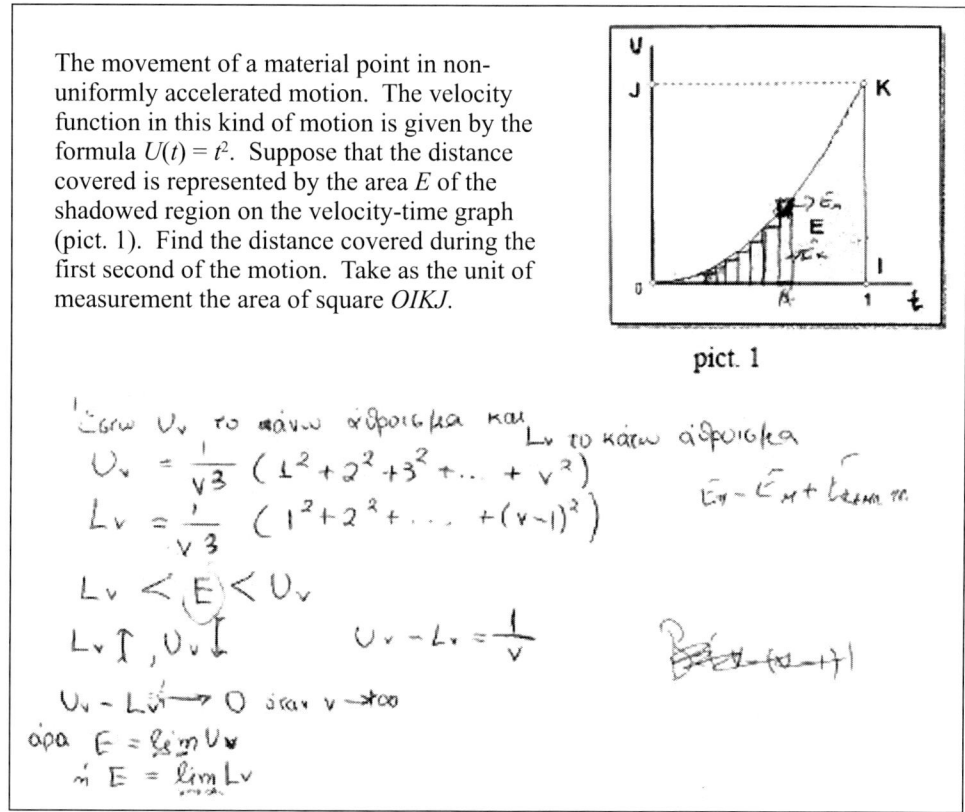

Figure 14.12.

Comment: Stelios and Dimitra divided the time axis up without any explanations (pict. 1) on the worksheet. During the interview, Stelios completed his answers to produce the worksheet as it appears below pict. 1.

(b) *An excerpt from the interview with Stelios:*

(1) **Researcher:** What are you doing here? (pointing at the worksheet)

(2) **Stelios:** We wanted to calculate the area of the parabolic region... I drew a partition here (*pointing at the worksheet*) using rectangles... (*pause*)... the lower sums... (*pause*)... (*he picks up the worksheet and a pencil*).

(3) **R.:** Do you want to write something down?
(4) **S.:** Yes, it is better to speak that way, using symbols…
(5) **R.:** Then…
(6) **S.:** According to Riemann…
(7) **R.:** You learned the Riemann theory after these activities.
(8) **S.:** Yes, …I had partitioned [the region] into elementary rectangles.
(9) **R.:** Why rectangles? Why not something else?
(10) **S.:** I thought the rectangles…. were the same as…the accelerated motion in which we had partitioned [the region] into elementary rectangles.
(11) **R.:** The step functions?
(12) **S.:** Yes, the sum of the elementary rectangles. The limit of the sum, as Δt approximates to zero, is the area of the shadowed region.
(13) **R.:** Here, there is a curve. It is not a line segment as it was in the previous case.
(14) **S.:** It also works in the case of the curve.

(*Then Stelios asked for some time to complete his worksheet*).

(15) **R.:** Have you written U_v for the upper sum and L_v for the lower sum?
(16) **S.:** I bound the area between…this is bigger than the small, meaning the lower, and smaller than the upper. By abstraction, $U_v - L_v$ equals $1/v$. Thus, $U_v - L_v$ approximates to zero as v approximates to infinity. That is to say, the bigger v gets (*he indicates the points of the partition on the time axis*), the more negligible the area becomes (*he shows the region on the graph bounded by the upper and lower sums of the rectangles*). Hence, the area is equal to the limit of the U_v or the L_v.

(c) *Content analysis & observations:*

(a) Stelios divided the region using rectangles. This choice is closely connected to the 4th activity in the case of uniform acceleration motion [(10)]. He understands that the model of the family of step functions can be generalized ("*It also works in the case of the curve*") [(14)]. The step functions come to the fore as models of situations, while these graphs later develop into models for formal mathematical reasoning about calculus [8].

(b) He precisely describes the process for calculating the parabolic region area using symbols and mathematical reasoning in order to justify his initial intuitive choice during the activity [(15)–(16)]. Stelios has thus connected the formal theory with the activity and made the transition from the embodied to the proceptual mathematical world [16]. The mathematical process he describes above is integrated into the definition of the definite integral concept. He uses it as a tool for solving a motion problem [3].

(d) *An excerpt from the interview with Dimitra*

(1) **Researcher:** You worked with Stelios on the 11th activity. I see that you drew rectangles on the graph. What were you trying to calculate?
(2) **Dimitra:** The area of the region. We wanted the displacement. We considered this area (*she indicates the area on the worksheet*) to represent the displacement which we can approach by dividing it into very small pieces.
(3) **R.:** Small pieces…and I see that you drew rectangles under the curve, while here (he shows the position A on the graph of the Figure 14.12) you drew a rectangle above the curve. Why did you do that?
(4) **D.:** ……Because the rectangles under the curve are ultimately an approximation to what we wanted. The rectangle above the curve is something more.
(5) **R.:** Why do you need the rectangles above and below the curve?
(6) **D.:** I don't know if I need these.…… (*pause*)….. I would only use the rectangles under the curve, but I don't think that's right or effective.

14.3. Designing Didactic Activities Inspired by History of Mathematics

(7) **R.:** Something is missing....

(8) **D.:** Yes.

(9) **R.:** What if you made the bases of the rectangles smaller?

(10) **D.:** Yes. While we would never reach the area required, we would get a satisfactory approximation.

(11) **R.:** If the number of rectangles approaches infinity, meaning that you continuously minimize their bases?

(12) **D.:** I would get a better approximation. But I still can't connect the above rectangles.

(e) *Content analysis & observations:*

Dimitra identifies the displacement with the area of the parabolic region, speaking about the division of the region [(2)]. She tries unsuccessfully to describe a process for calculating the area. It seems that she faces considerable difficulty understanding various concepts such as the limited approximation and the definite integral [(9)–(12)]. Dimitra cannot connect the activity with formal mathematics. She cannot make the transition from the *embodied* to the *proceptual* mathematical world [16], and cannot develop the model of the family of step functions into a model for mathematical reasoning [8].

14.3.3 The activities (III)

In the next activity (12th activity/Figure 14.13), a moving particle changes direction at some instant. This means that the sign of the velocity changes and the displacement of the particle and the distance covered are not equal throughout the time interval. In the commentary of this activity we discuss the relations between displacement, distance covered and area of regions on the velocity-time graph.

The 13th activity (Figure 14.14) is a guided activity aiming at a proof of the fundamental theorem of Calculus in the case of a nonnegative, continuous and increasing velocity function concerning a non-uniformly accelerated motion, using the velocity-time graph.

We shall now present the worksheet and an excerpt from the interview with Panagiotis.

Panagiotis' worksheet; his interview; an analysis of its content

(a) *The worksheet:*

The worksheet, cont.

(b) *An excerpt from the interview with Panagiotis:*

(1) **Researcher:** Here is your worksheet from the 13th activity. What is the process here? How have you answered with regard to "important observation" at the end of the worksheet?

(2) **Panagiotis:** It is the Fundamental Theorem of Calculus. It is the same thing we proved in the lecture (*he means the lecture which followed the activities during the subsequent formal Calculus study program*).... We proved it in the case of the uniform acceleration motion in a previous activity.

(3) **R.:** And what about the "important observation"?

(4) **P.:** It's what we said at the beginning. The velocity is the first derivative of the position function. Here we have a graph (*he indicates it on the graph*) plotting both the velocity and the displacement of the moving particle.

(5) **R.:** What is the relationship between the velocity and the area of the region under the velocity graph?

(6) **P.:** It value changes quickly, meaning it is the rate of change of the area.

(7) **R.:** What does that mean?

(8) P.: We have the graph of the rate of change of a continuous function, and below the graph we have the function itself, which is altered.

(c) *Content analysis & observations:*

(a) Panagiotis immediately connects the 13th activity with: (a) the 4th activity (the case of uniform acceleration motion), and (b) the Fundamental Theorem of Calculus, which was taught during the formal program of Calculus studies [(2)].

> The velocity-time graphical representation of a moving particle is given below. Assume that the velocity $V(t)$ is a positive, continuous and increasing function so that $V: [0, 1] \to R$, $V(0) = 0$.
> $E(t)$ is the function of the area of the region which is enclosed between the $V(t)$ graph, the time axis and the line $t = t_0$, for every $t_0 \in [0, 1]$.
>
>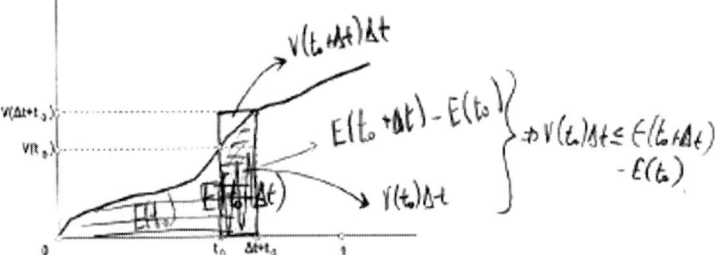
>
> 1. Determine on the graph the values $E(t_0)$ and $E(t_0 + \Delta t)$, as $0 < t_0 < t_0 + \Delta t < 1$.
>
> 2. Determine on the graph the difference $E(t_0 + \Delta t) - E(t_0)$.
>
> 3. Using the graph, prove that: $V(t_0) \Delta t \leq E(t_0 + \Delta t) - E(t_0) \leq V(t_0 + \Delta t)\Delta t$.
>
> Note that: $V(t_0) \leq \dfrac{E(t_0 + \Delta t) - E(t_0)}{\Delta t} \leq V(t_0 + \Delta t)$.
>
> 4. Does the $\lim\limits_{\Delta t \to 0} \dfrac{E(t_0 + \Delta t) - E(t_0)}{\Delta t}$ exist? Justify your answer.
>
> 5. Conclude that $E'(t_0) = V(t_0)$, and
>
> $\forall t_0 \in [0, 1]$, $\mathbf{E'(t) = V(t)}$.   (1)

Figure 14.13.

(b) Can we say that Panagiotis understands the interrelation between the functions on the graph? He explicitly states that the velocity function is the rate of change of the area function of the region below the graph [(6), (8)]. He has thus created a cognitive scheme which exploits the successive activities (from 4th to 13th) and the formal theory underpinning the recognition, generalization and formulation of the essence of the concepts inherent in the fundamental theorem. He uses the graphical representation to gave "body" to the theorem as he shifts to the "embodied mathematical world" [16]. As the worksheet and the interview make clear, Panagiotis easily makes the move into the "proceptual mathematical world" of processes and symbols [16] in which expressions like "the first derivative of the position function" [(4)] or "the rate of change of the area" [(6)] acquire meaning. He mentions the formal proof of the theorem [(2)], which allows us to conclude that he has approached the "formal/axiomatic mathematical world" introduced by Tall [16]. According to Piaget [12], Panagiotis has interiorized, coordinated and generalized concepts and processes (function, graph, area, derivative). He has also understood the two-way relation between a function and its rate of change–as reversed processes in the same context [4]–that lies at the heart of the fundamental theorem.

6. It is known, from the introductory definition, that the velocity function $V(t)$ of a moving particle is the derivative of the position function $S(t)$. Thus:

$$S'(t) = V(t), \; t \in [0, 1]. \quad (2)$$

Using equalities (1) and (2) and the graph, determine the relation between the functions $E(t)$ and $S(t)$.

$$\left.\begin{array}{l} E'(t) = V(t) \\ S'(t) = V(t) \end{array}\right\} \Rightarrow E'(t) = S'(t)$$

$$E(t) = S(t) + C$$

Επειδή για $t = 0 \Rightarrow \left.\begin{array}{l} E(0) = 0 \\ \text{κ' } S(0) = 0 \end{array}\right\} \Rightarrow c = 0$

AN IMPORTANT OBSERVATION:

In all previous activities the velocity-time graph gave a close interrelation between the velocity function and the distance covered which was represented by the area of the region below the graph in a time interval.

Can you describe the relation of these functional magnitudes?
Can you come in a general conclude?

Figure 14.14.

In this guided activity, almost all the students easily proved that the position function $S(t)$ is equal to the area function $E(t)$, as they did in the 4th activity. However, most of them failed to provide a satisfactory answer to the questions in the last part of the activity relating to the interrelation between the velocity function and distance covered. Most of the students interviewed failed to recognize the Fundamental Theorem of Calculus in this activity.

Let us refer to a theoretical issue concerning the relationship between rates and totals.

14.4 The Multiple Linked Representations Between Rates and Totals

Kaput [9, p. 9], states that:

> Situations or phenomena admitting of quantitative analysis almost always have two kinds of quantitative descriptions, one describing the total amount of the quantity at hand with respect to some other quantity such as time, and the other describing its rate of change with respect to that other quantity.... The understanding of the two-way relations between totals and rates descriptions of varying quantities (and the situations that they describe) is a fundamental aspect of quantitative reasoning. It is exactly this relationship that is at the heart of the Fundamental theorem of Calculus, and indeed, at the heart of Calculus itself.

Kaput [9, p. 10] illustrated the relations between the representations of total and rates as follows (Figure 14.15).

Through these connections between rates and totals we take advantage of linked representations, so that not only can we connect graphs and formulas, but also we can cross-connect, for example, a rate graph to a totals formula.

As we mentioned, the velocity-time graph plays a central role in the activities we presented. We call this representation **holistic** because of two important reasons: (1) the holistic representation allows the three functional variables to be represented differently on the same graph (velocity and time are represented by lines in the Oresmian sense and distance covered by the area of a figure), and (2) the representation of the distance covered by an area, and the interrelation of velocity with distance on the same graph, constitutes the students' first contact with the **definite integral** of the velocity function in a time interval, and the **fundamental theorem** of Calculus in this case. Generally, according to Kaput, we can say that a *holistic* graph represents simultaneously the "total quantity at hand with respect to some other quantity such as time, and its rate of change with respect to that other quantity."

Taking into account that the *holistic* graphs connect the representations of Rates and Totals in a common 'region' we reconstructed this two-way relation (Figure 14.16). Thus this representation in the same context of the two different

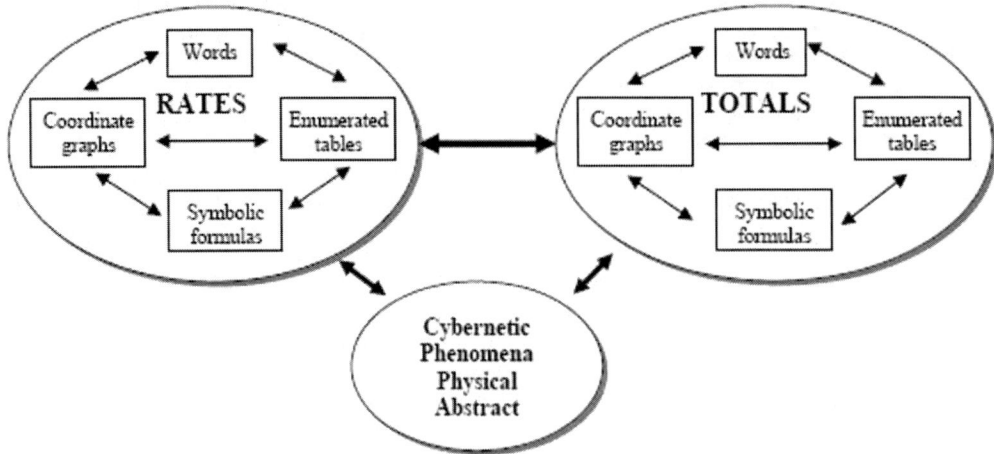

Figure 14.15.

quantitative descriptions may lead the students to a better understanding of the two-way relations between totals and rates (Figure 14.16).

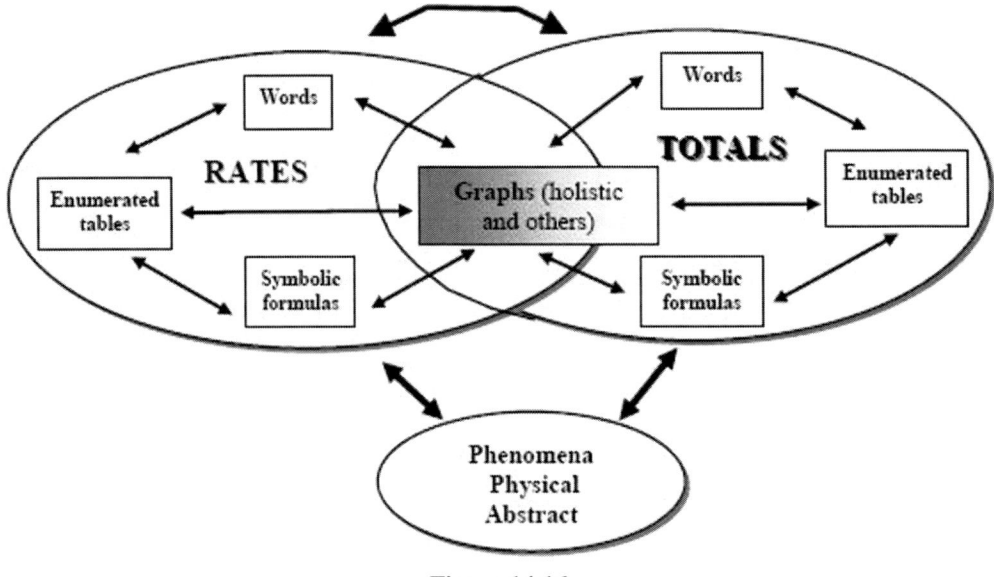

Figure 14.16.

In particular, in our case, the above scheme is formulated as follows (Figure 14.17).

14.5 Analysis of the Data Collected–Results

We based the evaluation of our didactic approach mainly on the interviews' content analysis. We investigated the mental operations of the students, the difficulties and the understanding of the mathematical concepts under consideration, using various appropriate theoretical perspectives.

In particular, concerning the definite integral concept we connected and interrelated, in a scheme, elements of different perspectives on the learning of mathematics, such as: (a) *the three worlds of mathematics* [16], (b) The reflective abstraction [12], (c) The RME theory [8], and (d) a mathematical concept as a "tool" and an "object," and their relation [3], as mentioned in the Section 14.3.

This scheme functions as follows: We want the students to approach the definite integral concept. Initially, the students make the transition from real life situations (motion problems) to the *embodied* mathematical world [16], using the velocity-time graph in which the concept appears as an area of a figure. They create *models of* solving

14.5. Analysis of the Data Collected–Results

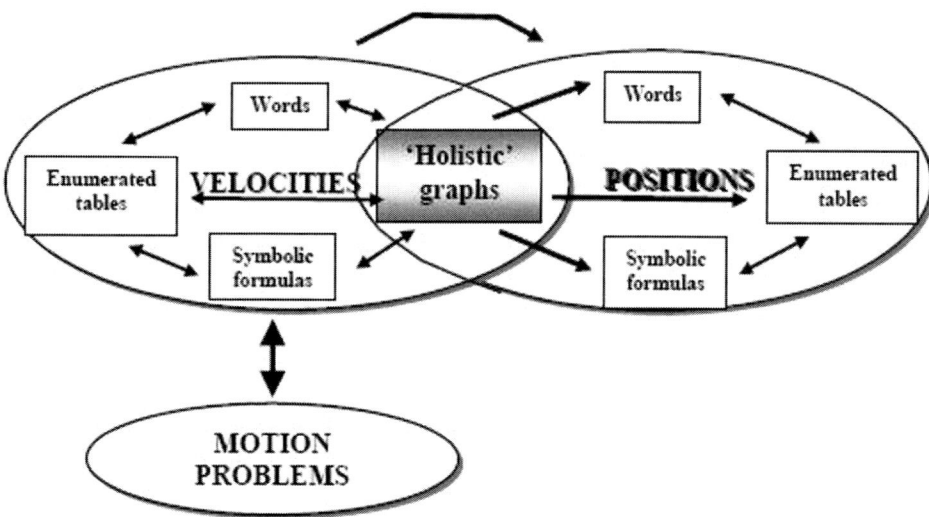

Figure 14.17.

particular problems, which evolve into *models for* mathematical reasoning [8] in the *proceptual* mathematical world of symbols and processes [16]. The students can also make the transition from motion problems to the proceptual mathematical world using previous knowledge from Algebra and Calculus. They act on mathematical objects such as function, limit and graph, by the mental operations of the *reflective abstraction* [12], for the construction of the definite integral concept as a *tool* [3] for calculating areas of curvilinear regions. Then, by generalization they make the transition to the *formal-axiomatic* mathematical world, where the definite integral concept is given by the formal definition [16]. We argue that the mathematical concept of the definite integral 'connects' the proceptual and the formal mathematical worlds in a common region [10, 11]. Schematically (Figure 14.18):

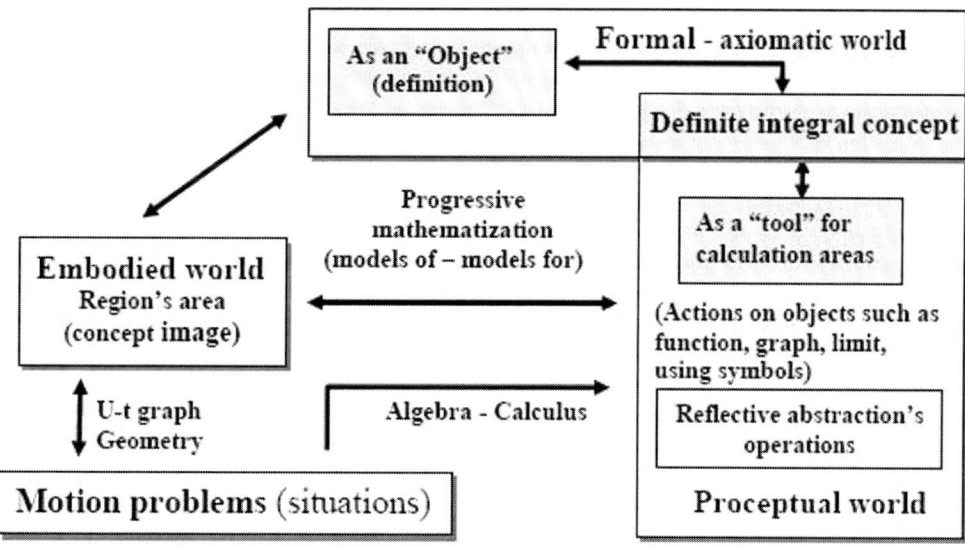

Figure 14.18.

The analysis of the data collected (pre-test, worksheets, interviews, post-test), according to the theoretical perspectives which guided our research, led to four different categories concerning the students' mental operations [10].

Category A: The students make the transition from real life situations, to the *embodied mathematical world* [16] using the velocity-time graph and Euclidean geometry, exploiting their experience and intuition. They take into account the assumptions and constraints of the activities. They create models of solving particular problems, which evolve into models for mathematical reasoning [8] in the *proceptual mathematical world* [16]. The students act on mathematical objects such as function, limit, graph, by the mental operations of the reflective abstraction (*interiorization, coordination, encapsulation* and *generalization of mental schemata*), [12], for the construction of the definite integral concept as a tool [3] for calculating areas. The students also approach the fundamental theorem of Calculus by coordination of the differentiation and integration processes, as a means of constructing a process which consists of reversing another one, by exploiting the graphical context [4]. They are able to justify their initial intuitive choices in the activities using statements, theorems and proofs in the context of the formal mathematical world. Schematically (Fig. 14.19):

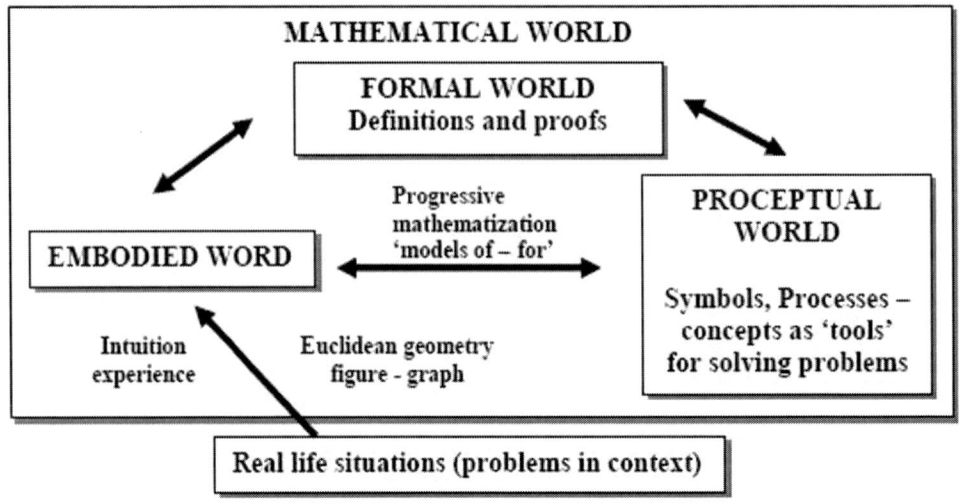

Figure 14.19.

The sum total of the data collected (worksheets, pre and post-tests, interviews) and their analysis, allow us to include Peter's and Panagiotis' mental operations in Category A.

Category B: The students in the initial activities use previous knowledge from Physics without taking the assumptions into account. Then, they make the conversion in the proceptual world using symbols and formulas. However, they quickly make the transition to the embodied mathematical world using the velocity-time graph in accordance with the activities. They create models of solving particular motion problems, which evolve into models for mathematical reasoning in the proceptual world. The students act on mathematical objects, in the same manner as Category A, for the construction of the definite integral concept as a tool for calculating areas of curvilinear regions. However, they cannot see the definite integral concept as an object through generalization in the context of the formal mathematical world. The students extend their mathematical justification to give explanations concerning their initial choices in the activities, but they can not express satisfactory statements of the formal mathematical theory and recognize theorems that are implicit in the activities. They have not made the passage to the formal world (Figure 14.20).

We have included Vangelis, Christos and Stelios' mental operations in this category.

Category C: The students, without using concepts and formulas from Physics, as in the previous category, act in the same manner as the students in Category B. They create models of the solution of motion problems, which extend to models for mathematical reasoning, only in the case of the construction of the definite integral concept as a tool for calculation of areas. They cannot generalize, nor recognize elements of the theory in the activities or express statements and definition of the formal theory.

Category D: The students make the transition to the embodied mathematical world using the v-t graph and Euclidean geometry. They face many difficulties when trying to pass to the proceptual world of symbols and processes: difficulties in translating v-t graphs to algebraic formulas of velocity, difficulties which are connected with the understanding of basic mathematical concepts such as limit and limit approximation, etc. The students are not able to construct the definite integral concept as a tool for calculating areas in the context of the activities. They cannot construct models for mathematical reasoning, since they are constrained in an intuitive action strictly in the context of the activities. There is no evidence that the students have approached the formal mathematical world.

14.5. Analysis of the Data Collected–Results

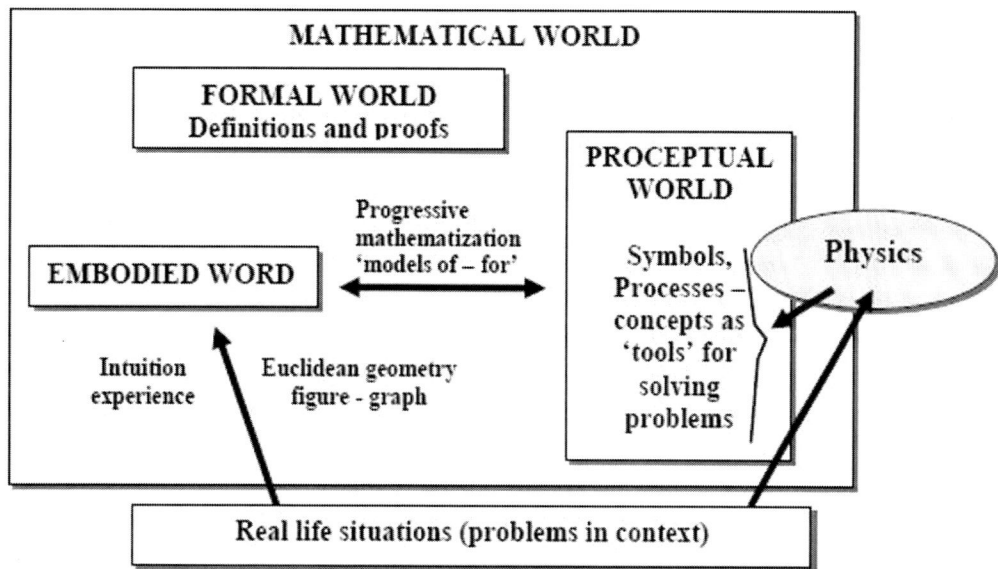

Figure 14.20.

We have included Dimitra's mental operations in Category D.

Bibliography

[1] Brousseau, G., 1997, 'Theory of Didactic Situations in Mathematics', (ed. & tr N. Balacheff et al.), *Mathematics Education Library*, Kluwer, Dordrecht.

[2] Clagett, M., 1959, *Science of Mechanics in the Middle Ages*, The University of Wisconsin Press, Madison.

[3] Douady, R., 1991, 'Tool, Object, Setting, Window: Elements for Analysing and Constructing Didactical Situations in Mathematics', in A. J. Bishop et al. (eds), *Mathematical Knowledge: Its Growth Through Teaching*, Kluwer Academic Publishers, The Netherlands, 117–130.

[4] Dubinsky, Ed, 1991, 'Reflective Abstraction in Advanced Mathematical Thinking', in D. Tall (Ed.), *Advanced Mathematical Thinking*, Kluwer Academic Publishers, Dordecht.

[5] Farmaki, V. and T. Paschos, 2007a, 'Employing genetic 'moments' in the history of Mathematics in classroom activities', *Educational Studies in Mathematics*, Vol. 66 (1), 83-106.

[6] ——, 2007b, 'The interaction between intuitive and formal mathematical thinking. A case study', *International Journal of Mathematical Education in Science and Technology*, 38 (3), 353–365.

[7] Freudenthal, H., 1991, *Revisiting Mathematics Education*, Kluwer Academic Publishers, Dordrecht.

[8] Gravemeijer, K. and M. Doorman, 1999, 'Context problems in realistic mathematics education: A calculus course as an example', *Educational Studies in Mathematics*, 39, 111–129.

[9] Kaput, J., 1999, 'Representations, Inscriptions, Descriptions and Learning: A Kaleidoscope of Windows', "*Representations and the Psychology of Mathematics Education: Part II*," Goldin and Janvier, eds. *Journal of Mathematical Behavior*, 17(2), 265–281.

[10] Paschos, T., 2007, 'The integration of genetic moments of the History of Mathematics in the didactical approach to basic concepts of Calculus' (Doctoral thesis), Department of Mathematics, University of Athens, Greece.

[11] Paschos, T. and V. Farmaki, 2007, 'The Integrating of genetic moments in the History of Mathematics and Physics in the designing of didactic activities aiming to introduce first-year undergraduates to concepts of Calculus', in Barbin, Stehlicova & Tzanakis (Eds), *History and Epistemology in Mathematics*, Proceedings of the 5th European Summer University, Prague, pp. 297–310.

[12] Piaget, J., 1972, *The Principles of Genetic Epistemology*, London, Routledge & Kegan Paul.

[13] ——, 1980, *Adaptation and Intelligence*, (S. Eames, trans.), University of Chicago Press, Chicago (original published 1974).

[14] ——, 1985, *The Equilibration of Cognitive Structures* (T. Brown and K. J. Thampy, trans.), Harvard University Press, Cambridge MA (original published 1975).

[15] ——, and R. Garcia, 1989, *Psychogenesis and the history of science*, (H. Feider, trans.), New York: Columbia University Press.

[16] Tall, D. O.,2004, 'Thinking through three worlds of mathematics', *Proceedings of the 28th PME*, Bergen, Norway, 4, 281-288.

[17] Tzanakis, C. and A.Arcavi, 2000, 'Integrating history of mathematics in the classroom: An analytic survey', in J. Fauvel & J. van Maanen (Eds.), *History in Mathematics Education: The ICMI Study*, Dordrecht, The Netherlands: Kluwer Academic Publishers, pp. 201–240.

[18] Treffers, A., 1987, *Three Dimensions. A Model of Goal and Theory Description in Mathematics Education: The Wiskobas Project*, D. Reidel, Dordrecht.

About the Authors

Theodorus Paschos received his B.Sc. in Mathematics in 1981, his M.Sc. in Methodology and Didactics of Mathematics in 1999, and his Ph.D. in Didactics of Mathematics, all from the University of Athens, Greece. His dissertation was entitled "Integration of Genetic Moments of the History of Mathematics in the Didactic Approach of Basic Concepts of Mathematical Analysis." He is currently a teacher of Mathematics in Secondary Education at the Experimental High School in Athens, and his research interests are in the didactics of mathematics, especially in the use of the history of mathematics in the teaching of mathematics. Many research articles of his and V. Farmaki's, concerning the integration of history of mathematics in classroom activities, have been published in various journals and proceedings of conferences.

Vassiliki Farmaki is Professor of Mathematics at Athens University, Athens, Greece. He received his B.Sc. in Mathematics in 1977 and his Ph.D. in Mathematics in 1984, both from Athens University. His research interests include Infinitary Ramsey Combinatorics, Banach space theory, Topology, and Mathematics Education.

15

History in a Competence Based Mathematics Education: A Means for the Learning of Differential Equations

Tinne Hoff Kjeldsen
Roskilde University, Denmark

15.1 Introduction

In a series of papers, Michael N. Fried has discussed a dilemma in historical approaches to mathematics education arising because "mathematics educators are *committed* to teaching *modern* mathematics ..." and he continues "However, when *history* is being *used* to justify, enhance, explain, and encourage distinctly modern subjects and practices, it inevitably becomes what is "anachronical" [...] or "Whig" history" [6, p. 395, italics in the original]. Whig history refers to the kind of history that is written from the present, i.e., a reading of the past in which one tries to find the present. On account of the mathematics teacher, Fried phrased the dilemma as follows:

> if one is a mathematics educator, one must choose: either (1) remain true to one's commitment to modern mathematics and modern techniques and risk being Whiggish, i.e., unhistorical in one's approach, or, at best, trivializing history, or (2) take a genuinely historical approach to the history of mathematics and risk spending time on things irrelevant to the mathematics one has to teach. [6, p. 398].

In Fried [7, p. 203], he emphasizes that this should not be understood as if history has no place or role to play in mathematics education, but was meant to point out "that a dilemma arises when the traditional commitments of mathematics education are assumed."

The purpose of the present paper is to argue that this dilemma can be resolved by adopting both (1) a competence based view of mathematics education, and (2) a multiple-perspective approach to the history of the practice of mathematics. Hereby, a genuinely historical approach to the history of mathematics can be taken, in which the study of original sources is also relevant to the mathematics one has to teach. A student directed project work on the influence of physics on the development of differential equations will be analyzed in order to present some empirical evidence for the claim that, within the proposed methodology, history can be used as a means for the learning of (parts of) traditional core curriculum mathematics without turning "Whiggish." The project belongs to a cohort of mathematics projects made over the past 30 years by students at Roskilde University, Denmark. Only one project is analyzed in the present paper, but the reflections and discussions brought forward in the present paper are based on knowledge about and experiences from supervising many of those projects.

First, mathematical competence and the role of history in a competence based mathematics education are presented. Second, a multiple-perspective approach to a history of the practice of mathematics is introduced. Third, the chosen project work is analyzed and discussed with respect to specific potentials for the learning of differential equations within the proposed methodology. Finally, the paper ends with some conclusions and critical remarks.

15.2 Mathematical Competence and the Role of History

The Danish so-called KOM-project (2000–2002) was initiated by the Ministry of Education. Its purpose was to describe mathematics curricula on all levels in the Danish system based on mathematical competencies instead of on a catalogue of subjects, notions, and results [20]. In this context mathematical competence means the ability to act appropriately in response to mathematical challenges of given situations. In the KOM-report it is argued that mathematical competence in general can be spanned by eight main competencies [20]. Half of them involve asking and answering questions in and with mathematics: (1) to master modes of *mathematical thinking*; to be able to formulate and solve problems in and with mathematics, i.e., (2) *modeling competency* and (3) problem solving, resp.; and (4) to be able to reason mathematically. The other half concerns language and tools in mathematics: (5) to be able to handle different *representations* of mathematical entities; (6) to be able to handle *symbols and formalism* in mathematics; (7) to be able to *communicate* in, with, and about mathematics; and (8) to be able to handle *tools and aids* of mathematics. In the discussion below of the chosen project work, the possible learning outcomes of reading sources are analyzed with respect to these competencies.

History of mathematics is not included as one of the main competencies in this scheme for the teaching and learning of mathematics. It is integrated, though, in the description of a person competent in mathematics, who, besides mastering the eight competencies, should also possess three kinds of *overview and judgment* regarding mathematics as a discipline. The first concerns actual applications of mathematics in other areas, the second, historical development of mathematics in culture and societies, and the third, the nature of mathematics as a discipline [20]. The second one explicitly requires knowledge about history of mathematics, not as an individual discipline (the goal is not to educate competent historians of mathematics), but enough to exercise overview and judgment regarding the historical development of mathematics, based on concrete examples.

The KOM-understanding of the role of history in mathematics education has the honesty to have history as an intrinsic part. The objective of the present paper is to discuss in what sense such an understanding of history can be implemented in situations where the curriculum does not include history and does not assign time to teach history. Under such circumstances, history of mathematics is most likely going to play no role at all in the learning and teaching of mathematics unless it can also be used as a means to learn and teach core subjects in the syllabus.

15.3 A Multiple Perspective Approach to History of Mathematics

How can we understand and investigate mathematics as a historical product? One way is to think of mathematics as a human activity and of mathematical knowledge as created by mathematicians. This has been the foundation for many recent studies in the history of the practice of mathematics, in which internal as well as external factors that have altered the course of mathematics have been identified ([4]), [15].

To study the history of the practice of mathematics involves asking why mathematicians situated at a specific place in space and time, in a certain society, and/or in a particular intellectual context, decided to introduce specific definitions and concepts, to study the problems they did, in the way they did it. In this line of thinking, mathematics is viewed as a cultural and social phenomenon, despite its universal character. Hence, studying the history of mathematics then also involves searching for explanations for historical processes of change, such as, but not limited to, changes in our perception of mathematics as such, in our understanding of mathematical notions or objects, in our idea of what counts as legitimate arguments for mathematical statements and so on.

A way of answering such questions and providing such explanations is to adopt a multiple perspective approach [11] to history of mathematics where episodes of mathematical activities are analyzed from multiple points of observations [13]. The perspectives can be of different kinds and the mathematics can be looked upon from different angles, such as sub-disciplines, techniques of proofs, applications, philosophical positions, other scientific disciplines, institutions, personal networks, beliefs, and so forth.

The education of historians of mathematics is not the objective of mathematics education, so how can this approach be brought into play to ensure the honesty to history, in a teaching situation where the teacher wants to use history as a means for students to learn a specific mathematical topic or concept? It can be implemented on a small scale, by having students read selected pieces of original mathematical texts where they focus explicitly on perspectives that address research approaches or the nature and function of specific mathematical entities (problems, concepts, methods, arguments), in order to uncover, discuss, and reflect upon the differences between how these approaches and entities are presented in their text book and the former way of conceiving and using them. In such teaching settings, the students have to read the mathematical content of the original text as historians, using the "tools" of historians, and answering historians' questions about the mathematics.

One such tool is the methodological framework of epistemic objects, techniques, and configurations [21, 5]. In short, *epistemic objects* refer to the mathematical objects about which new knowledge is searched for. *Epistemic techniques* refer to the methods and mathematical techniques that are used to investigate the mathematical concepts in question and *epistemic configuration of mathematical research* refers to the total of intellectual resources present in a specific episode. These terms are not intrinsically given, but are bound to concrete episodes of mathematical research. They are to be understood functionally, they change during the course of mathematical research and they might shift place. They are excellent tools for analyzing the production of knowledge and the understanding of mathematical entities in historical texts, because they are constructed to differentiate between how problem-generating and answer-generating elements of specific research episodes functioned and interacted [12].

Through activities where students work with historical texts guided by historical questions, connections between the students' historical experiences of the involved mathematics and their experiences from having been taught the textbook's version, can be created in the learning process. When students read historical texts from the perspectives of the nature and function of specific mathematical entities, they can be challenged to use other aspects of their mathematical conceptions in new situations and discussions of mathematics. Therefore, from a didactical perspective it is of interest to analyze historical episodes of mathematical research with respect to their potential to challenge students' mathematical conceptions.

In situations where the teacher has the opportunity to choose historical episodes determined by the perspective(s) he or she wants the students to reflect upon, without being constrained by pre-decided mathematical topics, it is of didactical interest to investigate if students are able to reflect upon and engage in historical discussions of mathematics from the chosen perspective(s) in a way that connects to the mathematics in the historical episode(s) [9]. For further discussions of these two scenarios of integrating history in mathematics education, see [10].

15.4 A Student Project on Physics' Influence on the History of Differential Equations

In the following, the student directed project work on the influence of physics on the development of differential equations will be analyzed with respect to how and in what sense the students' work with original sources provided potentials for the learning of differential equations—without losing sight of history.

15.4.1 The educational context of problem oriented student directed project work at RUC

The project report on physics' influence on the development of differential equations was written by five students enrolled in the mathematics programme at Roskilde University (RUC). All programmes at RUC are organized such that in each semester the students spent half of their time working in groups on a problem oriented, student directed project supervised by a professor. The projects are not described by a traditional curriculum, but are constrained by a theme [19, 1].

The requirement, or theme, for this project was that the students should work with a problem that deals with the nature of mathematics and its "architecture" as a scientific subject such as its concepts, methods, theories, foundation etc., in such a way that the status of mathematics, its historical development, or its place in society gets illuminated. Among the cohort of project reports, constrained by these objectives, this particular project was chosen, because the students happened to investigate differential equations, which are included in the core curriculum of advanced high school mathematics and mathematics and science studies in universities. Hence, the project work could be analyzed with respect to the issues addressed in the present paper.

15.4.2 Analysis of the project work with respect to learning outcomes and the competencies

The students formulated the following problems that guided them through their project work:

> How did physics influence the development of differential equations? Was it as problem generator? Did physics play a role in the formulation of the equations? Did physics play a role in the way the equations were solved? (Paraphrased from [18, p. 8]).

On the one hand, these are fully legitimate research questions within history of mathematics. They address issues about an episode in the history of mathematics seen from the perspective of how another scientific discipline influenced mathematicians' formulation of problems as well as the methods they used to solve the problems. On the other hand, these questions can only be answered by analyzing the details of original sources that deal with this particular episode in the history of mathematics, studying how the differential equations were derived from the problems under investigation, how the equations were formulated, why they were formulated in that particular way, how they were solved and with which methods—issues which are also relevant for the learning and understanding of the subject of differential equations. Based on readings of three original sources from the 1690s, the students discussed these issues within the broader social and cultural context of the involved mathematicians, critically evaluating their own conclusions within the standards for research in history of mathematics. Hence, in this way of working with history in mathematics education, history is neither Whiggish nor trivialized. The students did not explicitly use the methodology of epistemic objects, techniques, and configurations, but that would have been an excellent framework for their investigations, because—as we will see below—they did study and analyze the original sources with respect to which objects were under investigation and what kind of techniques the involved mathematicians used.

I will discuss three instances where the students—qua the historical work—were forced into discussions in which they came to reflect on issues that enhanced their understanding of certain aspects of differential equations in particular and of mathematics in general. The discussion will end with a short presentation of some of the learning outcomes with regard to the eight main mathematical competencies.

1. Johann's differential equation of the catenary problem. The catenary problem is to describe the curve formed by a flexible chain hanging freely between two points. The students read the solution that Johann Bernoulli presented in his lectures on integral calculus to the Marquis de l'Hôpital, supported by English translations of extracts [2]. Bernoulli formulated five hypotheses about the physical system that, as he claimed, follow easily from statics. For the students, of whom none studied physics, to derive these assumptions was the first mathematical challenge in reading Bernoulli's text: "we had to derive most of them ourselves. We use 18 pages to explain what Johann Bernoulli stated on a single page" [18, p. 19].

Below is one of the extracts of Bernoulli's text [2, p. 36] that the students read. As can be seen from the text, Bernoulli used the five hypotheses to describe the catenary and the infinitesimals dx and dy of the curve geometrically and derived the equation $dy : dx = a : s$ between the differentials. The figure was produced by the students and is similar to a figure in Bernoulli's lectures, except for the sine-cosine circle. [Beware that the letter a is used to denote two different entities in the following extract.] In their report, the students went through Bernoulli's text and filled in all the arguments. They were not familiar with this way of setting up differential equations from scratch so to speak, so the mathematization of the physical system was a major challenge for which they needed to consult some textbooks on statics and to combine the physics with mathematical results about triangles and the sine-cosine relations.

Bernoulli's arguments do not meet modern standards of rigor, and that created cognitive hurdles for the students. Didactically, it is important to identify such hurdles because they create situations where the students, during their struggle with understanding the mathematical content of the original text, can be challenged to reflect upon the differences between our modern understanding and the one presented in the source, thereby enhancing their own understanding of the concept of, in this case, differential equations and the mathematical techniques and concepts underneath. A concrete example of this is Bernoulli's use of the infinitesimal triangle. In the text above, Bernoulli used similar triangles to argue that $EL : AL = AH : Ha = dx : dy$, but, as the students pointed out in their report, a does

Assuming these results, we go on to find the common Catenary as follows: Let BAa be the required curve, with its lowest point at B; its axis, the vertical line through B, is BG. The tangent to the curve at its lowest point is the line BE, which will be horizontal. Let the tangent at any other point A be AE. We draw the ordinate AG and a line EL parallel to the axis. Let $BG = x$, $GA = y$, $Gg = dx$ and $Ha = dy$.

Because the weight of the chain is distributed evenly along its length we may put that weight equal to the length of the curve BA which we call s. Thus since an equal and constant force will always be required at the point B (by hypothesis (iv)) whether the chain BA is made longer or shorter, let this force be of magnitude a, expressed by the straight line C. Let us now imagine the weight of the chain AB to be concentrated and suspended at the point E where the tangents AE and EB meet. Then (by hypothesis (ii)) the same force is required at B to support the weight E as was previously required to support the chain BA. Indeed (by hypothesis (v)) the ratio of the weight E to the force at B is the same as the ratio of the sine of the angle AEB, or of its complement angle EAL, to the sine of the angle AEL, that is the ratio of EL to AL. Therefore, whatever position on the curve is taken for the fixed point A (the curve, by hypothesis (iii), being always the same) the ratio of the weight of the chain AB to the force at B is the same as the ratio of EL to AL, that is $s:a = EL:AL = AH:Ha = dx:dy$, and inversely $dy:dx = a:s$.[1]

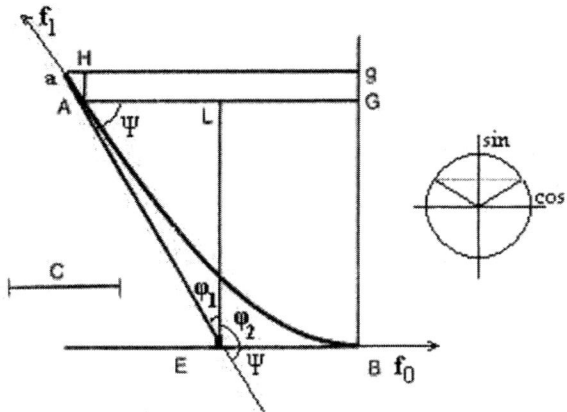

Figure 15.1.

not lie on the tangent, but on the catenary. Bernoulli also used the infinitesimal triangle later in the lecture, when he reformulated the differential equation derived above, using the relationship $ds = \sqrt{dx^2 + dy^2}$. Again—as pointed out by the students—ds is a part of the catenary, not the hypotenuse of a right angled triangle.

This mixed use of geometrical arguments and infinitesimals in deriving and reformulating the differential equation was very different from the students' textbook experiences of differential equations. The fact that Bernoulli's method worked in this particular case, despite its lack of rigor, provoked a discussion among the students and their supervisor (the author) about Bernoulli's use of the infinitesimal triangle and his use of the infinitesimals, dx and dy, as actual infinitely small quantities. This made the students focus more systematically on the differences between now and then, questioning, at first, why we need to define a differential quotient as the limit (in case it exists) of difference quotients, then analyzing the situation again to understand why Bernoulli's method worked fine for the catenary, and trying to picture situations where it would go wrong. This is an incidence where connections were created between the students' historical experiences and their experiences from modern mathematics which challenged them to examine their own understanding of the involved concepts. Through these discussions, the students built up intuition about infinitesimals and awareness about the reasons behind the construction of our modern concepts. Major differences were the lack, in the seventeenth century, of the concept of a function, of a limit, and the formalized concept of continuity. In this project work, the historical texts provided a framework for discussions among the students and with their supervising professor, about what constitutes the concept of a differential equation, and how we can read meaning into such an equation. Through these discussions, which were triggered by the historical texts, the students came to reflect upon the concept of a differential quotient and the meaning of a differential equation on a structural level that went beyond mere calculations and operational understanding of the concepts. This is an example of what Jahnke et al. [8] calls a reorientation effect of studying original sources.

2. Johann's solution of the catenary differential equation. Through some further manipulations Bernoulli reached the following formulation of the equation for the catenary $dy = adx/\sqrt{x^2 + 2ax}$ which he used to construct the

Let the normals AK and GH be drawn, meeting in B. Take $BA = a$ and describe an equilateral hyperbola BC with vertex B and centre A. Now construct a curve DI with the property that everywhere BA is the middle proportional between KC and KD, that is such that $KD = aa:\sqrt{(2ax + xx)}$. Now draw a parallel AF and take the rectangle AG equal to the area $HBKDI$. Prolong DK and FG, then their intersection point E will be on the required curve.[1]

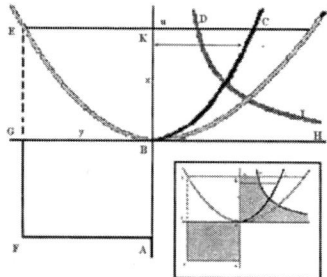

Figure 15.2.

curve geometrically. This puzzled the students and initiated discussions about, what it means to be a solution to a differential equation. As can be seen in Figure 15.2 [2, p. 41], Bernoulli interpreted the integral geometrically as the area below a curve. The students added an illustration of this in their figure, as can be seen above, with the two shadowed areas which are not present in Bernoulli's figure. This way of solving the equation by constructing the curve forced the students into discussions about conceptual aspects of solutions to differential equations. It made them articulate what constitutes a solution in our modern understanding, an articulation that does not automatically manifest itself from solving differential equation exercises from modern textbooks. In order to follow Bernoulli's construction, the students were challenged to think about and use integration differently than they would normally do when solving differential equations analytically. They were also forced to use the properties of the curve represented geometrically which they felt as a challenge. They were used to using the direct relationship between the analytical expression of a function and the coordinate system, to produce a graph. Here they went "the other way" and had to think of the curve as being represented by its graph instead of its analytical expression. Historically, they realized that what is understood by a solution to a differential equation has changed in the course of time.

3. Different solution methods of the brachistochrone problem. The brachistochrone problem is to describe the curve of fastest descent between two points for a point only influenced by gravity. Jacob and Johann Bernoulli published different solution methods to the problem in 1697. Johann Bernoulli interpreted the point as a light particle moving from one point to another. By using Fermat's principle of refraction, he derived an equation for the brachistochrone, i.e. the cycloid, involving the infinitesimals dx and dy. Jacob Bernoulli considered the problem as an extremum problem noting that, since the brachistochrone gives the minimum in time, an infinitesimal change in the curve will not increase the time.

The differences between Johann's and Jacob's solution of the brachistochrone illustrated for the students the power of mathematics. Johann's solution was tied to the physical conditions of the problem and could not be generalized beyond the actual situation, whereas Jacob's solution was independent of the physical situation and could be used on different kinds of extremum problems. Through their work with the historical texts on the solution of the brachistochrone, the students experienced the characteristics of the nature of mathematics that makes it possible to generalize solution methods of particular problems. Thereby, they were able to understand why Jacob's method could generate new kinds of questions that eventually led to a new research area in mathematics, the calculus of variations, and why Johann's could not[1].

15.4.3 Development of mathematical competencies

In the discussions above of episodes where the students through their work with the original sources used other aspects of their mathematical conceptions in new situations and discussions, some learning potentials regarding differential equations and the mathematical concepts underneath have already been emphasized, especially in the discussion of the students' work with Johann Bernoulli's text on the catenary. A more systematic analysis of the students' report with respect to the KOM-report showed that the students, in their work with the historical texts, actually were challenged within all of the eight main competencies. The students' awareness of the special nature of *mathematical thinking* (1) was especially enhanced in their comparison of Johann's and Jakob's solutions of the brachistochrone as discussed above. The students' *problem solving* (2) skills were trained extensively and in different areas of mathematics. As mentioned in the discussion of their work with Johann's solution of the catenary problem, the students had to fill in a lot of gaps in order to understand Johann's results. Each of these gaps required that the students derive intermediate results on their own about similar triangles using trigonometry, and that they solve mathematization problems. Through their work with understanding the Bernoulli brothers' mathematization of the physical problems, parts of the students' *modeling* competency (3) were developed. The competency to *reason* (4) in mathematics was developed in all those parts of the project work where the students tried to make sense of the original sources by means of their own mathematical training and knowledge. (5) *Representations*: as exemplified in the discussion of the students' work with Johann Bernoulli's construction of the solution to the differential equation of the catenary, the students were challenged to work with a representation of the solution to the differential equation that is different from the analytical representation given in modern textbooks. In the report, the students also solved the differential equation analytically

---

[1] For a presentation of the historical problem of the brachistochrone in a didactical perspective, see [3].

and compared the analytical representation with Bernoulli's geometrical one. During their mathematization of the five hypotheses from statics that Bernoulli took for granted, the students were trained both in working with different representations and in using the mathematical language of *symbols and formalism* (6). This competency was especially developed in the students' work with the two original sources on the brachistochrone problem in their struggle to understand Johann's mathematization of the path of the light particle and Jakob's use of the minimizing property of the brachistochrone. The writing of the report (90 pages) in which the students, through a thorough presentation and analysis of the original sources, answered their problems for their project work within the historical context, developed their competency to *communicate* (7) in, with, and about mathematics in ways that go far beyond what normal exercises in solving differential equations requires. The students wrote their report using the mathematical editor programme TeX and they used other computer software programmes to draw their own versions of Bernoulli's figures. These activities show engagement by the students in *tools and aids* (8) of mathematics, especially in their reproduction and modification of Johann Bernoulli's figure of the construction of the solution to the differential equation of the catenary.

15.5 Some Conclusions and Critical Remarks

Based on their studies of the original sources and relevant secondary literature, the students concluded that physics did function as problem generator in the early history of the development of differential equations and played a decisive role in the derivations of the equations describing the catenary and the brachistocrone. They further concluded that physics played a significant role for Johann's solutions of both the catenary and the brachistochrone problem, but not for Jacob's solution of the brachistochrone problem. Jacob's arguments were not linked to the physical system; hence his method could be transferred to other problems of that type. This became the beginning of the calculus of variations. The students did not move beyond this in their project, but it is interesting to notice that the calculus of variation later became central in physics, providing an important feedback in the opposite direction.

The analysis of the chosen project has shown that, if we adopt a competence based view of mathematics education and evaluate learning outcomes not with reference to standard procedures and lists of concepts and results, but with respect to how and which mathematical competencies the students have been challenged to invoke, and thereby develop, and if we let the students work with the history of the practice of mathematics studied from specific perspectives that address significant issues regarding the mathematics in question, then history can be used as a means to teach and learn core curriculum subjects without losing sight of history. Another point, which has not been taken up in the present paper, but has been developed in [17], is that interdisciplinary work with history and mathematics in the sense outlined in the present paper seems to provide special opportunities for students to investigate meta-discursive rules of mathematical discourse. According to Sfard's [22] theory of learning as communicating, the development of proper, meta-discursive rules are necessary for the learning of mathematics, and, as has been illustrated in the analysis of the students' work with physics' influence on the development of differential equations, by working with history in mathematics education within the framework outlined in the present paper, meta-discursive rules of mathematical discourse can be exposed and turned into objects of reflection, precisely because they can be investigated at the object-level of history discourse. This implies an important role for history in the learning of mathematics, because is suggests that history can facilitate meta-level learning of mathematical discourse.

The above claims are further supported through analyses of other historically oriented mathematics projects that have been performed by students at Roskilde University either in the two year introductory science programme or in the mathematics master's programme. A project on the history of mathematical biology, where the students read an original source of Nicholas Rashevsky on a mathematical model for cell division is treated in [16, 14] and analyzed with respect to learning outcomes regarding deriving and understanding the general differential equation of diffusion, the students' understanding of the integral concept, development of the students' modeling competency, and its contribution to general educational goals, respectively. Other examples of projects with substantial learning outcomes of core mathematics, in university mathematics education, are "Paradoxes in set theory and Zermelo's III axiom," "What mathematics and physics did for vector calculus," "Generalisations in the theory of integration," "Infinity and "integration" in Antiquity," "Bolzano and Cauchy: a history of mathematics project," "The real numbers: constructions in the 1870s," and "D'Alembert and the fundamental theorem of algebra."

In the present paper focus has been on how history can be used for the learning of core curriculum mathematics without either trivializing it or using a Whiggish approach to history. The learning outcome of the above history projects can also be analyzed with respect to *mathematical awareness*, as explained by Tzanakis and Arcavi [23], which includes aspects related to the intrinsic and the extrinsic nature of mathematical activity. The above mentioned projects can then also be seen as empirical evidence for some of the possibilities history offers as referred to by Tzanakis and Arcavi [23, p. 211]. With respect to the KOM-report these aspects relate to the three kinds of overview and judgment.

It can be raised as a criticism that only certain perspectives of the history are considered, and that *e.g.*, to gain insights into historical processes of change, episodes from different time periods need to be studied. In the above project work, the students did not experience the historical process of change, but they did experience that the understanding of the involved mathematics in the 17th century was different from our understanding. The students solved the differential equations related to their project work, but they did not learn analytical techniques of solving differential equations extensively during their historical studies, and they did not learn to distinguish between different types of differential equations.

Acknowledgments I wish to thank Constantinos Tzanakis for helpful comments and suggestions on an earlier version of this paper which was first presented at the CERME 6 in Lyon in January 2009.

Bibliography

[1] Blomhøj, M. and T.H. Kjeldsen, 2009, "Project organised science studies at university level: exemplarity and interdisciplinarity," *ZDM Mathematics Education, Zentralblatt für Didaktik der Mathematik*, 41, 183–198.

[2] Bos, H.J.M., 1975, "The Calculus in the Eighteenth Century II: Techniques and Applications," *History of Mathematics Origins and Development of the Calculus 5*, Milton Keynes: The Open University Press.

[3] Chabert, J-L., 1997, "The Brachistochrone Problem," in *History of Mathematics, Histories of Problems*, E. Barbin (ed.), Editions Ellipses, 183–202.

[4] Epple, M., 2000, "Genies, Ideen, Institutionen, mathematische Werkstätten: Formen der Mathematikgeschichte," *Mathematische Semesterberichte*, 47, 131–163.

[5] ——, 2004, "Knots Invariants in Vienna and Princeton during the 1920s: Epistemic Configurations of Mathematical Research," *Science in Context*, 17, 131–164.

[6] Fried, M.N., 2001, "Can Mathematics Education and History of Mathematics Coexist?," *Science & Education*, 10, 391–408.

[7] ——, 2007, "Didactics and history of Mathematics: Knowledge and Self-Knowledge," *Educational Studies in Mathematics*, 66, 203–223.

[8] Jahnke, N.H. et al., 2000, "The use of original sources in the mathematics classroom," in *History in Mathematics Education: The ICMI study*, J. Fauvel & J. van Maanen (eds.), Dordrecht: Kluwer, pp. 291–328.

[9] Jankvist, U.T., 2009, "Evaluating a Teaching Module on the Early History of Error Correcting Codes," in *Proceedings 5. International Colloquium on the Didactics of Mathematics*, M. Kourkoulos & C. Tzanakis (eds.), Rethymnon: The University of Crete

[10] Jankvist, U. T. and T. H. Kjeldsen, 2011, "New Avenues for History in Mathematics Education: Mathematical Competencies and Anchoring," *Science and Education*, 20, 831–862.

[11] Jensen, B.E., 2003, *Historie—livsverden og fag*, Copenhagen: Gyldendal.

[12] Kjeldsen, T.H., 2009a, "Egg-forms and Measure-Bodies: Different Mathematical Practices in the Early History of the Modern Theory of Convexity," *Science in Context*, 22, 1–29.

15.5. Some Conclusions and Critical Remarks

[13] ——, 2009b, "Abstraction and application: new contexts, new interpretations in twentieth-century mathematics," in Robson, E., & Stedall, J. (eds.): *The Oxford Handbook of the History of Mathematics*(pp. 755–778). New York: Oxford University Press.

[14] ——, 2010, "History in mathematics education - why bother? Interdisciplinarity, mathematical competence and the learning of mathematics." *Interdisciplinarity for the 21st Century: Proceedings for the 3th International Symposium on Mathematics and its Connections to the Arts and Sciences*. Bharath Sriraman and Viktor Freiman. Information Age Publishing, pp. 17–48.

[15] Kjeldsen, T.H., S.A. Pedersen, and L.M. Sonne-Hansen, 2004, *New Trends in the History and Philosophy of Mathematics*, Odense: SDU University Press, 11–27.

[16] Kjeldsen, T.H. and M. Blomhøj, 2009, "Integrating history and philosophy in mathematics education at university level through problem-oriented project work," *ZDM Mathematics Education, Zentralblatt für Didaktik der Mathematik*, 41, 87–104.

[17] ——, forthcoming. "Beyond Motivation—History as a method for the learning of meta-discursive rules in mathematics," Submitted June, 2010, *Educational Studies in Mathematics*, accepted for publication, November 2010.

[18] Nielsen, K.H.M., S.M. Nørby, T. Mosegaard, S.I.M. Skjoldager, and C.M. Zacho, 2005, *Physics' influence on the development of differential equations and the following development of theory*. (In Danish). IMFUFA, Roskilde University.

[19] Niss, M., 2001, "University mathematics based on problem-oriented student projects: 25 years of experiences with the Roskilde model," in *The teaching and learning of mathematics at university level: an ICMI study*, D. Holton (ed.), Dordrecht: Kluwer, pp. 405–422.

[20] ——, 2004, "The Danish "KOM" project and possible consequences for teacher education," in *Educating for the Future*, R. Strässer, G. Brandell, B. Grevholm & O. Helenius (eds.), Göteborg: The Royal Swedish Academy, pp. 179–190.

[21] Rheinberger, H-J., 1997, *Towards a History of Epistemic Things: Synthesizing Proteins in the Test Tube*, Stanford: Stanford University Press.

[22] Sfard, A., 2008, *Thinking as Communicating*. Cambridge: Cambridge University Press.

[23] Tzanakis, C. and A. Arcavi, 2000, "Integrating history of mathematics in the classroom: an analytic survey," in *History in Mathematics Education: The ICMI study*, J. Fauvel & J. van Maanen (eds.), Dordrecht: Kluwer, pp. 201–240.

About the Author

Tinne Hoff Kjeldsen is associate professor in history and didactics of mathematics at IMFUFA, NSM, Roskilde University, Denmark. Her main research areas are within the history of pure and applied mathematics in the twentieth century and the integration of modelling and of history and philosophy of mathematics in mathematics education. Her main interests include history of twentieth century mathematics, history and philosophy of mathematical modeling, historiography, and didactics of mathematics, especially regarding modeling competencies and the use of history in mathematics and science education. She has been an invited lecturer at numerous international conferences, workshops, mathematics departments, and history of science departments in Europe, North America, and Asia. Her research results are published in books and major journals of history, philosophy, and didactics of mathematics and science.

16

History of Statistics and Students' Difficulties in Comprehending Variance

Michael Kourkoulos and Constantinos Tzanakis
University of Crete, Greece

16.1 Introduction

Although variation is central to statistics (*e.g.*, [28, 29]), before the end of the 90's not much attention was given to it in didactical research, as Bakker [1, p. 16], Reading [33] and others have remarked. Only recently have there been some systematic studies on the development of students' conception of variation (*e.g.*, [43, 33, 34, 3, 14, especially pp. 382–386]). However, although didactical research on variance and standard deviation (s.d.) is very limited, it still points out that students face important difficulties in understanding these concepts. Mathews, Clark and their colleagues [25, 5, 14, pp. 383, 385–386, 388] examined students in four tertiary USA institutions, shortly after they had completed their introductory statistics course with an A. The majority of the students had not understood even the basic characteristics of the s.d.: (i) one third of them considered the s.d. simply as the outcome of an algorithm to be given and performed, did not understand its meaning and were unable to calculate it if the formula was not given; (ii) some confounded the s.d. with the Z-values; (iii) others believed that the s.d. of an ordered set of numbers depends on the distance between the successive values of this set and expresses a kind of mean value of these distances. Although these students got an A, they had not even formed the simplified idea that the s.d. is a kind of an average distance from the mean. In other recent studies it was found that after the conventional introductory teaching of the s.d. in higher education, many students continued to face serious difficulties in understanding and coordinating even the simplest of the underlying foundational concepts associated with the s.d. ([27], [6, pp. 95–97], [15, pp. 128–135]). Moreover, these difficulties were often stubborn [15, pp. 135–140].

16.2 Classroom Observations

We have designed and implemented two experimental introductory statistics courses with two groups of prospective primary school teachers (24 and 52 students, respectively) at the Department of Education of the University of Crete [21, 24]. Students worked in small groups (3–5 students) and performed *guided research work* [13, 20, 16, 41] in which they not only had to work out problems given by the teacher, but also had to get involved in forming the research questions, and gradually pose their own research questions and problems (closed and open questions, conjectures, etc). With the exception of certain examples initially provided by the teacher, students had to find and produce examples of data necessary in their investigations of the properties of statistical aggregates. They looked for examples in different textbooks on introductory statistics and on the Internet. The examples they chose to elaborate on, as well as those that they produced, relate to every-day life situations and educational phenomena with which they felt acquainted (weight and height of human populations, students' grades, income, wedding data, etc); they also elaborated on purely numerical examples. In fact, the setting of the examples they chose to elaborate on is common in introductory statistics textbooks and courses [4]. Besides the purely numerical ones, the examples used mainly (or almost exclusively) refer to social phenomena (distributions of students' grades, income, employment, etc), or phenomena of human biology, also (strongly) related to social factors (weight and height distributions of human populations, number of children per adult or per family, disease and death rate distributions, etc), whereas, meaningful examples of situations from other fields, like physics or geometry, are absent. The use of similar examples is particularly frequent in introductory statistics textbooks and courses addressed to high school students and students of human and social sciences (*e.g.*, see [4, 24, pp. 11, 13–14])

The feeling of familiarity was the dominant criterion that led our students to confine the non-purely numerical examples they chose to work with to those that are related to social factors. This choice, however, did not facilitate their understanding of different aspects of statistical aggregates, *e.g.*, the mean as an equilibrium point (indeed, important insights on this can be provided by physical examples, like a bar with weights attached to it, or a simple model of springs having a common attachment point, [42]. Nevertheless, despite working on this restricted set of examples, our students succeeded in examining, at a rather satisfactory level, the basic properties of aggregates of central tendency (mode, median and mean) and the simplest aggregates of variation (range, interquartile range and mean absolute deviation). To this end, however, they had to do considerable experimental work and in some cases reduce the treated situations to the corresponding purely numerical situations and/or graphical representations. This was due to the fact that in these cases, the examined properties were better understood if considered numerically, rather than in the context of the concrete situations to which the examples of data referred. The analysis of students' behavior and difficulties points out that a different collection of the examined examples—in particular the use of adequate physical models—could significantly facilitate their work [21, 24, 22].

The difficulties in the understanding of variance were much more prominent than those in the understanding of central tendency. From the beginning of the course, many students could not understand its meaning and were reluctant to accept it. One main reason for these difficulties was that in all the situations students used to obtain examples of distributions for elaboration, variance had an unclear, or even problematic, meaning. In most cases, the sum of squares was dimensionally meaningless (squares of weights, grades, money, etc). In fact, with the exception of purely numerical examples, the only cases that the sum of squares was not dimensionally meaningless for the students were those involving distribution of lengths (*e.g.*, students' height, travel distances). However, even in these cases, the squares of lengths and their sums had unclear meanings in the context of the corresponding situations.

Some of the students' comments that are eloquent concerning their difficulties in understanding and their reluctance in accepting this concept appear below:

> "The squares of the heights of the students...I understand the height, or the weight of a student, but what does the square of a student's height mean?...I have never heard anybody talk about the square of the height of people."

Referring to the distribution of distances of bus trips, a student remarked:

> "I don't understand why we use this sum of squares to measure the dispersion...I see that these squares of distances are certain areas, but what do they have to do with the trips, or buses?...I mean, there are many quantities really related to the distances of the bus trips—the traveling time, the fuel, the cost of the trip, but the square of the distances of the trips?...I don't see what they have to do with the trips, or the buses...I mean that it's only a mathematical artifice; it has nothing to do with the real trip."

Another student remarked on the same case:

> "I see that when the distances of bus trips are more dispersed, the sum of the squares of the distances from the mean increases, and so...somehow, this sum indicates the dispersion. But why use these squares of distances that don't really mean anything for the trips, while we can use the sum of the absolute distances from the mean, or the interquartile range that are simple and we do understand what they measure?"

16.2. Classroom Observations

What the third student said is characteristic of the conceptual difficulties of many other students. They can understand that when the dispersion around the mean increases, the sum of the squares of the distances from the mean also increases, and, therefore, the variance is a parameter that does express the dispersion. However, they are reluctant to accept this way of measuring dispersion, which involves quantities that have unclear meaning in the context of the examined situations and, in most cases, are dimensionally ill-defined[1]. Their resistance against variance was reinforced by the fact that they already used other parameters measuring dispersion (mean absolute deviation from the mean (MAD), interquartile range) that are easier to understand and conceptually simpler than the variance.

Some students have pushed their arguments on the subject even further, *e.g.*:

> "If it is permitted to use such strange quantities, like the squares of weights, or squares of the students' height, why not use the **cubes** or the **fourth powers** of the weights to find the dispersion?...What I want to say is, why use these strange quantities and not use the absolute differences from the average and the **mean absolute deviation**, which directly measures the dispersion from the average?"

In response to the students' reluctance and objections, the teacher said that the essential reasons for using variance as a dispersion parameter are that it possesses important computational properties and that in most cases it is easier to manipulate than MAD. However, for the students to understand this, they need to work to find its properties. In this context, thinking mainly by analogy to questions they had posed and investigated for the previously examined aggregates, students were able to pose interesting questions on variance, *e.g.*: *if we want to calculate the variance of the union of two populations (or of two sets of values), what information do we have to know and how will we calculate it?*

An additional cause for difficulty in the students' work, concerning both the initial meaning of variance and its properties, was that in the usual graphical representations of statistical data, they did not find interpretative elements to help them significantly in their investigations[2]. Since, neither in the context of the often examined real or realistic examples of data, nor in their graphical representations were students able to find interpretative elements that were of significant help, their productive work on the properties of variance was mainly confined to the corresponding purely numerical situations. As the students failed to find such interpretative elements, their difficulties in investigating the properties of variance were significantly increased. However, despite this, they succeeded in finding some of its basic properties and understanding others explained by the teacher[3]. By comparing them with those of MAD, they realized some of its important computational advantages and that in most cases it was significantly easier to manipulate than was MAD. For example, they considered it an important advantage that the variance of the union of two or more mutually disjoint subsets of a population can be obtained only from the variance, the mean, and the total frequency of each subset[4], whereas, the corresponding information is generally insufficient for the calculation of MAD. Moreover, they remarked that this was not unique and that similar phenomena appear for other properties, since they found that: (a) Generally, less information is needed in order to calculate the change of variance due to the change of values of a subset of the total population than that of MAD; (b) to calculate the second moment of a statistical variable X about a value x_0, they needed only the distance from x_0 to the mean value of X and the variance of X, whereas, this distance and MAD are not sufficient for the calculation of the mean absolute deviation of the variable X about x_0[5].

Properties of this type and their comparisons convinced students that variance is a useful dispersion parameter and that there are reasons to prefer it to MAD. Nevertheless, many students still felt that they had not satisfactorily understood the meaning of variance, especially in the context of the real or realistic situations they had examined.

In the second group, students raised the question of the existence of a relation between variance (s^2) and MAD (μ), hoping that since μ was better understood, such a relation could help them understand s^2. After persistent research

---

[1] Here the teacher used a common argument; using the s.d. instead of variance, at least partially resolves the dimensional problem, since the s.d. is dimensionally meaningful, given that it has the same dimensions as the variable itself. But students argued that this was not at all satisfactory because the s.d. was obtained through the use of meaningless quantities.

[2] Since such representations were helpful for understanding the statistical parameters that students had studied already, they looked for the graphical representation of variance, or the sum of squares of distances; however, their efforts were not fruitful, in this case.

[3] Students in the first group, who were weak in algebra, faced even more important difficulties than those in the second. It was difficult for them to find the answers to the research questions, and the teacher had to give considerable help, or even give and explain the answer.

[4] Excluding the discussion of cases with an infinite number of subsets, or subsets of infinite total frequency.

[5] All these aspects of variance and their difference from those of MAD, come from the same fundamental fact: the product of variance with the corresponding frequency n has properties of an extensive quantity, while that of n with MAD has not. However, this point was not discussed with the students in such a general perspective.

work, they succeeded in finding first, that there is a "Pythagorean"–like relation ($s^2 = \mu^2 + s_R^2 = \mu^2 + s_L^2$) for symmetric distributions (for which $s_R = s_L$) and then, that $s^2 = \mu^2 + (s_R^2 + s_L^2)/2$ when the mean equals the median; and finally that $s^2 = \mu^2/(4p_1p_2) + p_1 s_R^2 + p_2 s_L^2$ in the general case [21][6]. The proof of these relations is outlined in the Appendix.

The majority of the students considered these relations as a satisfactory answer to the initial question of linking MAD to s^2. Nevertheless, others pursued their investigation further. By iterating the general formula, they tried to find an approximation to s^2 using only absolute deviations[7].

Besides the didactical interest of students' research work on this subject, the extended work and the considerable effort that this investigation demanded show the students' great interest in finding relations between MAD and s^2. This interest was motivated by their feeling of insufficient understanding of the meaning of s^2 and their hope that linking it to MAD—considered to be simpler and better understood—would permit them to understand it. Students' work on this subject helped them significantly improve their understanding of variance and become more willing to accept it as a basic dispersion measure. Nevertheless, many of them continued to believe that MAD was conceptually simpler and with a clearer meaning in the context of the realistic situations that they examined. Therefore, a significant number of them continued to express their preference to MAD.

In conclusion, our students' behavior demonstrates that restricting the set of non-purely numerical examples of data used to introduce variance to examples referring to social phenomena and phenomena of human biology creates important difficulties as far as its understanding is concerned and may result in activating important resistance to accepting it as a dispersion measure. Such phenomena are often related to everyday life situations that give the impression of simplicity, ignoring that they are in fact very complex; our students' behavior points out that this impression is deceptive. This resistance is widely activated if, as in our case, the teaching approach requires the students to carefully and coherently examine the meaning of the newly introduced aggregate. In the context of such examples, the meaning of variance appears unclear to them and, what is worse, it is very often dimensionally meaningless. In fact, by working exclusively with such examples, students got the impression that the only examples to which variance appeared not to be a questionable quantity were the purely numerical ones. Moreover, even at the purely numerical level students had difficulties seeing the **rationale** behind the use of the sum of **squared** deviations as a measure of dispersion.

As has already been mentioned, very often in introductory statistics textbooks and courses the non-purely numerical examples mainly (or almost exclusively) refer to the aforementioned types of phenomena. However understandable the interest in elaborating examples of data referring to social phenomena and/or phenomena of human biology may be, the aforementioned elements of students' behavior point out the need for the introductory teaching of variance to be enriched and enlarged with examples of situations in which:

- Variance has a clear meaning in-context;
- Its properties can be interpreted satisfactorily;
- It is possible to understand the significance of variance as an important dispersion parameter.

16.3 Looking for Solutions: History Enters the Scene

A didactically oriented historical analysis of the emergence and evolution of the concept of variance [24], as well as a similar analysis of the interactions between relevant concepts in statistics and physics [42] revealed the following crucial point: The use of sums of squared deviations from a center (central point, line, or surface) as a measure of

---

[6]s_R is the s.d. of the "right subpopulation" (i.e., that part of the population with values of the variable greater than the mean of the distribution). s_L -the s.d. of the "left subpopulation"—is defined similarly; p_1, p_2 are the relative frequencies of the right and left subpopulation. This is students' terminology; their research work was limited to distributions of finite populations (e.g., peoples' income, students' height & weight, etc). To formulate these properties, it is also assumed that the frequency exactly on the mean is zero, or negligible.

[7]At the conceptual level, the first and second relation pointed out to the students that, for the examined distributions, variance is equal to μ^2 plus an additional quantity expressing the dispersions of the right and left subpopulation around their own means. The third (general) relation was considered to express almost the same thing, but with a correction due to the unbalance between the size of the right and left subpopulations. Furthermore, students found empirically that in most of the cases they examined, if in the general relation s_R^2, s_L^2 are replaced by μ_R^2, μ_L^2 (μ_R, μ_L being the MAD of the right and left subpopulations) then the calculated value was close to that of the variance ($s^2 \approx \mu^2/(4p_1 p_2) + p_1 \mu_R^2 + p_2 \mu_L^2$). Thus, many students had the idea that for a large class of distributions, they had succeeded in approximating variance by a rather simple relation that involves only μ, μ_R and μ_L. Students who found this approximation unsatisfactory continued with the aforementioned iteration.

16.3. Looking for Solutions: History Enters the Scene

dispersion around this center, together with the relevant statistical methods and aggregates (Method of Least Squares (MLS), variance, residual sum of squares, etc) resulted in serious conceptual barriers and was very difficult to understand in treating problems in the field of social sciences. In contrast, their conception and evolution was by far easier in the treatment of measurements in geodesy and astronomy[8]. There were three interrelated main reasons:

(i) Social phenomena and phenomena of human biology influenced by social factors are very complex, since they depend on a very large number of factors.

(ii) Such phenomena are very rarely subject to experimental treatment. In most cases it is not possible for some factors to vary (while keeping the others constant) in order for their importance and influence to be evaluated. This was not only an obstacle in the creation and evolution of social theories, but also an additionally significant difficulty in elaborating and evaluating statistical methods of treating social data.

(iii) There were no theories of these phenomena that satisfactorily incorporate the influence of all pertinent factors (or at least those which, individually or collectively, exert considerable influence) [39, ch. 5, in particular pp. 163–169, 195–201].

These reasons did not apply to the examined astronomical and geodesic phenomena, which were much simpler and could more easily be subject to empirical check and (since the 18th century) had a solid theoretical background—namely classical mechanics—permitting their efficient modelization and interpretation. This background infused statistical objects and methods with meaning, inspired and oriented their development, and permitted the interpretation of their results. Furthermore, it offered a reliable a priori expectation, which was a critical element in the evaluation of these methods [39, chs. 1, 3, 4]. As a consequence, statistical aggregates of central tendency and dispersion in astronomical and geodesic problems usually corresponded to clear (real, or conceptual) objects of key importance for the situation examined. For example, the statistical centre of gravity of a multidimensional distribution of measurements was often the approximation of the position of a celestial body (and accordingly a mean was often the approximation of a coordinate of one such position); a line of best fit to measurements was often an approximation to the trajectory of a celestial body; and the variance was a measure of the inaccuracy of observations. In contrast, in the elaboration of data related to social phenomena, statistical aggregates were (and still are) in most cases only data tendencies, with a meaning that was much more difficult to construct. Moreover, the absence of reliable *a priori* expectations, due to the absence of adequate and efficient theories for these phenomena, made the evaluation of applied statistical methods difficult. Of course, the combination of all this greatly enhanced the difficulties encountered ([39, pp. 358–361], [24], Section 2 and references therein).

Furthermore, it is important to note that crucial information on the fundamental significance of variance came from other domains of physics as well, more specifically, from the kinetic theory of gases and statistical mechanics in the 19th century, through the works of Krönig (1856), Clausius (1857–58), Maxwell (1860, 1867), Boltzmann (1868, 1872), Gibbs (1902) and others, and from Einstein's theory of Brownian motion, and Einstein and Debye's early quantum theory of solids in the first decade of the 20th century ([2, §§1.11, 1.13], [19, §§60, 119], [24, pp. 8, 9, 20], [26, §§I.6, I.7], [31, pp. 110–128], [37, §§2.II.1-3], [42, pp. 287, 290, 291]).

Conventional introductory statistics textbooks and courses do not consider this imposing historical reality, and this omission allows for the phenomenon stressed in Section 16.1; namely, that very often in such textbooks and courses, the non-purely numerical examples are mainly (or almost exclusively) examples referring to social phenomena or phenomena of human biology, whereas meaningful examples of situations from other domains are absent. Both the historical development and students' behavior presented in Section 16.2 underline the substantial difficulties they have in understanding the use of squared deviations as a measure of dispersion in the context of examples referring to such phenomena. On the other hand, however, history points out that there are other domains—in particular, physics—that provide examples of situations in the context of which variance has a clear meaning; hence, they are valuable in

---

[8]The MLS was conceived and developed at the beginning of the 19th century in the context of geodesic and astronomical problems (initially by Legendre (1805), Gauss (1809, 1816, 1821, 1823), Laplace (1810–1812, 1818) and others; see [39, chs. 1, 3, 4], [35, ch. 9], [18, chs. 6, 7, §§12.3], [24] and footnote 10 here). However, it was difficult to adequately transfer the statistical methods of measuring and treating dispersion developed in geodesy and astronomy to biological and, even less, to social sciences. To this end, it was necessary to overcome important conceptual barriers during a complex scientific evolution that lasted for almost a century. Pioneering work by Galton, Edgeworth, Pearson, Yule and others was necessary to elaborate a new conceptual framework adequate for considering regression and correlation in these sciences ([39, chs. 5, 8–10], [31, part 3], [40, part I]). It is characteristic of the conceptual difficulties of treating such problems statistically that, only in 1897, did Yule—relying on theoretical arguments—propose a generalized method of linear regression based on the MLS for problems in the social sciences [39, chs. 9, 10].

facilitating its understanding and that of related concepts, thus enriching an introductory teaching of statistics. Looking for such examples in the rich reservoir of history allowed us to identify such models. According to their content, these models may be classified into:

- *Physical models* in which variance expresses:

    (a_1) the dispersion energy of a physical system, in the form of kinetic energy (the model of moving particles);

    (a_2) the dispersion energy of a physical system, in the form of potential energy (the model of springs);

    (b) the moment of inertia of a system of masses.

- *Multi-dimensional geometric models*: A sample of size N of a statistical variable x, with values on its elements $x_1, x_2, \ldots x_N$, is represented by an N-dimensional vector X. The norm of $X - \overline{X}$ is $N^{1/2}$ times the s.d. s, where the vector $\overline{X}$ is along the "mean" direction with all its coordinates equal to the mean $\overline{x}$ of x for this sample and the coefficient of variation, $C_v \equiv s/|\overline{x}|$, turns out to be equal to $|\tan \varphi|$, φ being the angle of X and $\overline{X}$.

- *Stochastic models*: These are models based on simple discrete (jump), or continuous (diffusion) Markov processes, describing intuitively clear physical situations, *e.g.*, a random walk, or the motion of a Brownian particle with, or without friction (corresponding to a Wiener, or Ornstein-Uhlenbeck process, respectively).

As a first step, we experimentally investigated the didactical use of models (a_1), (a_2), since, on the one hand, they have important interpretative virtues concerning variance and its properties and on the other hand, they are the simplest ones. Only rudiments of elementary physics are prerequisite, so they could be understood by students like ours (prospective elementary school teachers). The results were very encouraging concerning both the initial understanding of variance and its properties[9].

Moreover, the extension of these models in two and three dimensions can be productively used in introductory teaching of other important relevant statistical concepts, like the MLS, linear regression, residual sum of squares, Pearson's correlation coefficient, etc (for results of a relevant empirical investigations see [23]).

16.4 An Important Predecessor

When Legendre proposed the MLS (1805), methods to treat inconsistent observations in astronomy and geodesy had been developed, using simpler measures of deviations (errors). These included first-order relative and absolute deviations (Boscovich's method, presented in his work of 1757, 1760, 1770), as well as weighted first-order relative and absolute deviations (Laplace's "method of situation" of 1799, which was an elaborated version of Boscovich's method). An earlier but widely used method was that of Mayer, developed in 1750 and amended by Laplace in 1787. According to this method, when the number of unknowns is smaller than that of the initial (linear) equations, which are, in addition, inconsistent (because they follow from observed values with measurement errors), the latter were weighted in a simple way (multiplying each equation by 1, 0, or −1) and then added, to get an aggregate equation. The final solution was found by solving a system of such aggregate equations (for these methods, see [39, pp. 16–55], [11, ch. 2, 3], [18, ch. 6]). These influential methods were part of the conceptual background that allowed Legendre to conceive the MLS. Thus, its emergence appears as a natural evolution of already existing methods of data treatment, rather than a jump or discontinuity in their evolution due mainly to one man's genius. Seen in the conceptual context of these methods, the MLS appears as another way of weighting errors (in which the weighting coefficient is equal (or proportional) to the error), whose important advantages were supported, initially by Legendre using theoretical and practical arguments and then by the Gauss-Laplace synthesis (1809–1812) and the accumulated experience from its use.[10]

There is some rough similarity worth noting between our students' behavior and the historical development of the treatment of measurement errors (in geodesy and astronomy). In both cases the use of sums of first-order absolute deviations appears as an important conceptual predecessor and competitor to the use of the sums of squared deviations. More specifically:

---

[9]For further discussion on the epistemological origins of these models and the results of our experimental investigation, see [42, 22, 24]. On the epistemological origins and the interpretative virtue of the springs' model see also [11, ch. 12], [12, ch. 4].

[10]That at about the same period Gauss conceived the method independently is an additional indication that the MLS was the outcome of a natural evolution of pre-existing methods of data treatment in geodesy and astronomy; Legendre was the first to publish it in 1805, but, from 1806 onwards, Gauss claimed that he had used the method since 1795 ([30], [39, pp. 15, 145–146], [40, ch. 17], [35, §§9.1.4]).

16.4. An Important Predecessor

(i) Initially, our students considered MAD as conceptually simple and clear, whereas they considered variance as difficult and/or obscure.

(ii) By comparing the properties of MAD and variance, they appreciated the computational advantages of the latter. Thus they understood an important, albeit operational, argument in favor of its use. By searching and finding relations between MAD and variance, they tried to improve their understanding of the meaning of variance through linking it to the conceptually simpler MAD; to a significant extent they did so and had the feeling that it was so.

(iii) Although they accepted variance as an important dispersion parameter, a significant number of students continued to believe that MAD was conceptually simpler and clearer and therefore, they continued to prefer it to the variance.

On the other hand, historically:

(a) The use of squared deviations and their sums emerged as an advantageous way of weighting deviations (errors) in an intellectual environment where weighting errors and equations (using simpler weights) was a common practice in obtaining aggregate equations. Among the methods of weighting deviations, previously used, a principal one was the use of first-order absolute deviations.

(b) Open discussions on the characteristics of the method that uses sums of squared deviations in comparison to others were essential for the appreciation of both its potential advantages and disadvantages. One of its main advantages was the important computational properties of the sums of squared distances from a center, which yield the easiness and generality of its application.[11]

(c) Though the MLS had, already by the end of 1825, become a standard and widely used method in geodesy and astronomy, it was not universally accepted. A still existing competitor was the use of the sum of first-order absolute deviations. As late as 1832, Bowditch was recommending Boscovich's method, over the MLS, because it attributes less weight to defective observations [39, p. 55].[12]

It is also worth noting Laplace's preference of minimizing the mean absolute deviation as a fundamental criterion for finding a best estimate for an unknown parameter sought, versus Gauss's preference (initially expressed in his works of 1821 & 1823 and maintained afterwards) of minimizing the mean squared deviation as such a criterion (*e.g.*, [17, pp. 222–229], [24, pp. 5–6], [36, §§11.2, 11.3]).

In the historical development the use of sums of first-order absolute deviations was an earlier and conceptually simpler way to measure and treat dispersion that was important for understanding and appreciating the use of the sums of squared deviations. For our students' learning trajectory, MAD—through its comparison and link to variance—was a key element that helped them to improve their understanding and acceptance of variance as a dispersion measure. On the contrary, conventional introductory textbooks and teaching of statistics pay very little (if any) attention to the

---

[11] On these issues, the second supplement to Laplace's *Théorie Analytique des Probabilités* (1818) is particularly worth mentioning, because of its innovative and fruitful ideas. For a large number of observations and one unknown parameter sought, Laplace compared the accuracy of the estimate provided by the Least Squares (LS) and that provided by the "method of situation", in which a Least Absolute Deviation (LAD) criterion is used. He derived an accuracy comparison criterion, which, in modern terminology, is analogous to the asymptotic relative efficiency of the estimates. Following this criterion, the LS estimate is more accurate if the errors follow a normal distribution, but for other error distributions the "method of situation" provided more accurate estimates ([11, §§9.7], [§§12.3][18]). Laplace did not provide any specific example of distribution for the second case, but Edgeworth did so; using Laplace's criterion he pointed out in 1888 that if the error distribution is a mixture of two normal distributions with very different variances, then the LAD estimate of the unknown parameter sought is more accurate than that provided by the LS—[11, §§9.7, ch. 15]. (For other relevant works see also [7].) Laplace took a further remarkable step in this work; he examined if a linear combination of the LS and LAD estimates could provide a more accurate one and succeeded in determining the coefficients of such a linear combination for certain cases in which the error distribution is not normal [38, 18, §§12.3]. In a way closely similar to Laplace's work of 1818—but without knowing it in advance and definitely in different circumstances—R.A. Fisher in 1920 compared the sample s.d. and the sample mean absolute deviation as estimates of the dispersion of the involved error distribution. It is in the context of this work that Fisher first conceived the concept of *sufficiency*, which is very important for modern theoretical statistics [38, 18, §§12.3, 18.4]. The approaches in these pioneering works flourish in modern research on regression: the elaboration, or the improvement of fitting methods and processes based on different normed function spaces (mainly L^2 (LS), L^1 (LAD) and L^{∞} (Minimax principle)) and the comparison of their performances (mostly concerning accuracy, robustness and/or computational properties and characteristics) constitute an important trend in modern research in this domain. Moreover the elaboration and evaluation of methods combining the use of more than one fitting criteria (*e.g.*, LAD and LS) constitute another important such trend. (*e.g.*, see [8, 9, 10, 32, 44]).

[12] Additionally, the Mayer-Laplace method, which is simpler and demands much less labor than the MLS, enjoyed popularity until the mid 19th century, even though it is less accurate than the MLS [39, pp. 38–39].

mean absolute deviation. For example, Chalepaki [4], analyzing 15 widely used introductory statistics textbooks in Greek universities as well as the chapters on statistics of the official Greek high school textbook, found that in many of these books the mean absolute deviation is not even mentioned; in others only a definition or description is given and none proposes any systematic work on it nor on its comparison to variance.

The preceding historical considerations and the results of our experimental teaching point out that in introductory teaching of statistics, one should work more thoroughly with the mean absolute deviation. Because of its relative conceptual simplicity, such a work can facilitate students' introduction to the concept of dispersion measures around a centre. Furthermore, with adequate linking and comparison, the use of the mean absolute deviation can substantially facilitate the understanding of variance and s.d.

16.5 The Normal Distribution and the Central Limit Theorem

Historically, the use of the normal distribution (directly, or, mainly, as an approximate distribution) was an influential argument for using variance, s.d., and more generally, sums of squared deviations from a center (central point, or central line). However, this argument became important only gradually, as the understanding of the scientific community deepened on the role of the normal distribution and the Central Limit Theorem (CLT) in different domains, including measurement errors in geodesy and astronomy, physics, natural variation, and the social sciences ([31, ch. 4, 5], [39, ch. 4, 8–10], [24]).

There are introductory teaching approaches to statistics in which there is only a very first, or superficial, approach to the normal distribution and the CLT. For example, in the senior year of Greek high school, the graph of the normal distribution and the probability corresponding to an interval of one, two or three s.d. around the mean are presented, but the formula for the normal distribution is not! Hence, the normal distribution and the CLT cannot be convincing for the use of variance and the s.d.; such conceptually complex objects cannot be justified and understood through linking them to another, even more complex object, like the normal distribution, when the latter is very incompletely understood. In these or other teaching approaches, where variance and s.d. are introduced long before the examination of the normal distribution and/or the CLT, other arguments, such as those mentioned previously, should be used to explain and justify variance and s.d. as principal dispersion parameters. Otherwise, it is possible that, at least for a long time, students will have the feeling that the s.d. and its use are unjustifiably imposed by the teacher, while there are other efficient dispersion parameters with clearer meaning, and/or that s.d. is another statistical object whose reason of existence remains obscure.

It is also worth noting that the Chebychev inequality can be taught soon after the introduction of variance and the s.d., offering interesting insights to students. It allows for a quick, albeit partial, answer to some of the students' reserves about the s.d., showing that the s.d. offers information on the corresponding probability of a range of values centered around the mean, when the radius of this range is greater than the s.d. Of course, this information is not as precise as the information deduced from the s.d. in case the normal distribution can be used as an approximation, but it can be applied more generally, and it leads to a simple and direct way to the proof of the (weak) law of large numbers.

Bibliography

[1] Bakker, A., 2004, *Design research in statistics education: On symbolizing and computer tools*, Utrecht: The Netherlands, CD-Beta Press

[2] Brush, S.G., 1983, *Statistical Physics and the Atomic Theory of Matter*, Princeton: Princeton University Press.

[3] Canada, D., 2006, "Elementary pre-service teachers' conceptions of variation in a probability context," *SERJ* 5(1), 36–63

[4] Chalepaki, M., 2009, *Didactical analysis of introductory statistics textbooks for secondary and tertiary education*, unpublished Master degree dissertation, Department of Education, University of Crete (in Greek)

[5] Clark, J., G. Karuat, D. Mathews, and J. Wimbish, 2003, *The Fundamental Theorem of Statistics: Classifying Student understanding of basic statistical concepts*, unpublished paper,
    www1.hollins.edu/faculty/clarkjm/stat2c.pdf

16.5. The Normal Distribution and the Central Limit Theorem

[6] delMas, R. and Y. Liu, 2007, "Students' conceptual understanding of the standard deviation," in M.Lovett & P.Shah (Eds), *Thinking with Data*, NJ: Lawrence Erlbaum, pp. 87–116.

[7] David, H.A., 1998, "Early sample measures of variability," *Statistical Science* 13(4), 368–377

[8] Dielman, T.E., 2005, "Least absolute value regression: recent contributions," *Journal of Statistical Computation and Simulation*, 75(4), 263–286

[9] Dodge Y. and J. Jurecková, 1987, "Adaptive combination of least squares and least absolute deviations estimators," in Y. Dodge (Ed), *Statistical Data Analysis Based on L^1,-Norm and Related Methods*, Amsterdam: North-Holland, pp. 275–284.

[10] ——, 2000, *Adaptive Regression*, NY: Springer

[11] Farebrother, R.W., 1999, *Fitting linear relationships: A history of calculus of observations 1750–1900*, NY: Springer

[12] ——, 2002, *Visualizing statistical models and concepts*, NY: Marcel Dekker

[13] Freudenthal, H., 1991, *Revisiting mathematics education—China lectures*, Dordrecht: Kluwer

[14] Garfield, J. and D. Ben-Zvi,, 2007, "How Students Learn Statistics Revisited," *International Statistical Review* 75(3), 372–396

[15] Garfield, J., R. delMas, and B. Chance, 2007, "Using students' informal notions of variability to develop an understanding of formal measures of variability," in M.Lovett & P.Shah (Eds), *Thinking with Data*, NJ: Lawrence Erlbaum, pp. 117–148.

[16] Goos, M., 2004, "Learning Mathematics in a Classroom Community of Inquiry," *JRME* 35(4), 258-291

[17] Gnedenko, B., and O.B. Sheinin (Sheynin), 1992 "The theory of probability," in A. N. Kolmogorov and A. P Yushkevich (Eds) *Mathematics of the 19th Century*, vol I, Basel: Birkhäuser, pp. 211–288.

[18] Hald, A., 2007, *A History of parametric statistical inference from Bernoulli to Fisher 1713–1935*, NY: Springer.

[19] Jeans, J., 1954/1904, *The dynamical theory of gases*, NY: Dover (first published in 1904, Cambridge: Cambridge University Press)

[20] Legrand, M, 1993, "Débat scientifique en cours de mathématiques et spécificité de l'analyse," *Repères*, 10, 123–158

[21] Kourkoulos, M. and C. Tzanakis, 2003, "Introductory Statistics with problem-solving activities and guided research work, assisted by the use of EXCEL" in T. Triandafyllidis, C. Hadjikyriakou (eds) *Proceedings of the 6th International Conference on Technology in Mathematics Teaching* (ICTMT6), Athens: New Technologies Publications, pp.109–117.

[22] Kourkoulos, M., E. Mantadakis, and C. Tzanakis, 2006, "Didactical models enhancing students understanding of the concept of variance in Statistics" in D. Hughes-Hallett, I. Vakalis, H. Arikan (eds) *Proceedings of ICTM3*, Istanbul: Turkish Mathematical Society, Pub. N.Y. : John Wiley & Sons, on CD-ROM, Paper–151.pdf

[23] Kourkoulos, M. & Tzanakis, C., 2008a, "Enhancing students understanding on the Method of Least Squares: An interpretative model inspired by historical and epistemological considerations," in E. Barbin, N. Stelikova, C. Tzanakis (Eds), *Proceedings of ESU5* (pp. 271–283), Prague: Vydavatelsk

[24] ——, 2008b, "Contributions from the study of the history of statistics in understanding students' difficulties for the comprehension of the Variance," in R. Cantoral, F. Fasanelli, A. Garciadiego, B. Stein, C. Tzanakis (Eds), *Proceedings of the HPM2008* (on CD-ROM), Mexico City

[25] Mathews, D. and J. Clark, 1997, "Successful Students' Conceptions of Mean, Standard Deviation and the Central Limit Theorem," *Paper presented at the Midwest conference on teaching statistics*, Oshkosh, WI. `www1.hollins.edu/faculty/clarkjm/stats1.pdf`

[26] Mehra, J. and H. Rechenberg, 1982, *The historical development of Quantum Theory*, NY: Springer, vol.I.

[27] Meletiou-Mavrotheris, M. and C. Lee, 2005,"Exploring introductory statistics students' understanding of variation in histograms," *Proceedings CERME4*, Spain http://cerme4.crm.es/Papers%20definitius/5/MeletiouLee.pdf

[28] Moore, D., 1990, "Uncertainty," in L. Steen (Ed.), *On the shoulders of giants*, Washington: National Academy Press, pp. 95–138.

[29] ——, 1997, "New pedagogy and new content: The case for statistics" *International Statistical Review*, 65(2), 123-165

[30] Plackett, R.L., 1972, "The discovery of the method of least squares," *Biometrika*,59, 239–251

[31] Porter, Th.M., 1986, *The rise of statistical thinking: 1820–1900*, Princeton: Princeton University Press

[32] Portnoy S. and R. Koenker, 1997, "The Gaussian Hare and the Laplacian Tortoise: Computability of Squared-Error versus Absolute-Error Estimators," *Statistical Science*, 12(4), 279–296

[33] Reading, C., 2004, "Student description of variation while working with weather," *SERJ*, 3(2), 84–105, http//www.stat.auckland.ac.nz/serj

[34] Reading, C. and J.M. Shaughnessy, 2004, "Reasoning about variation," in D.Ben-Zvi & G.Garfield (eds), *The Challenge of Developing Statistical Literacy, Reasoning and Thinking*, Dordrecht: Kluwer, pp. 201–226.

[35] Sheynin, O.B., 2005, *Theory of probability. A historical essay*, Berlin: NG Verlag

[36] ——, 1977, "P.S. Laplace's theory of errors," *Arch. Hist. Ex. Sci.*, 17(1), 1–61.

[37] Sklar, L., 1993, *Physics and Chance: Philosophical issues in the foundations of Statistical Mechanics*, Cambridge: Cambridge University Press

[38] Stigler, S.M., 1973, "Laplace, Fisher, and the discovery of the concept of sufficiency". *Biometrika*, 60, 439–445

[39] ——, 1986, *The History of Statistics: The measurement of uncertainty before 1900*, Cambridge (MA): Harvard University Press

[40] ——, 1999, *Statistics on the table: The history of statistical concepts and methods*, Cambridge (MA): Harvard University Press.

[41] Stonewater, J., 2005, "Inquiry Teaching and Learning: The Best Math Class Study," *School Science and Mathematics*, 105(1), 36–47

[42] Tzanakis, C. and M. Kourkoulos, 2006, "May history and physics provide a useful aid for introducing basic statistical concepts?" *Proceedings of the HPM2004 & of the 4th Summer University on the History and Epistemology in Mathematics Education*, revised edition, F. Furinghetti, S. Kaijser & C. Tzanakis (eds), Greece: University of Crete, pp 284–295.

[43] Watson, J.M., B.A. Kelly, R.A. Callingham, and J.M. Shaughnessy, 2003, "The measurement of school students' understanding of statistical variation," *IJMEST*, 34, 1–29

[44] Wilson, H. G., 1978, "Least-squares versus minimum absolute deviations estimation in linear models" *Decision Sciences* 9, 322–335

Appendix

(a) Below is an outline of the justification of the 3rd (most general) formula in Section 16.2, constructed by the students with the teacher's help, in the notation of Footnote 6. $s^2 = \mu^2/(4p_1 p_2) + p_1 s_R^2 + p_2 s_L^2$.

From their previous work, students knew the following property concerning the variance of the distribution of a quantitative characteristic of a finite population: If the population is considered to be divided into two or more mutually disjoint subsets, then the variance of the total population equals the variance of the means of the subsets plus the mean of their variances. This can be expressed formally as $s^2 = \sum_i p_i(\overline{x_i} - \overline{x})^2 + \sum_i p_i s_i^2$, where $\overline{x_i}, s_i, p_i = n_i/n$ are respectively, the mean, the s.d. and the relative frequency of the i-th subset (n_i, n being the absolute frequencies of the i-th subset and the total population) and $\overline{x}$, s are respectively the mean and the s.d. of the total population[13]

As mentioned in Footnote 6, students assumed for simplicity that the frequency of the distribution exactly on the mean is zero, or negligible. Hence, for such distributions they considered the total population to be separated into two subsets, the right subpopulation and the left subpopulation in the terminology of Footnote 6. Applying the previous relation, we have $s^2 = p_1(\overline{x_R} - \overline{x})^2 + p_2(\overline{x_L} - \overline{x})^2 + p_1 s_R^2 + p_2 s_L^2$, (where $\overline{x_R}, s_R, p_1$ are the mean, the s.d. and the relative frequency of the right subpopulation and $\overline{x_L}, s_L, p_2$, are similarly defined for the left one).

From their previous work on the mean and the mean absolute deviation about the mean (μ), students also knew that $p_1(\overline{x_R} - \overline{x}) = p_2(\overline{x} - \overline{x_L}) = \mu/2$. Hence, putting $\overline{x_R} - \overline{x} = \mu/(2p_1)$ and $\overline{x_L} - \overline{x} = -\mu/(2p_2)$ in the previous equality, they obtained the result.

(b) The first and second formulae in Section 16.2 are special cases of the third one. However, students proceeded conversely. After persistent empirical research work, they found the first relation (for symmetric distributions). Then, with the teacher's assistance, they obtained its proof, which was close to that above. This allowed them to find rather easily the more general second relation. Further generalization to obtain the third relation was more demanding for them, but they finally succeeded in doing so[14] [15].

About the Authors

Michael Kourkoulos graduated from the mathematics department of the University of Athens, Greece, and holds a master's degree and a Ph.D. in Mathematics Education from the Université Louis Pasteur de Strasbourg, France. His research interests and publications are in Mathematics Education, especially the teaching and learning of statistics, elementary algebra and number theory, as well as the integration of history and epistemology in teaching, collaborative learning and guided research work on teaching activities. He is currently Assistant Professor of Mathematics Education at the Department of Education of the University of Crete, Greece. Together with Constantinos Tzanakis, he has been organizing the International Colloquium on Didactics of Mathematics at the University of Crete, Greece for several years.

Constantinos Tzanakis graduated from the mathematics department of the University of Athens, Greece, holds a master's degree in Astronomy from the Department of Astronomy, Sussex University, UK and a Ph.D. in Theoretical Physics from the Université Libre de Bruxelles, Belgium. His research interests and publications are in Statistical Physics, Relativity Theory and related areas, as well as in mathematics and science education, especially the integration of history and epistemology in teaching, the relation between mathematics and physics and its didactical implications, and the didactics of statistics. He has been the chair of the ICMI affiliated International Study Group on the Relations

---

[13] This property is the analogue of the *law of total variance* for distributions in finite populations, where it is evidently much simpler conceptually.

[14] Due to space limitations, we omit the details of this part of students' research work.

[15] The three formulas in Section 16.2 can be slightly transformed, so that they are valid when the frequency of the distribution exactly on the mean is not zero, or negligible. Moreover the three relations hold for univariate random variables having finite variance and a probability density function that is a continuous real function.

between the History & Pedagogy of Mathematics (the HPM Group) from 2004 to 2008 and co-organizer of several meetings of this group, including the European Summer University on the History and Epistemology in Mathematics Education (ESU). He is currently professor of mathematics and physics at the department of Education of the University of Crete, Greece. Together with Michael Kourkoulos, he has been organizing the International Colloquium on Didactics of Mathematics at the University of Crete, Greece, since 1998.

17

Designing Student Projects for Teaching and Learning Discrete Mathematics and Computer Science via Primary Historical Sources[1]

Janet Heine Barnett, Jerry Lodder, David Pengelley, Inna Pivkina and Desh Ranjan

Colorado State University - Pueblo and *New Mexico State University, United States of America*

17.1 Introduction

A discrete mathematics course often teaches about precise logical and algorithmic thought and methods of proof to students studying mathematics, computer science, or teacher education. The roots of such methods of thought, and of discrete mathematics itself, are as old as mathematics, with the notion of counting, a discrete operation, usually cited as the first mathematical development in ancient cultures [7]. However, a typical course frequently presents a fast-paced news reel of facts and formulae, often memorized by the students, with the text offering only passing mention of the motivating problems and original work that eventually found resolution in the modern concepts. This paper describes a pedagogical approach to teaching topics in discrete mathematics and computer science intended to place the material in context and provide direction to the subject matter via student projects centered around actual excerpts from primary historical sources. Much has already been written about teaching with primary historical sources [6, ch. 9]. Here we focus on a list of specific pedagogical goals and how they can be achieved through design of student projects based on primary sources.

Our interdisciplinary team of mathematics and computer science faculty has completed a pilot program funded by the US National Science Foundation, in which we have developed and tested over a dozen historical projects for student work in courses in discrete mathematics, graph theory, combinatorics, logic, and computer science. These projects have appeared in print [1], and are presently available through the web resource [3].

We are now in the second year of a four-year NSF expansion grant through which additional projects based on primary sources are being developed, tested, evaluated, revised, and published. The expansion will support classroom testing by faculty at twenty other institutions, careful evaluation of their effectiveness, and provide training in teaching with these projects to graduate students. Projects created to date under the expansion grant are available at our new web resource [2], along with descriptions of projects yet to be developed. We welcome instructors who would like to collaborate in testing or writing projects.

---

[1]Based on a workshop presented at the HPM Group Satellite Meeting of ICME 11 (HPM 2008) in Mexico City, Mexico, 14–18 July 2008.

Here we will briefly describe the pedagogical goals which guide project design, illustrate these principles with excerpts from two particular projects, discuss how such historical projects can be implemented in the classroom, and present preliminary evaluation evidence of their effectiveness.

17.2 Pedagogical Goals and Design Principles

A central goal of our pedagogical approach is to recover motivation for studying particular core topics by teaching and learning these topics *directly* from a primary historical source of scientific significance. Primary source authors represented in the collection include Archimedes, Cantor, Euler, Hamilton, Leibniz, Pascal, Shannon, Turing, Veblen, and von Neumann, writing on topics such as mathematical induction, finite sums of powers, graph theory, transfinite arithmetic, binary arithmetic, combinatorics, computability, and decidability.

Designed to capture the spark of discovery and motivate subsequent lines of inquiry, each historical project is built around primary source material which serves either as an introduction to a core topic in the curriculum, or as supplementary material to a textbook treatment of that topic. Through guided reading of the selected primary source material and by completing a sequence of activities based on these excerpts, students explore the science of the original discovery and develop their own understanding of the subject. Each project also provides a discussion of the historical exigency of the piece and a few biographical comments about the author to place the source in context.

The following pedagogical goals further guide our selection of primary source material and the design of a project. The next section will illustrate how these goals have been implemented in two particular projects.

Fifteen Pedagogical Goals Guiding the Development of Primary Source Based Projects

1. Hone students' verbal and deductive skills through reading.

2. Provide practice moving from verbal descriptions of problems to precise mathematical formulations.

3. Promote recognition of the organizing concept behind a procedure.

4. Promote understanding of the present-day paradigm of the subject through the reading of an historical source which requires no knowledge of that paradigm.

5. Promote reflection on present-day standards and paradigm of subject.

6. Draw attention to subtleties, which modern texts may take for granted, through the reading of an historical source.

7. Promote students' ability to equally participate, regardless of their background or capability.

8. Offer diverse approaches to material which can serve to benefit students with different learning styles through exposure to multiple approaches.

9. Provide a point of departure for students' work, and a direction for their efforts.

10. Encourage more authentic (versus routine) student proof efforts through exposure to original problems in which the concepts arose.

11. Promote a human vision of science and of mathematics.

12. Provide a framework for the subject in which all elements appear in their right place.

13. Promote a dynamical vision of the evolution of mathematics.

14. Promote enriched understanding of subject through greater understanding of its roots, for students and instructors.

15. Engender cognitive dissonance (dépaysement) when comparing a historical source with a modern textbook approach, which to resolve requires an understanding of both the underlying concepts and use of present-day notation.

17.3 Incorporating Pedagogical Design Goals: Two Sample Projects

In this Section we provide excerpts from two student projects in order to give a flavor of their nature. For each, we display selected excerpts from the primary historical source in the project, from our own narrative in the project, and also from student assignment questions, and discuss how these elements promote our pedagogical goals. To set project excerpts apart from our writing in this paper, all project excerpts are indented relative to our main text. Moreover, note that within the project excerpts, the primary historical source material is set in sans serif font. Our specific pedagogical goals are referred to by the numbers used to enumerate them in Section 17.2.

17.3.1 *Treatise on the Arithmetical Triangle*: Blaise Pascal

The project *Treatise on the Arithmetical Triangle* is intended for an introductory level discrete mathematics and proofs course, and presents the concept of mathematical induction from the pioneering work written by Blaise Pascal [8] in the 1650s.

After arranging the figurate numbers in a table, forming "Pascal's triangle," Pascal observes several patterns in the table, which he would like to claim continue indefinitely. Pascal offers a condition for the persistence of a pattern, stated verbally in the proof of his "Twelfth Consequence," a condition known today as mathematical induction. In fact, Pascal's treatise is the first place in the mathematical literature where the principle of mathematical induction is enunciated so completely and generally (Goals 11, 12, 13, 14, 15). Moreover, Pascal's Twelfth Consequence results directly in the modern factorial formula for combination numbers and binomial coefficients, allowing him to proceed to algebra and probability. In this project, students learn first-hand about the issues involved in proofs by iteration, generalizable example, and mathematical induction.

Students begin by reading Pascal's defining description of his triangle, which is highly verbal, entirely labeled using letters, and rotated in comparison to how most people view it today. We display excerpts from a few of his critical definitions here. Such a verbal approach, to a triangle all students think they are already familiar with, but without modern indexing notation, and geometrically tilted from the modern view, challenges students' skill at translating to modern mathematical descriptions, and places all students on the same unfamiliar footing (Goals 1, 2, 7, 15).

The idea of having students learn by being placed on unfamiliar footing is expressed in French as *dépaysement*, a condition in which one must approach things from unaccustomed points of view, and pay great attention to subtleties (Goals 3, 4, 5, 6, 8, 15). One might translate it as 'being thrown off guard', 'being bewildered', 'being taken out of one's element'. It is one of the great strengths of learning from primary historical sources, since they were often written in dramatically different context and time, thus providing a very distinct, and often extremely valuable, point of view. By viewing a topic from a point of view different from the standard modern one, a broader, but also deeper and more flexible, understanding is gained (Goal 14).

Treatise on the Arithmetical Triangle

Definitions

> ... Through each of the points of section and parallel to the sides I draw lines whose intersections make little squares which I call *cells*.
>
> Cells between two parallels drawn from left to right are called *cells of the same parallel row*, as, for example, cells G, σ, π, etc., or φ, ψ, θ, etc.
>
> Those between two lines drawn from top to bottom are called *cells of the same perpendicular row*, as, for example, cells G, φ, A, D, etc., or σ, ψ, B, etc.
>
> Those cut diagonally by the same base are called *cells of the same base*, as, for example, D, B, θ, λ, or A, ψ, π....
>
> Cells of the same base equidistant from its extremities are called *reciprocals*, as, for example, E, R and $B, \theta, \ldots$
>
> Now the numbers assigned to each cell are found by the following method:

The number of the first cell, which is at the right angle, is arbitrary; ... Each of the others is specified by a single rule as follows:

The number of each cell is equal to the sum of the numbers of the perpendicular and parallel cells immediately preceding. Thus cell F, that is, the number of cell F, equals the sum of cell C and cell E, and similarly with the rest.

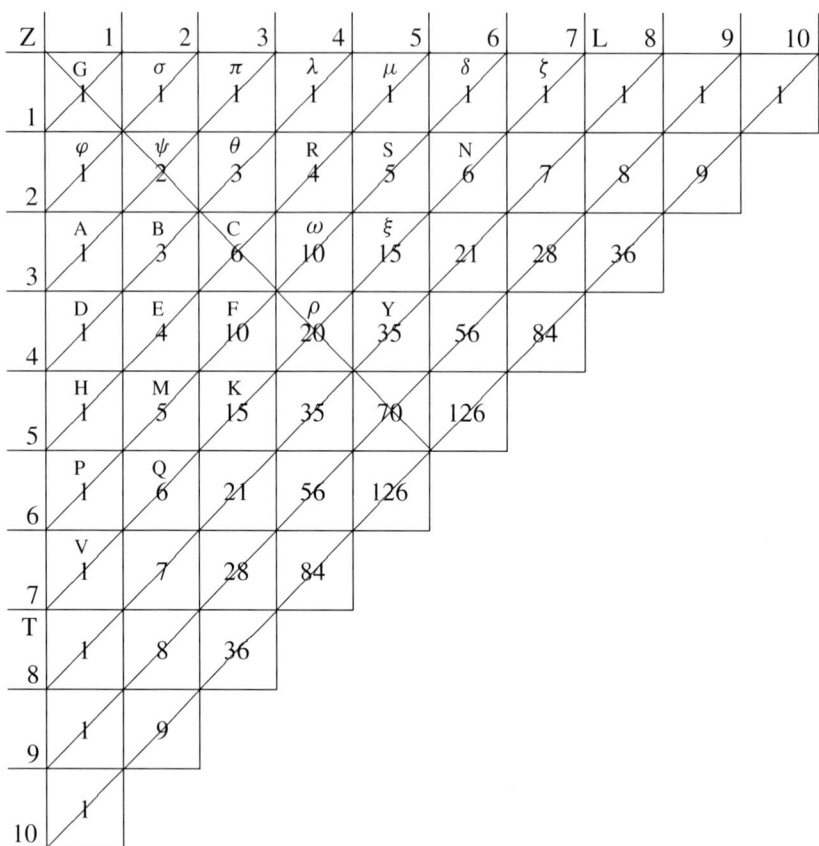

Formalizing Pascal's entire description using modern indexing notation is the first big challenge to students, through exercises like the following (Goals 1, 2, 4, 5). Sophisticated ideas like double-indexing arise immediately and naturally (Goal 6), and students will also learn from this project, in relevant context, about summation and product notations, and recurrence relations (Goals 4, 5, 8).

Project Question 1. Pascal's Triangle and its numbers

(a) Let us use the notation $T_{i,j}$ to denote what Pascal calls the number assigned to the cell in *parallel row i* (which we today call just *row i*) and *perpendicular row j* (which we today call *column j*). We call the i and j by the name *indices* (plural of *index*) in our notation. Using this notation, explain exactly what Pascal's rule is for determining all the numbers in all the cells. Be sure to give full details. This should include explaining for exactly which values of the indices he defines the numbers.

After Pascal proves some simple properties of the triangle essentially iteratively, mathematical induction first arises naturally but implicitly in demonstrating its symmetry. In addition, Pascal proves his claim by "generalizable example," largely because he has no indexing notation to deal conveniently with arbitrary elements. Having students make all this precise in full generality with modern notation enables them to begin to think in terms of induction before it is formally introduced, and to powerfully appreciate the efficacy of indexing notation (Goals 3, 4, 5, 10).

Fifth Consequence

In every arithmetical triangle each cell is equal to its reciprocal.

For in the second base, $\phi\sigma$, it is evident that the two reciprocal cells, φ, σ, are equal to each other and to G.

In the third base, A, ψ, π, it is also obvious that the reciprocals, π, A, are equal to each other and to G.

In the fourth base it is obvious that the extremes, D, λ, are again equal to each other and to G.

And those between, B, θ, are obviously equal since $B = A + \psi$ and $\theta = \pi + \psi$. But $\pi + \psi = A + \psi$ by what has just been shown. Therefore, etc.

Similarly it can be shown for all the other bases that reciprocals are equal, because the extremes are always equal to G and the rest can always be considered as the sum of cells in the preceding base which are themselves reciprocals.

Project Question 3. Symmetry in the triangle: first contact with mathematical induction

> Write the Fifth Consequence using our index notation. Use index notation and the ideas in Pascal's proof to prove the Consequence in full generality, not just for the example he gives. Explain the conceptual ideas behind the general proof.

Only after students grapple with and explain Pascal's Fifth Consequence in their own way are they expected to read about the principle of mathematical induction. The crowning consequence in Pascal's treatise is the Twelfth, in which Pascal derives a formula for the ratio of consecutive numbers along a base in the triangle. From this he will obtain an elegant and efficient "closed" formula for all the numbers in the triangle, a powerful tool for much future mathematical work. And it is right here that Pascal enunciates the general proof principle we call induction. Again we ask students to translate Pascal's proof by generalizable example into a modern and completely general proof. This is far from trivial, and even involves an understanding of a property of proportions that is largely lost today. (This single rich source excerpt and its tasks for students encompasses all fifteen of our pedagogical goals.)

Twelfth Consequence

In every arithmetical triangle, of two contiguous cells in the same base the upper is to the lower as the number of cells from the upper to the top of the base is to the number of cells from the lower to the bottom of the base, inclusive.

Let any two contiguous cells of the same base, E, C, be taken. I say that

E	:	C	::	2	:	3
the lower		the upper		because there are two cells from E to the bottom, namely E, H.		because there are three cells from C to the top, namely C, R, μ.

Although this proposition has an infinity of cases, I shall demonstrate it very briefly by supposing two lemmas:

The first, which is self-evident, that this proportion is found in the second base, for it is perfectly obvious that $\varphi : \sigma :: 1 : 1$;

The second, that if this proportion is found in any base, it will necessarily be found in the following base.

Whence it is apparent that it is necessarily in all the bases. For it is in the second base by the first lemma; therefore by the second lemma it is in the third base, therefore in the fourth, and to infinity.

It is only necessary therefore to demonstrate the second lemma as follows: If this proportion is found in any base, as, for example, in the fourth, $D\lambda$, that is, if $D : B :: 1 : 3$, and $B : \theta :: 2 : 2$, and

$\theta : \lambda :: 3 : 1$, etc., I say the same proportion will be found in the following base, $H\mu$, and that, for example, $E : C :: 2 : 3$.

For $D : B :: 1 : 3$, by hypothesis.

Therefore $\underbrace{D + B}_{E} : B :: \underbrace{1 + 3}_{4} : 3$

$E : B :: 4 : 3$

Similarly $B : \theta :: 2 : 2$, by hypothesis.

Therefore $\underbrace{B + \theta}_{C} : B :: \underbrace{2 + 2}_{4} : 2$

$C : B :: 4 : 2$

But $B : E :: 3 : 4$

Therefore, by compounding the ratios, $C : E :: 3 : 2$. Q.E.D.

The proof is the same for all other bases, since it requires only that the proportion be found in the preceding base, and that each cell be equal to the cell before it together with the cell above it, which is everywhere the case.

Project Question 6. Pascal's Twelfth Consequence: the key to our modern factorial formula

(a) Rewrite Pascal's Twelfth Consequence as a generalized modern formula, entirely in our $T_{i,j}$ terminology. Also verify its correctness in a couple of examples taken from his table in the initial definitions section.

(b) Adapt Pascal's proof by example of his Twelfth Consequence into modern generalized form to prove the formula you obtained above. Use the principle of mathematical induction to create your proof.

From his Twelfth Consequence Pascal can develop a "formula" for the numbers in the triangle, which can then be used in future work on combinatorics, probability, and algebra. We have students follow Pascal's generalizable example to do so (Goals 1, 2, 5, 8, 9, 14, 15) in modern form.

Problem

Given the perpendicular and parallel exponents of a cell, to find its number without making use of the arithmetical triangle. . . .

Project Question 7. Pascal's formula for the numbers in the Arithmetical Triangle

(a) Write down the general formula Pascal claims in solving his "Problem." Your formula should read $T_{i,j} =$ "some formula in terms of i and j." Also write your formula entirely in terms of factorials.

(b) Look at the reason Pascal indicates for his formula for a cell, and use it to make a general proof for your formula above for an arbitrary $T_{i,j}$. You may try to make your proof just like Pascal is indicating, or you may prove it by mathematical induction.

The project can continue on perfectly naturally to integrate combinatorics, the binomial theorem, Fermat's Theorem (proof by induction on the base using Pascal's formula for the binomial coefficients and uniqueness of prime factorization), and end with the RSA cryptosystem. This goes far beyond the historical source, but shows how the source serves as foundation for important modern topics (Goals 9, 14).

17.3.2 *The solution of a problem relating to the geometry of position*: Leonhard Euler

The project *Treatise on the Arithmetical Triangle* just discussed illustrates how rich individual small excerpts from a primary source can be in terms of promoting the pedagogical goals of our approach to using history in discrete mathematics courses. Next we illustrate how specific pedagogical goals can be emphasized in a project by using the project narrative and student tasks to carefully frame a sequence of excerpts from a primary source. The project we consider, *Early Writings on Graph Theory: Euler Circuits and The Königsberg Bridge Problem*, is suitable for a beginning-level discrete mathematics course, or for a 'transition to proof' course.

17.3. Incorporating Pedagogical Design Goals: Two Sample Projects

In the paper [5] on which this project is based, Leonhard Euler (1707–1783) provides a mathematical formulation of the Königsberg Bridge Problem. Considered today to be the starting point of modern graph theory, this foundational paper offers a unique window on a dynamical vision of the evolution of mathematics (Goal 13). The first Euler excerpt which students read in the project conveys this vision in his own words.

Solutio Problematis Ad Geometriam Situs Pertinentis

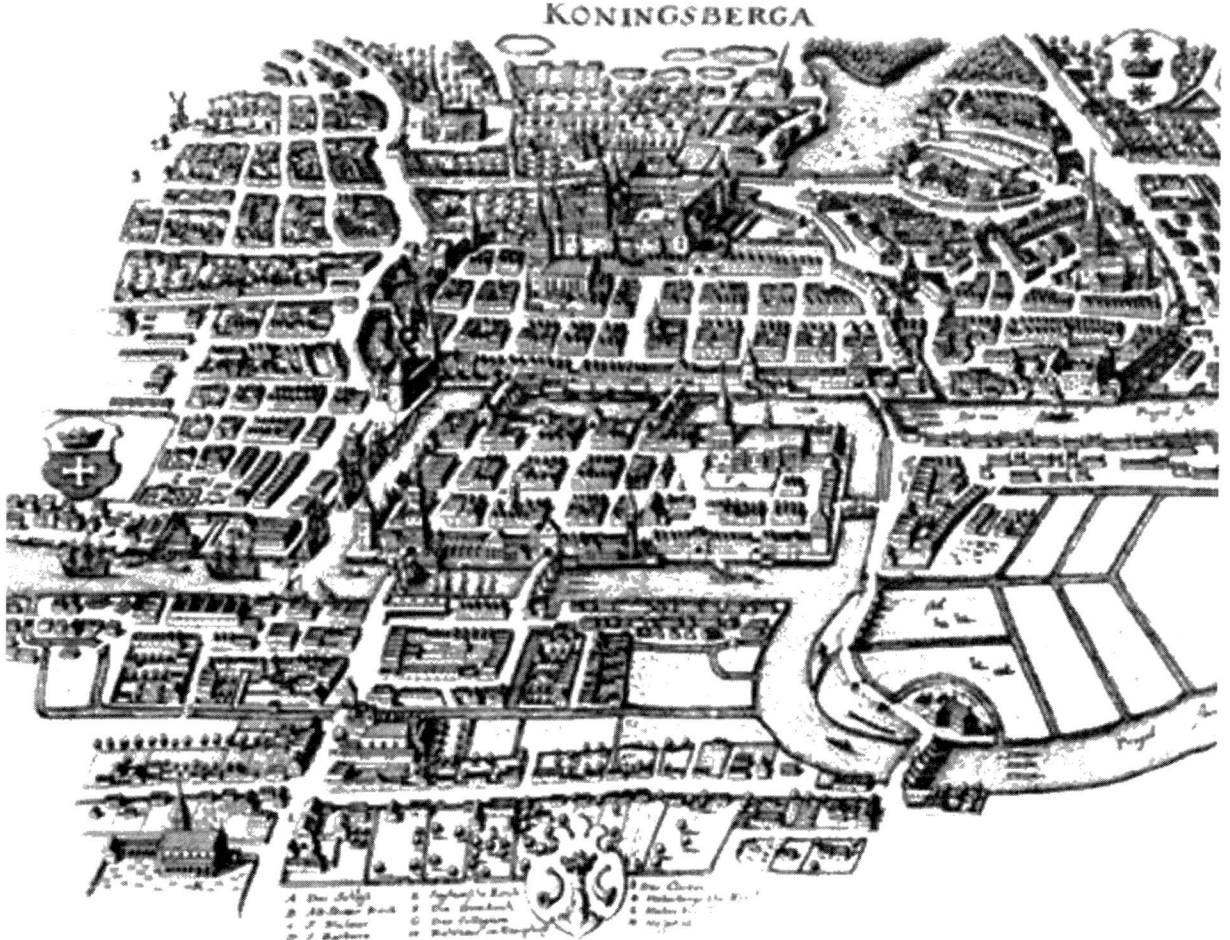

Figure 17.1.

1. In addition to that branch of geometry which is concerned with magnitudes, and which has always received the greatest attention, there is another branch, previously almost unknown, which Leibniz first mentioned, calling it the *geometry of position*. This branch is concerned only with the determination of position and its properties; it does not involve measurements, nor calculations made with them. It has not yet been satisfactorily determined what kind of problems are relevant to this geometry of position, or what methods should be used in solving them. Hence, when a problem was recently mentioned, which seemed geometrical but was so constructed that it did not require the measurement of distances, nor did calculation help at all, I had no doubt that it was concerned with the geometry of position — especially as its solution involved only position, and no calculation was of any use. I have therefore decided to give here the method which I have found for solving this kind of problem, as an example of the geometry of position.

The project places further emphasis on this evolutionary vision of mathematics in two ways. First, the project's introduction places Euler's paper in a historical context with a first-hand account[2] from Gottfried W. Leibniz (1646–1716) concerning why he felt the need for a 'geometry of position,' followed by a later account[3] of the state of this new field in 1833 from C. F. Gauss (1777–1855). Secondly, the project explicitly introduces the more abstract terminology and notation of modern graph theory in parallel with Euler's analysis, as illustrated in the following project excerpt.

2. The problem, which I am told is widely known, is as follows: in Königsberg in Prussia, there is an island A, called *the Kneiphof*; the river which surrounds it is divided into two branches, as can be seen in Fig. 17.2, and these branches are crossed by seven bridges, a, b, c, d, e, f and g. Concerning these bridges, it was asked whether anyone could arrange a route in such a way that he would cross each bridge once and only once. I was told that some people asserted that this was impossible, while others were in doubt: but nobody would actually assert that it could be done. From this, I have formulated the general problem: whatever be the arrangement and division of the river into branches, and however many bridges there be, can one find out whether or not it is possible to cross each bridge exactly once?

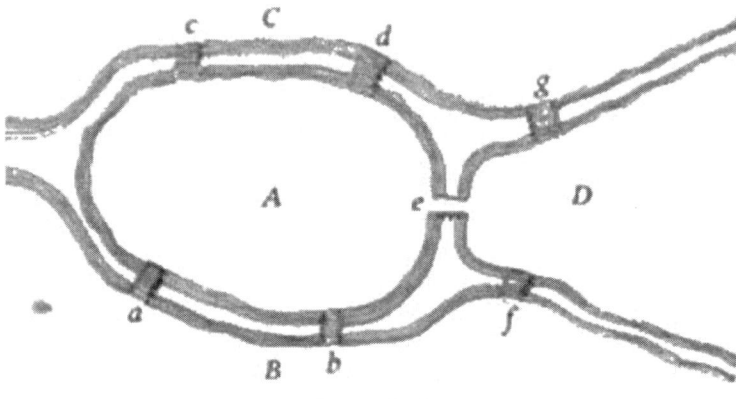

Figure 17.2.

Notice that Euler begins his analysis of the 'bridge crossing' problem by first replacing the map of the city by a simpler diagram showing only the main feature. In modern graph theory, we simplify this diagram even further to include only points (representing land masses) and line segments (representing bridges). These points and line segments are referred to as *vertices* (singular: vertex) and *edges* respectively. The collection of vertices and edges together with the relationships between them is called a *graph*. More precisely, a graph consists of a set of vertices and a set of edges, where each edge may be viewed as an ordered pair of two (usually distinct) vertices. In the case where an edge connects a vertex to itself, we refer to that edge as a *loop*.

Project Question 1. Sketch the diagram of a graph with 5 vertices and 8 edges to represent the bridge problem in Fig. 17.3.

Through this interweaving of primary source excerpts, project narrative and project questions, students are thus prompted to connect Euler's writing with the modern theory which (eventually) evolved out of it from the start of the project. Additional definitions are introduced and analyzed as they arise in connection with Euler's analysis of the bridge crossing problem. This strategy is also one way to promote the pedagogical goal of building students' understanding of the subject in its current form without requiring prior knowledge of its present-day paradigm to read the historical sources (Goal 4). In the *Arithmetical Triangle* project, this is done by referring students to a current

---

[2] This appeared in an 1670 letter from Leibniz to Christian Huygens (1629–1695), quoted in [4, p. 30],) as follows: 'I am not content with algebra, in that it yields neither the shortest proofs nor the most beautiful constructions of geometry. Consequently, in view of this, I consider that we need yet another kind of analysis, geometric or linear, which deals directly with position, as algebra deals with magnitude.'

[3] As quoted in [4, p. 30], Gauss reported 'Of the geometry of position, which Leibniz initiated and to which only two geometers, Euler and Vandermonde, have given a feeble glance, we know and possess, after a century and a half, very little more than nothing.'

17.3. Incorporating Pedagogical Design Goals: Two Sample Projects

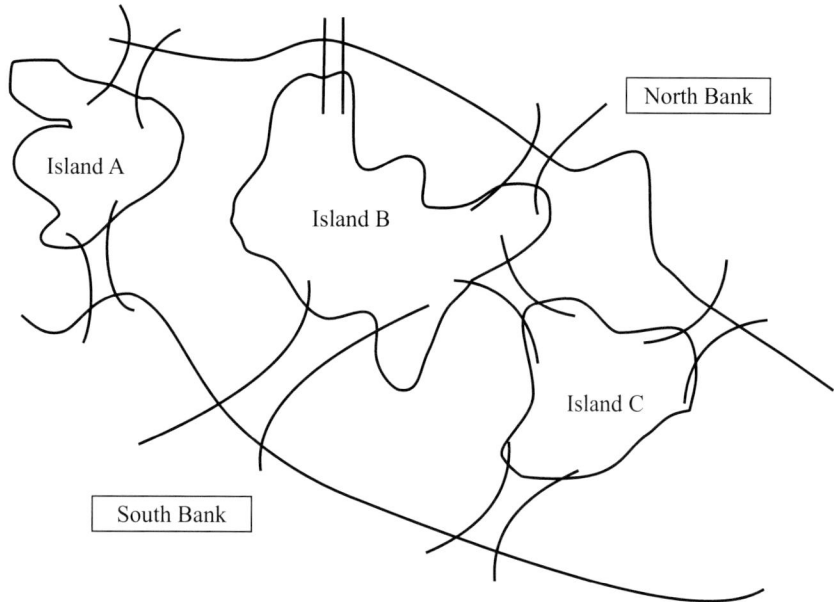

Figure 17.3.

textbook at appropriate points; other projects promote this goal through the use of concluding sections which connect the mathematical concepts developed in the primary source to its current terminology and notation.[4]

The strategy employed in the *Euler Circuits* project of developing the modern theory in parallel with its original formulation also enhances students' ability to more equally participate, regardless of their background or capability (Goal 7). This goal is further advanced by the primary source selections themselves. For example, Euler develops two procedures in his paper for determining whether a given bridge crossing problem admits of a solution, the first of which is quite distinct from the usual process presented in a current textbook for determining whether a given graph contains an Euler path.[5] Thus, even students who have studied graph theory in another course are exposed to an algorithm which is not only different from the one they are likely to have encountered previously, but also considerably more concrete. The concrete nature of Euler's preliminary algorithm has further advantages, including that of benefiting students with different learning styles through exposure to multiple approaches (Goal 8).

Euler's derivation of his first algorithm for determining if a bridge crossing problem is solvable (i.e., if a graph contains an Euler path) occupies several pages in his paper, following which he gives two examples of its use. He then states the following two observations which he will use to develop his second (simpler) procedure.

> 16. In this way it will be easy, even in the most complicated cases, to determine whether or not a journey can be made crossing each bridge once and once only. I shall, however, describe a much simpler method for determining this which is not difficult to derive from the present method, after I have first made a few preliminary observations. First, I observe that the numbers of bridges written next to the letters A, B, C, etc. together add up to twice the total number of bridges. The reason for this is that, in the calculation where every bridge leading to a given area is counted, each bridge is counted twice, once for each of the two areas which it joins.
>
> 17. It follows that the total of the numbers of bridges leading to each area must be an even number, since half of it is equal to the number of bridges. This is impossible if only one of these numbers is odd, or if three are odd, or five, and so on. Hence if some of the numbers of bridges attached to the letters A, B, C, etc. are odd, then there must be an even number of these. Thus, in the Königsberg problem, there were odd numbers attached to the letters A, B, C and D, as can be seen from Paragraph 14, and in the last example (in Paragraph 15), only two numbers were odd, namely those attached to D and E.

---

[4] See, for example, *Origins of Boolean Algebra in the Logic of Classes: George Boole, John Venn and C. S. Peirce*, author Janet Heine Barnett, available on the web resource [2].

[5] As we will see below, the second procedure which Euler develops and justifies is the more streamlined procedure of checking the parity of the vertex degrees which is standard in current textbook treatments.

As students work through Euler's various arguments, project questions such as the following also prompt them to pay attention to his style of argumentation, and to reflect upon how it differs from that used in a modern textbook, again emphasizing an evolutionary vision of mathematics while engendering cognitive dissonance through the comparison of a historical source with a modern textbook approach (Goals 13, 15).

> **Project Question 6.** The result described in Paragraph 16 is sometimes referred to as 'The Handshake Theorem,' based on the equivalent problem of counting the number of handshakes that occur during a social gathering at which every person present shakes hands with every other person present exactly once. A modern statement of the Handshake Theorem would be: *The sum of the degree of all vertices in a finite graph equals twice the number of edges in the graph.* Locate this theorem in a modern textbook, and comment on how the proof given there compares to Euler's discussion in paragraph 16.
>
> **Project Question 7.** The result described in Paragraph 17 can be re-stated as follows: *Every finite graph contains an even number of vertices with odd degree.* Locate this theorem in a modern textbook, and comment on how the proof given there compares to Euler's discussion in paragraph 17.

The theme of reflection on present-day proof standards (Goal 5) apparent in these questions is also raised in the introduction of the project, where the computer-assisted resolution of the Four Color Problem by Appel and Haken is discussed. The project's emphasis on this theme culminates with a set of final exercises in which students are asked to 'fill in the gaps' of a modern proof of Euler's final theorem. Students first read Euler's own proof of this theorem, which appears in the paragraphs leading up to its statement. The following project excerpt gives his statement of the theorem, along with portions of the associated project narrative and exercises that continue the theme of reflection on present-day standards, while also drawing students' attention to subtleties (*e.g.*, connectedness) which modern texts often take for granted (Goal 6).

> 20. So whatever arrangement may be proposed, one can easily determine whether or not a journey can be made, crossing each bridge once, by the following rules:
>
>> If there are more than two areas to which an odd number of bridges lead, then such a journey is impossible.
>>
>> If, however, the number of bridges is odd for exactly two areas, then the journey is possible if it starts in either of these areas.
>>
>> If, finally, there are no areas to which an odd number of bridges leads, then the required journey can be accomplished starting from any area.
>
> With these rules, the given problem can always be solved.

A complete modern statement of Euler's main result requires one final definition: a graph is said to be *connected* if for every pair of vertices u, v in the graph, there is a walk from u to v. Notice that a graph which is not connected will consist of several components, or subgraphs, each of which is connected. With this definition in hand, the main results of Euler's paper can be stated as follows:

Theorem *A finite graph G contains an Euler circuit if and only if G is connected and contains no vertices of odd degree.*

Corollary *A finite graph G contains an Euler path if and only if G is connected and contains at most two vertices of odd degree.*

> **Project Question 8.** Illustrate why the modern statement specifies that G is connected by giving an example of a disconnected graph which has vertices of even degree only and contains no Euler circuit. Explain how you know that your example contains no Euler circuit.
>
> **Project Question 9.** Comment on Euler's proof of this theorem and corollary as they appear in paragraphs 16–19. How convincing do you find his proof? Where and how does he make use of the assumption that the graph is connected in his proof?

The various excerpts we have examined from this project demonstrate another emphasized pedagogical goal: that of providing practice moving from verbal descriptions to precise mathematical formulations (Goal 2). In fact, Euler's entire paper provides students with a model for this, as he moves from a detailed map of Königsberg (Euler's Fig. 17.1), to a simpler diagram showing only the main features of the problem (Euler's Fig. 17.2), to a reformulation of the problem in terms of sequences of letters (vertices), to his first algorithm for determining if a solution exists, to the final theorem with which he concludes his paper. Again, project questions interspersed between the source excerpts in which Euler makes these moves provide students with opportunities to reflect upon these reformulations, and apply them in specific cases.

Finally, we note that a deliberate effort was made while designing this project to include questions focused on developing students' verbal and deductive skills through reading (Goal 1), by prompting them to interact with Euler's text in the way which a mature mathematical reader naturally approaches a new text. For example, returning to an earlier part of Euler's paper in which he is developing a procedure for determining how many times a vertex must appear in the representation of a route for a given bridge problem, the following project question prompts students to more deeply reflect on the passage they have just read by applying Euler's rule to a particular example; they are then able to immediately compare their conclusion concerning that example to Euler's own conclusion in the subsequent paragraph of his paper.

> **Project Question 3.** In paragraph 8, Euler deduces a rule for determining how many times a vertex must appear in the representation of the route for a given bridge problem for the case where an odd number of bridges leads to the land mass represented by that vertex. **Before reading further**, use this rule to determine how many times each of the vertices A, B, C, and D would appear in the representation of a route for the Königsberg Bridge Problem. Given Euler's earlier conclusion (paragraph 5) that a solution to this problem requires a sequence of 8 vertices, is such a sequence possible? Explain.

By incorporating such a diversity of question types—some aimed at developing students' skills in reading and proof writing, others aimed at comprehension of Euler's analysis and its relation to modern graph theory, and still others aimed at promoting reflection on present-day proof standards—this and other projects in the collection naturally lend themselves to the use of multiple approaches to their implementation which can be of particular benefit to students with different learning styles (Goal 8). For instance, the first part of this particular project (in which students are required to read and understand Euler's analysis of the problem) is well suited for small group and whole class discussion, while the final exercises (in which students complete a modern proof of Euler's main theorem) are ideally suited for individual practice in proof writing, but could also be completed in small groups. In the next section, we consider other issues related to classroom implementation of the projects.

17.4 Implementation

Time spent working on the project is time for explanation, exploration, and discovery, for both the instructor and the student. Instructors are encouraged to adapt a project to their particular course by adding or rephrasing some questions, or deleting others to reflect what is actually being covered.[6] Each project also has guiding notes for the instructor on its use in teaching.

Our own experience with implementation suggests that it is important to be familiar with all details of a project before implementing it with students. We have also found it important to begin early in the course with a discussion of the relevance of the historical piece, its relation to the course curriculum, and how modern textbook techniques owe their development to problems often posed centuries earlier. A comparison with modern techniques could begin as soon as the students have read the related historical passages, or be postponed until after the project is completed.

For use in the classroom, instructors should allow one to several weeks per project, depending on the project(s) selected. Some projects also work well when implemented in several parts spread over the course of the academic term. For certain course topics, the project can simply replace other course activities for a time, with the main course

---

[6]The source files for projects developed under the pilot grant can be downloaded and edited from the web resource [3]. Source files for new projects being developed under the expansion grant can be obtained by contacting their authors; contact information is located on the web resource [1].

topics learned directly through the project. Each assigned project should count for a significant portion of the course grade (about 20%) and may take the place of an in-class examination, or be assigned in pieces as homework.

While a project is being implemented, several class activities are possible. Students could be encouraged to work on the project in class, either individually or in small groups, as the instructor monitors and assists their progress and explores meaning in language from time past. Many instructors also assign select project questions for students to complete based on their own reading, prior to a discussion in class. Whole class discussions or brief lectures may also be appropriate at certain junctures.

Some type of student writing or presentation is recommended in conjunction with an historical project. Again, instructors have considerable flexibility in terms of how this is done. One option is to assign and collect written responses to project questions in installments, either before or after class work. Other instructors elect to assign a final detailed paper based on the project, perhaps asking students to compare how the topic is treated in the primary source to how it is treated in a current textbook. Another possibility is to require that each student complete an historical project related to the course curriculum on an independent basis, either individually or in groups, and report on their topic to the class in an oral presentation.

17.5 Conclusion

Evaluation data from the initial pilot study based on student performance in later course work suggests our approach to using history to teach mathematics is effective in promoting students' understanding of the present-day paradigm of the subject. Of course, there are other factors at play that could explain these data, including differing entering preparation for varying groups of students in different courses and semesters, and individual instructor experience and pedagogical style. Efforts to compensate for these confounding variables are part of the more extensive evaluation now underway with our expansion grant.

Following completion of a course using historical projects, students' own perceptions of the benefits of learning from primary sources echo our pedagogical goals. We close with a selection of their comments, allowing students to have the final word. "See how the concepts developed and understand the process." "It ties in better, links can be made." "Appropriate question building." "Helps with English-math conversion." "Gives meaning to problems." "You get an understanding of why you are doing something." "You understand it without the middle man." "It leads me to my own discoveries." "We learn from the best."

Acknowledgment Development of the materials described here has been partially supported by the National Science Foundation's *Course, Curriculum and Laboratory Improvement Program*, Grants DUE-0231113, DUE-0715392 and DUE-0717752. All opinions, findings and conclusions are those of the authors and do not necessarily reflect the views of the NSF.

Bibliography

[1] Barnett, J., G. Bezhanishvili, H. Leung, J.Lodder, D. Pengelley, and D. Ranjan, 2009, Historical Projects in Discrete Mathematics and Computer Science, in *Resources for Teaching Discrete Mathematics*, B. Hopkins (ed.), Washington, D.C.: Mathematical Association of America, MAA Notes volume 74.

[2] Barnett, J., G. Bezhanishvili, H. Leung, J.Lodder, D. Pengelley, I. Pivkina, and D. Ranjan, 2008, *Learning Discrete Mathematics and Computer Science via Primary Historical Sources*, www.cs.nmsu.edu/historical-projects/.

[3] Bezhanishvili, G., H. Leung, J. Lodder, D. Pengelley, and D. Ranjan, 2003, *Teaching Discrete Mathematics via Primary Historical Sources*, www.math.nmsu.edu/hist_projects/.

[4] Biggs, N., E. Lloyd, and R. Wilson, 1976, *Graph Theory: 1736–1936*, Oxford: Clarendon Press.

[5] Euler, L., 1758–59, Solutio Problematis ad Geometriam Situs Pertinentis, *Novi Commentarii Academiae Scientarium Imperialis Petropolitanque* 7, 9–28.

[6] Fauvel, J. and J. van Maanen, 2000, *History in mathematics education*, Dordrecht; Boston: Kluwer Academic Publishers.

[7] Katz, V., 1998, *A History of Mathematics: An Introduction*, Second Edition, New York: Addison-Wesley.

[8] Pascal, B., 1991, Treatise on the Arithmetical Triangle, in *Great Books of the Western World*, Mortimer Adler (ed.), Chicago: Encyclopædia Britannica, Inc.

About the Authors

Janet Heine Barnett holds a B.S. in Mathematics and Humanities from Colorado State University—Fort Collins, and an M.A. and Ph. D. in Set Theory from the University of Colorado—Boulder. She is a Professor of Mathematics at the Colorado State University—Pueblo where she has taught since 1990. A 1995–1996 fellow at the MAA Institute for History of Mathematics and Its Use in Teaching (funded by the NSF), her scholarly interests have long included mathematics history and its use both to promote mathematical understanding and as a vehicle for promoting teacher reflection on pedagogical issues. Through her current collaboration with faculty at New Mexico State University, she has developed seven original source projects for the teaching of discrete mathematics, including an extensive project on group theory based on papers by Lagrange, Cauchy, and Cayley which is now being developed into a full-length junior-level abstract algebra textbook. Other recent and on-going projects include a study of the historical relation of mathematics and war, a history of the hyperbolic functions in the eighteenth century, and the mathematical history of Paris (jointly with her dance and travel partner/husband George Heine).

Jerry Lodder has incorporated history into a variety of mathematics courses at both the undergraduate and graduate level. He is co-author of *Mathematical Masterpieces: Further Chronicles by the Explorers*, and a contributor to the volumes *From Calculus to Computers* (MAA Notes #68) as well as *Resources for Teaching Discrete Mathematics* (MAA Notes #74). He is currently involved in an NSF-supported project to teach discrete mathematics and computer science from historical sources with many of the historical modules posted at www.cs.nmsu.edu/historical-projects/. He received his Bachelor of Science degree from Wabash College and his Ph.D. from Stanford University.

David Pengelley's mathematical research is in algebraic topology and the history of mathematics, and he develops the pedagogies of teaching with primary historical sources, with student projects, and with student advance reading, writing, and active classroom work replacing lectures. He has developed a mathematics education graduate course on the role of history in teaching mathematics. He recently authored the articles "Dances between continuous and discrete: Euler's summation formula," "Did Euclid need the Euclidean algorithm to prove unique factorization?," "Voice ce que j'ai trouvé: Sophie Germain's grand plan to prove Fermat's Last Theorem," and the book *Mathematical Masterpieces: Further Chronicles by the Explorers*. David was the 2009 recipient of the Mathematical Association of America's national Haimo Award for Distinguished College or University Teaching of Mathematics.

Inna Pivkina is an associate professor of computer science at New Mexico State University (NMSU). Her research interests are in computer science education, knowledge representation and reasoning, logic programming, and data mining applications. Inna received an honors diploma in Mathematics from Novosibirsk State University in Russia, and a Ph.D. in Computer Science from the University of Kentucky where she was awarded a dissertation year fellowship. Dr. Pivkina was a co-PI on an NSF DUE CCLI Phase 2 grant "Collaborative Research: Learning Discrete Mathematics and Computer Science via Primary Historical Sources" and on an NSF BPC-DP grant "Linked Communities and Computing in Context: Empowering Southern New Mexico Women in Computing." She was a distinguished member of the NMSU Teaching Academy in 2003–2010 and a member of the NMSU Teaching Academy Advisory Board in 2004–2006. Inna served on the NMSU ADVANCE Faculty Development Committee in 2004–2009 and was the chair of the Committee in 2008–2009.

Desh Ranjan received his undergraduate degree in computer science from the Indian Institute of Technology, Kanpur (INDIA) (1987) and a Masters' and Ph.D. in Computer Science, with a minor in Mathematics, from Cornell University in Ithaca, New York in 1990 and 1992 respectively. After finishing his Ph.D., he spent one year (1992–93) as a post-doctoral fellow at the Max Planck Institute for Computer Science in Saarbrucken, Germany. He was a member of the faculty of the computer science department at New Mexico State University for 16 years and served as the chair of the department from 2004 to 2009. He was the founding director of the NSF-funded Center for Bioinformatics and Computational Biology at NMSU. He is currently an endowed Professor and the Chair of the Computer Science Department at Old Dominion University in Norfolk, Virginia. Desh's research interests include Algorithms, Bioinformatics and Computational Complexity. He is greatly interested in the use of original historical sources for teaching computer science and has devised projects to do the same.

18

History of Mathematics for Primary School Teacher Education Or: Can You Do *Something* Even if You Can't Do Much?

Bjørn Smestad
Oslo University College, Norway

18.1 Introduction

In this article, I will describe my context (Norwegian pre-service teacher education) and give some examples of different ways I work on history of mathematics. A major part of the paper will be spent discussing some of the materials I have used with students.

As a subtitle, I have chosen "Can you do *something* even if you can't do much?" In conferences in the HPM community, we get to see wonderful examples of how rich a resource the history of mathematics can be, but often I am left with the question "Will I have time to do this with my students?" A dedicated history of mathematics course would have been great for prospective mathematics teachers—but when they can't have that, what can they have?

18.2 Background

I teach a course in mathematics for prospective primary and lower secondary school teachers. The course lasts for two years, and is supposed to occupy a fourth of the students' time for that period. After doing this course, students are expected to be able to teach mathematics from grade 1 to 10 in the Norwegian school system—in itself an optimistic expectation.

There are certain important factors that have to be taken into account when planning such a course. When it comes to history of mathematics, students usually know very little in advance. The time is so limited that we can't dedicate time to an overview of the history of mathematics—everything we do on history of mathematics must be part of a broader treatment of mathematics. The mathematics we study is mostly at the level of lower secondary and below. Moreover, my students generally do not enjoy working in other languages than Norwegian. On the other hand, they are to become teachers, so they should be interested in anything that can enhance their teaching. (It could be noted that most of these factors are also present when teaching pupils in school.)

Nonetheless, I would like to include some history of mathematics, to try to reach the following goals: I want my students

- to see that problems that pupils face have often also been present in history (thus, facing problems does not mean that their pupils, or the students themselves, are stupid...)
- to get a general sense that mathematics has developed and give mathematics a human and cultural dimension
- to see different ways in which history of mathematics can be included in teaching (even as games!)
- to know that questions about the origin of mathematical words usually have an answer, often even an interesting one (etymology)

Even though my students do not get a course in history of mathematics, I want them to get a taste of history of mathematics and a wish to learn more.

Previous studies that I've done give me two important insights:

1. History of mathematics easily becomes just biography when prepared for the classroom. This is shown both in my analysis of Norwegian textbooks [13] and in my analysis of 638 mathematics lessons in 7 countries from the TIMSS Video Study [14]. Therefore, it is important for me to work on "real mathematics," not just to give anecdotes or biographies.

2. History of mathematics is sometimes seen as "taking time away from the mathematics" [16]. Therefore, it is important to show the prospective teachers how history of mathematics can add value to the mathematics teaching, also from a purely mathematical point of view[1].

I should add that I have no ambition of being original—except in a purely local sense. I am happy to pick up ideas from conferences and articles to enrich my teaching, as long as my students have not seen the material before. Therefore, many of the examples in this article may be familiar.

18.3 Ways of Working with History of Mathematics

In my teaching, I have included history of mathematics in several different ways. I have included historical information in lectures, I have been working on original sources, my students have done projects (in which the students have included the history of mathematics in activities for pupils), I have given my students tasks from history and I have also created an etymology game. I will give examples of all of these, but mainly, we will look at tasks I've given my students.

There are, of course, several other possibilities which I have not explored, for instance having historical/mathematical plays [10] or historical/mathematical exhibitions [1]. Many possibilities are also described in the ICMI Study [8].

18.3.1 As part of a lecture

I mention history of mathematics in many of my lectures, and spend some time on Al-Khwārizmī (see [9]), on the history of measurements, on Platonic solids, on Eratosthenes' calculation of the circumference of the Earth, on equations (for instance Tartaglia and Abel) and on Florence Nightingale and the use of statistics. However, here I will give an example from a lecture on the history of perspective drawing. In Norway, perspective drawing is a part of the mathematics curriculum in both primary and secondary school.

First, I show an example of Egyptian art, for instance from the 4000 year old tomb of Khnumhotep in Beni Hasan. The students will notice that although some parts of the painting are naturalistic, other parts are not—and it is clear that it was never the intention of the artist to depict the world exactly as it is. Two men have very different height, even though they are standing in the same boat. Obviously, this is not because of the artist's lack of skill, but because the artist wanted to show that one of the men was more important than the other. I show this example to make the students aware that it would be misguided to judge paintings based on what we think is "right" or "wrong."

Then I go on to a few other paintings, for instance "The Little Garden of Paradise" by an unknown artist ("Upper Rhenish Master"). This painting, which is in the Städel Museum in Frankfurt, was painted about 1410. By this time, painters were able to paint naturalistically. When asked if there is something odd about the painting, the students immediately say that there is something a bit "wrong" with the table, or with the well or with the walls. The table, for instance, is seen slightly from above, while the rest (including the glass on the table) is seen more from the side. It may be that the painter wanted this effect, but it is still a fact that perspective drawing had not been developed.

---
[1] See Siu [11] for more reasons *not* to use history of mathematics in the classroom.

18.3. Ways of Working with History of Mathematics

Figure 18.1. From the tomb of Khnumhotep

Figure 18.2. Upper Rhenish Master: The Little Garden of Paradise

Then, of course, we look at a few paintings which are "perfect," for instance Raphael's "The School of Athens" (1509–10) in The Vatican Museum. By this time, those of the students who don't know (or remember) how to draw in perspective, are intrigued. Therefore, I give them a copy of a painting (or a simpler perspective drawing), and ask them to figure out what is going on—which geometrical properties are the same in the drawing as in the real world it portrays, and which are not? This leads to a discussion on concepts such as lines, parallels, angles and so on. Even if the students already know the laws of perspective, they will probably see that this is a possible way of introducing it to pupils in school—to let the pupils "discover" the rules.

After looking at the rules a bit closer, seeing a few more examples and doing a few drawings, we go on to looking at some examples of later art. Picasso rejected the single viewpoint, and instead painted objects as seen from several points of view at the same time. Escher played with perspective to create "impossible" drawings. My favorite, however, is Max Beckmann's "Synagogue" from 1919 (which is also in the Städel Museum). Max Beckmann knew very well how to draw a building in perfect perspective. The whole point was, however, that after the terrible war, nothing was perfect anymore. The painter consciously breaks the rules to get the effect he wants. These examples illustrate another important point: to draw in perspective, you need certain skills. Having these skills, however, does not mean that you have to use them. The skills give you more choice, also to create new effects.

What is the point of including the history when teaching perspective? In my opinion, there are many points. For instance, the students see that this particular part of mathematics has developed over time as people met artistic and mathematical problems that they needed to solve. They also see that there are important connections between mathematics and arts. This last point may be particularly important for some of the pupils or students who feel that

Figure 18.3. The School of Athens

Figure 18.4. Max Beckmann: Synagogue

mathematics is too "sterile." Moreover, the pre-perspective paintings give a motivation to learn how to avoid those "mistakes"—to become "better than these ancient masters."

18.3.2 Working on original sources

As mentioned earlier, my students prefer texts in Norwegian, therefore there are not many authentic original sources to choose from. Moreover, working on original sources tends to take more time than we have available (even though I acknowledge the great value in doing it)[2]. Consequently, I don't do this very often. The original source I use most often is the beginning of Leonardo Pisano (Fibonacci)'s "Liber abaci" (1202). The text shows how Fibonacci had to explain the Hindu-Arabic numerals to great length to make them understandable to his contemporary public. Students tend to have a feeling that numerals and number systems are simple, perhaps because they have understood them for such a long time. Working on Fibonacci makes the students see that these topics indeed are difficult—teachers should not be surprised that pupils have to struggle to understand them. Moreover, students see how mathematics has developed. Fibonacci also gives a good opportunity (among many) to illustrate the central role of non-European cultures in the development of the mathematics we do in school.

I need to give a little warning on translations, which I will also come back to later: I give my students a translation which I have done from an English translation of the Latin original [12]. This is certainly not optimal, but the alternative for me is to wait for some scholar to do the translation into Norwegian. In this particular context, I think the main points will be kept in the translation, and that the outcome for the students is better than if I did not use Fibonacci at all.

18.3.3 Projects

I will not say much about projects. That is not because projects are not valuable—they are—but because I find them to be difficult in my context. At times, my students have done wonderful projects where they have connected the history of mathematics to teaching in imaginative ways, but it does take quite a lot of time and tends to involve colleagues in parallel classes and preferably also my students' pupils in schools. So in the spirit of the subtitle ("Can you do *something* even if you can't do much?"), projects may not be the place to start. Moreover, I have discussed my use of projects more fully in another paper [15].

18.3.4 Tasks

Most of the rest of this article will be devoted to tasks given to students as part of their everyday task work. The tasks all have mathematical content, which means that they can not be seen as "taking time away from the mathematics," and they give the students an opportunity to discuss the problems in groups. Here are some selected examples. Please note that the tasks have been translated from Norwegian. This may have added inaccuracies compared to the worksheets actually used with students.

Geometry and medieval lawyers Before this activity, students have been given a translation of Jan van Maanen's article "Teaching geometry to 11-year-old 'medieval lawyers'" [20], in which van Maanen describes pupils working on a juridical document from 1355. In this document, division of new land (for instance formed by alluvial deposits) is discussed, and the following general principle is established: New land belongs to the owner of the nearest old land. The article goes on to explain that the borders can be found by bisecting angles, and gives a few examples of tasks (see Figure 18.5).

Comment: This well-known task is good for several reasons. Many Norwegian pupils learn to bisect angles mechanically, as a procedure, without ever discussing what the properties of the points on the line that bisects the angle are. By working on this kind of problem, the properties are in focus. Problems such as these also show that even this kind of geometry (ruler and compass constructions) is useful, and they present possibilities for connections to other subjects. Even discussions on what is a "fair" division can be included.

---

[2] See Bekken et al [2] and Clark and Glaubitz [7] for more on using original sources in teaching.

18.3. Ways of Working with History of Mathematics

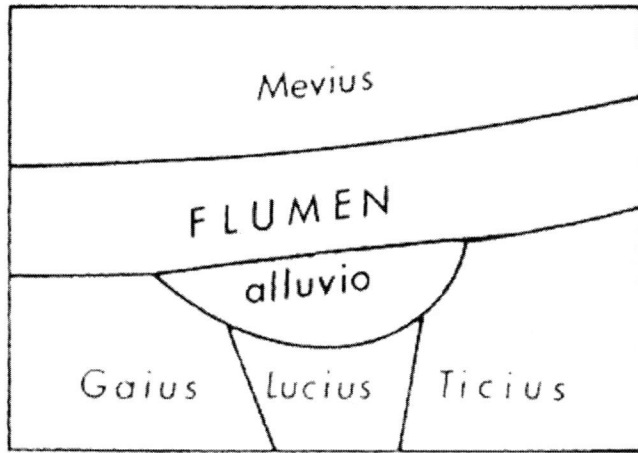

Figure 18.5. This figure is taken from the article by van Maanen. Mark the new borders in the new land (marked as "alluvio").

I also give a non-historical task about five farms scattered on an island, where the perpendicular bisector is useful for dividing the island between the farms in a "fair" way (by one definition of the word fair). Here, however, there is no historical component, and I see signs that the students find the problem a bit too artificial.

Algorithms for multiplication through time

Task: On this page are different algorithms for multiplication. For each algorithm, I want you to try to understand the procedure. Use the same algorithm to calculate 265 · 38. Try to understand why the algorithm gives the right answer, and what may be the advantages and disadvantages of the algorithm.

a) "Gelosia method": This method is found in *Lilavati* (by the Indian mathematician Bhaskara, who lived around 1150). The method came to Europe via Arab manuscripts, and was found in printed textbooks until the 1700s [17]. Here, 183 · 49 is calculated:

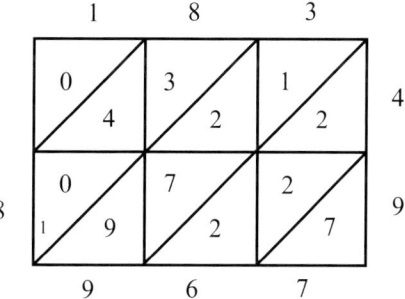

b) "Russian Peasant Multiplication": The method is called "Russian peasant multiplication" because it was used in rural communities in Russia all the way into modern times. But this is essentially the same method as the Egyptians used four thousand years ago [4]. Here, 183 · 49 is calculated:

Halves	Doubles
49	183
~~24~~	~~366~~
~~12~~	~~732~~
~~6~~	~~1464~~
3	2928
1	5856
	8967

c) Here is a third method, used by Eutocius of Ascalon (ca. 500 AD) [5]. Here, $534 \cdot 3$ is calculated:

$$
\begin{array}{rrr}
500 & 30 & \\
\times\ 3 & \times\ 3 & 4 \\
\hline
1500 & 90 & \times\ 3 \\
& & \hline \\
& & 12
\end{array}
$$

$$1590$$

$$1602$$

Comment: Many students enter my mathematics course with an opinion that there is only one way of doing multiplication—the algorithm they have been taught in school. It is essential that they learn to appreciate other methods and to understand the unfamiliar, as they will later meet pupils who are doing things in their own ways—either through their own invention or through education elsewhere. The first and the third algorithm also help us see more clearly why "our" algorithm is working. The second algorithm is interesting because it is not "basically the same" as ours. The algorithms also show how multiplication has been done in other cultures, and make it possible for us to discuss what characteristics of our mathematical culture makes our algorithm a good one for us (For Egyptians, doubling and division by two were the basic operations, which meant that their algorithm was good for them.)[3]

Problems from probability theory For the work on probability theory, I have written a booklet which includes historical notes and problems. I have chosen some problems from this booklet for discussion here. I chose to skip the "problem of points," which has been discussed in detail by Chorlay and Brin [6]. Many of the examples in this part are from the classic book by Isaac Todhunter [19].

Task 4–7:

"Supposing a tree fell down, Pooh, when we were underneath it?," Piglet asked. "Supposing it didn't," said Winnie-the-Pooh.

a) Is there anything so improbable that we don't worry about it, even though the probability is not equal to zero?

b) In his 1777 *Essai d'Arithmétique Morale*, Buffon argues that probabilities less than 1/10000 cannot be distinguished from a probability of 0. He argues that the probability that a 56 year old man will die in the course of one day (according to his tables) is about 1/10000, while such a man in reality regards the probability as 0. (A similar argument was given by d'Alembert earlier.) What do you think of such reasoning?

Task 4–12 (The St. Petersburg problem): A classical problem from probability theory is the following:

a) A throws a coin. If a head turns up on the first throw he gets one ducat from B, if a head does not turn up until the second throw, he gets 2 ducats from B, if a head does not turn up until the third throw, he gets 4 ducats from B and so on (getting 2^{n-1} ducats if a head doesn't turn up until the nth throw). Calculate the expected value for A.

(The problem is called the St. Petersburg problem simply because it was first published in St. Petersburg—by Daniel Bernoulli. It is, however, Nicolas Bernoulli who is credited for first posing the problem, in 1713.)

b) How much should A be willing to pay to play this game? Would you be willing to pay that amount?

c) What will the expected value be if we assume that B has only a limited sum to give to A—for instance 10.000 ducats?

d) Some of the "problem" with this problem is probably that we would rather not pay a very large amount of money to have a tiny chance of winning an incredible amount of money. Some (for instance Cramer) have tried to cut the knot by saying that for all practical purposes, winning ten million ducats is not ten times as valuable as winning one million ducats—I won't get that much happier by the additional nine million ducats. Do you see why this "cuts the knot"?

---

[3] For a pedagogically interesting explanation of Russian Peasant Multiplication, see mathforum.org/dr.math/faq/faq.peasant.html.

e) Others (for instance d'Alembert) have cut the same knot by saying that probabilities less than for instance 1/10000 can just as well be regarded as 0. Do you see how this "cuts the knot"?

f) Still others (for instance Buffon) argued that there are limits to how many throws you have the time for in the span of one life—therefore the number of throws must be limited. Do you see how this "cuts the knot"?

g) Why have so many good mathematicians used so much energy on "explaining away" the result in a), do you think?

Comment: Students often have an intuitive feeling that small probabilities are unimportant, and that probability theory is about deciding if something is "probable" or "not probable." These two problems show that small probabilities may be very important. Small probabilities are also part of the reason for the opposition to nuclear energy, for instance. However, the St. Petersburg problem also shows that there may at times be a gap between the theoretical and the practical (because of the formulation of the problem), and this gap, which leads to counterintuitive results, needs to be bridged. We see how mathematicians struggled to bridge it. More mathematically: when students really understand why each of the modifications of the original problem leads to a finite answer, they have surely understood important parts of the concept of expected value.

Task 2–22: In 1888, Joseph Bertrand published a paradox, the so-called "chord paradox" [3]. The question is simple: You have a circle of radius 1 cm, and choose a random chord. What is the probability that this chord has a length greater than $\sqrt{3}$ cm? Please try to answer before reading the following three alternative answers:

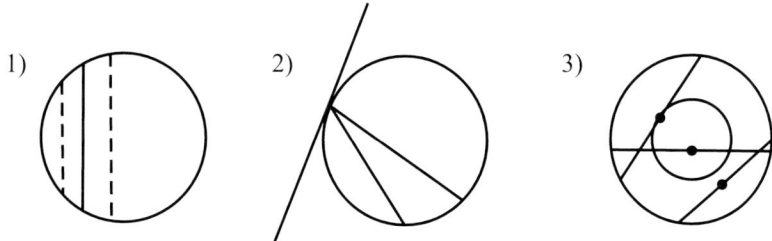

1) Because of symmetry, we can assume that the chord has a particular direction, for instance that it is vertical. With a little calculation, we see that only the chords which are less than 1/2 cm from the centre of the circle, have a length greater than $\sqrt{3}$ cm, while the ones that are more than 1/2 cm from the centre, will have a shorter length. Therefore, half of the distances give the length we want, so the probability is 1/2.

2) Because of symmetry, we can choose a point on the circle which the chord should touch. The question is then only which angle the chord should have to the tangent of the circle. A little calculation shows that only the chords which have angles greater than 60 degrees to the tangent, has a length greater than $\sqrt{3}$ cm. Out of 180 degrees, there is only a sector of 60 degrees that gives the length we want, so the probability is 1/3.

3) To choose a chord randomly is equivalent to choosing the midpoint M on the chord. The chord will only get a length of $\sqrt{3}$ cm (or more) if M is inside a circle with radius 1/2. This circle has only a fourth the area of the bigger circle. Therefore, the probability is 1/4.

Which of these alternatives is correct? (Or may all of them be correct?)

Comment: This problem is important in showing that "randomness" is not an easy concept—it must sometimes be carefully defined. It is not always obvious what it means to be "picked randomly."

Task 4–5 In the saga of Olaf the Holy, chapter 97, the following story is told: "Thorstein Frode relates of this meeting, that there was an inhabited district in Hising which had sometimes belonged to Norway, and sometimes to Gautland. The kings came to the agreement between themselves that they would cast lots by the dice to determine who should have this property, and that he who threw the highest should have the district (see Figure **??**). The Swedish king threw two sixes, and said King Olaf of Norway need scarcely throw. He replied, while shaking the dice in his hand,

Figure 18.6. Drawing by Erik Werenskiold

"Although there be two sixes on the dice, it would be easy, sire, for God Almighty to let them turn up in my favour." Then he threw, and had sixes also. Now the Swedish king threw again, and had again two sixes. Olaf king of Norway then threw, and had six upon one dice, and the other split in two, so as to make seven eyes in all upon it; and the district was adjudged to the king of Norway. We have heard nothing else of any interest that took place at this meeting; and the kings separated the dearest of friends with each other." [18]

a) What is the probability of getting a double six three times in a row, as is related here?

b) Do you think that the kings wanted chance to decide, or did they see some other significance in the throws of dice?

Comment: A basic assumption in our work on probability in school is that some events are random—such as throwing dice. Not everybody agrees in this. The outcome of a throw may be ascribed to gods or to what is perceived as "fair." (Even today, we talk of "fair dice," even though the outcomes often seem unfair to the loser.) Discussing such competing assumptions will be important for the students when they become teachers, so they should be made aware of them. Of course, Olaf the Holy's saga has the added benefit of being available in Norwegian.

18.3.5 A game

Teacher education (almost) always serves two purposes: to improve the students' content knowledge and to give insight into how teaching can be done. I try to provide my students with examples of different ways of including history of mathematics, even by making an etymology game. The game is quite simple, and the main point is that the player is given part of the etymology of a word, and is to guess which word it is. The reactions of students have been interesting: some students complain that they can't do it, because we have never studied etymologies, while others get fascinated and try to reason. I have no ambition that my students will learn the etymology of lots of words during my course, but I want them to remember that every word has a background, and many of these backgrounds cast light upon the concept—sometimes from a surprising angle. A teacher should not be uninterested in the origins of the words he or she is teaching.

For instance, students are interested to see that the word "trigonometry," which they mostly associate with abstract functions, comes from Greek words for "triangle" and "measure." The word "interval" comes from Latin and means something like "a place between the walls"—that is immediately understandable. That "asymptote" has its root in something meaning "not to meet" is also quite reasonable.

18.4 Discussion

When this paper was presented at the ESU5 in Prague, it was pointed out to me that one of the etymologies given in the game was wrong. That error is regrettable. But it also points back to the subtitle of this paper. For me as a teacher to make an etymology game, I have to rely on sources such as etymological dictionaries, which themselves have errors. For me to research every word would make the process too time-consuming. The same goes for almost everything else I do—when including history of mathematics in my teaching, I often do not go back to the primary sources, but instead rely on secondary sources (although preferably not only one). I think that trying to include history of mathematics, obviously with as few errors as possible, is better than not including it at all. When teachers show such attempts to each other, the materials will both be used more widely and, we might hope, corrected.

My answer to the question in the subtitle ("Can you do *something* even if you can't do much?"), is "yes." I do believe that it is possible to include history of mathematics that may light the interest of some students, even if you don't have the opportunity of doing everything you would have wanted. And it is possible to create resources based on others' ideas, removing the need to research everything from scratch. I have tried to show some of my work, and hope that the ideas that are available will multiply in the years to come, so that still more teachers who are not experts in the history of mathematics will feel able to enrich their teaching in this way.

Bibliography

[1] Abdounur, O.J., 2007, An Exhibition As a Tool to Approach Didactical and Historical Aspects of the Relationship Between Mathematics and Music. E. Barbin, N.a. Stehlíková, C. Tzanakis, eds. *European Summer University 5*. Vydavatelský servis, Prague, Czech Republic, 409.

[2] Bekken, O., E. Barbin, A. El Idrissi, F. Métin, and R. Stein, 2004, Panel: Original Sources in the Classroom. F. Furinghetti, S. Kaijser, C. Tzanakis, eds. *HPM2004 & ESU4*. Uppsala Universitet, Uppsala, Sweden, 185–191.

[3] Bekken, O. and H. Dahl, 1980, *Sannsynlighetsregning med litt statistikk*. Agder Distriktshøgskole, Kristiansand.

[4] Bunt, L.N.H., P.S. Jones, and J.D. Bedient, 1988, *The historical roots of elementary mathematics*. Dover publ., N.Y.

[5] Cajori, F., 1991, *A history of mathematics*. Chelsea Publ. Co., New York, N.Y.

[6] Chorlay, R. and P. Brin, 2007, Using Historical Texts in the Classroom: Examples in Statistics and Probability. E. Barbin, N.a. Stehlíková, C. Tzanakis, eds. *European Summer University 5*. Vydavatelský servis, Prague, Czech Republic, 193.

[7] Clark, K. and M.R. Glaubitz, eds., 2005, *Mini-Workshop on Studying Original Sources in Mathematics Education*, Oberwolfach.

[8] Fauvel, J. and J. Van Maanen, 2000, *History in mathematics education*. Kluwer Academic Publishers, Dordrecht.

[9] Glaubitz, M.R., 2007, The Use of Original Sources in the Classroom: Theoretical Perspectives and Empirical Evidence. E. Barbin, N.a. Stehlíková, C. Tzanakis, eds. *European Summer University 5*. Vydavatelský servis, Prague, Czech Republic, 373–381.

[10] Gonulates, F., 2007, Mathematics Theater, Mathematics on Stage. E. Barbin, N.a. Stehlíková, C. Tzanakis, eds. *European Summer University 5*. Vydavatelský servis, Prague, Czech Republic, 411–412.

[11] Siu, M.-K., 2004, "No, I do not use history of mathematics in my class. Why?" *HPM 2004*. Uppsala Universitet, Uppsala, Sweden.

[12] Smestad, B., 2000, Leonardo Pisanos Liber abbaci. *Tangenten*(2) 15–16.

[13] ——, 2003, Historical topics in Norwegian textbooks. O. Bekken, R. Mosvold, eds. *Study the Masters: The Abel-Fauvel Conference*. NCM, Kristiansand, 153–168.

[14] ——, 2004, History of mathematics in the TIMSS 1999 Video Study *HPM2004 & ESU5*, Uppsala, Sweden.

[15] ——, 2008a, A look at three years of projects with students *The International Conference on Mathematics Education*, Monterrey, Mexico.

[16] ——, 2008b, Teachers' conceptions of history of mathematics *HPM 2008*, Mexico City, Mexico.

[17] Smith, D.E., 1958, *History of mathematics*. Dover, New York.

[18] Sturlason, S., Heimskringla, Online Medieval and Classical Library.

[19] Todhunter, I., 1865, *A history of the mathematical theory of probability from the time of Pascal to that of Laplace*. Chelsea Publ. Co.

[20] van Maanen, J., 1992, Teaching geometry to 11 year old "mediaeval lawyers." *Mathematical Gazette* 76 (475), 37–45.

About the Author

Bjørn Smestad is a teacher educator at Oslo University College in Norway. He wrote his cand. scient. thesis on British attempts to develop a solid foundation for Newton's theory of fluxions. After becoming a teacher educator in 1998, his main research interests have been history of mathematics in mathematics education as well as technology in mathematics education. In particular, he has been interested in what place history of mathematics currently has in Norwegian schools. He has been a regular participant in and contributor to the ESU and HPM conferences since 2000, and he has been editor of the HPM Newsletter since 2004.

19

Reflections and Revision: Evolving Conceptions of a *Using History* Course

Kathleen Clark
Florida State University

19.1 Introduction

As with the construction of any secondary mathematics education course, a course on the history of mathematics for teaching can assume many different forms. For example, if the secondary mathematics education major resides in a Department of Mathematics, the course may tend to be more of a pure mathematics course instead of one with explicit attention to pedagogical ideas. Alternatively, if the course is a College of Education offering, it may shed some of its strict mathematical content and concentrate more on biographical, anecdotal, or pedagogical information. In recent years, what constitutes a history of mathematics course has become the subject of discussion for different audiences focused on undergraduate mathematics teaching [6]. Given the professional discussion that takes place about the content of history of mathematics courses in general, I conducted a study to investigate undergraduate mathematics education students' learning in the course, *Using History in the Teaching of Mathematics* (or, *Using History*), over four semesters. A natural consequence of the research has been to reflect on subsequent offerings of the course in order to revise topics and assignments for the purpose of fulfilling course objectives. Each of the objectives was designed to create opportunities for pre-service mathematics teachers (PSMTs) to consider using the history of mathematics in their future teaching.

As part of a broader inquiry, I began with the following research questions:

(1) In what ways does the study of the history of mathematics impact PSMTs' mathematical, historical, and pedagogical knowledge?

(2) What do PSMTs report as being significant to their engagement with and influential on their learning of the history of mathematics?

(3) What kinds of learning experiences are most promising for increasing critical knowledge (mathematical, historical, and pedagogical) of PSMTs?

The presentation given at the *5th European Summer University on the History & Epistemology in Mathematics Education* (ESU5) focused primarily on the first research question, in an effort to investigate and understand the impact of prescribed experiences that call for PSMTs to obtain or demonstrate historical knowledge of the topics they will be called upon to teach [1, 4]. Data collected subsequent to ESU5, however, revealed that PSMTs, given appropriate guidance, are able to explore their own ideas for how to present school mathematics topics using historical perspectives. Given less guidance however, PSMTs are more apt to consider *Using History* only for anecdotal additions in their teaching, as opposed to capitalizing on historical problems and methods of solution.

19.1.1 Perspectives for consideration

Both my own perspective about how prospective teachers will realistically consider the use of history of mathematics in teaching and the PSMTs' perspectives on why the use of history may be beneficial were made explicit before the start of each semester of *Using History*. Many students, when reflecting in their journal about taking the required course, stated that they did not understand why they needed to take such a course and more strongly, they asked why one would ever need to include the history in their teaching. One student shared the following:

> I heard something interesting in one of my classes today. I heard about a teacher in a local school who doesn't understand the proper placement of history in a math class. When he attempted to teach his students the Pythagorean Theorem, he first introduced them to Greece, then to Pythagoras, and on and on until he had completely lost his students. I don't think this is the place of history of math in the classroom. In the same sense, I don't know that I've placed it in the right place either. (That is, at the end of a class to catch the last of students' attention.) I think that I am really lost as to its real roots. (Sharon[1], Fall 2006).

So, even though Sharon was engaged with and positive toward studying the history of mathematics, she struggled with finding its "proper placement in a math class." Other students, however, were not so sure of the need to study the history of mathematics in the first place—either from their own perspective or that of their future students:

> Before taking this class, I did not understand why I needed to learn the history of mathematics for teaching. Do not get me wrong, I was very interested in it; however, I just did not know how it would make me a better teacher. (Kristie, Spring 2007)

> Upon entering this course, I had a hard time understanding how incorporating history into a math class is necessary for a student's education. (James, Spring 2007)

With the knowledge that PSMTs do not understand (or, in many ways appreciate) the requirement of a *Using History* course, I approached the course with the hypothesis that if they experience the benefit of learning mathematics through the study of the history of mathematics, prospective teachers can envision the use of an historical perspective in their future teaching. In planning the course, *Using History*, I designed activities and tasks that I hoped would provide prospective teachers with learning mathematics in ways that would motivate them to plan for the use of history in teaching.

19.2 Course Context

Using History is a required mathematics education course for all prospective middle grades (students aged 10-13) and high school (students aged 14-18) teachers in the secondary mathematics education program at Florida State University. In addition to *Using History*, the pedagogical preparation includes courses in using technology, how adolescents learn mathematics, instructional methods, classroom management and planning, and student teaching. In the last decade, *Using History* has most often been delivered in one of two formats. Most recently, the course has been conducted as more of a mathematics course, with an emphasis on the mathematical contributions of more prominent mathematicians (i.e., Archimedes, Euler, Pascal). Prior to this manifestation, the course included a combination of mathematics content with a culminating course project in which students developed or located a collection of classroom activities containing some historical significance. It is not clear (due to lack of institutional records), however, to what extent students either participated in or had modeled for them the various ways to engage in the study of the history of mathematics both for personal understanding of mathematics and potential instructional practice.

---

[1] All student names are pseudonyms.

The mathematical preparation of the students enrolled in *Using History* is a student contextual feature worth noting. Formerly, secondary mathematics education majors at Florida State University did not complete the same mathematics courses as mathematics majors—unlike other secondary mathematics education programs within the state. Instead, prospective middle grades teachers complete up through Calculus I and take three mathematics courses in the College of Education (courses in algebra, geometry, and problem solving). Students preparing to teach high school must complete through Calculus II, take four prescribed courses beyond the calculus requirement (Applied Linear Algebra, Modern Algebra, College Geometry, and an elective with Calculus II as the prerequisite), in addition to the three College of Education mathematics courses. Prospective middle grades teachers represented approximately one-third of the *Using History* enrollment each semester, creating a diversity of mathematical preparedness among the students taking the course. Indeed, each semester, one-half of the students pursuing middle grades mathematics certification claimed they did so because the undergraduate mathematics courses required for high school mathematics certification were too difficult. Furthermore, the variability of student experience with mathematics content courses may have impacted student participation in *Using History*, particularly with respect to completion of the capstone assignment in the course.

19.2.1 Course goals and foci

The goals and foci of the current course were developed from the philosophy that, "the beauty of the study of the history of mathematics is that it can give a sense of place…from which to learn mathematics, rather than merely acquiring a set of disembodied concepts" [5, p. 14]. The goals of the course ask students to engage in the study of the history of topics that prospective mathematics teachers are expected to teach in the content areas of number, algebra, geometry, precalculus, and calculus and to consider alternative perspectives when teaching mathematics. In addition, a significant aspect of the course is to provide students opportunities to gain ability in identifying and creating appropriate resources for the purpose of integrating an historical perspective in teaching mathematics. The three course foci include (1) working with mathematical ideas that evolved over time; (2) studying and discussing the historical and cultural influences on and because of the mathematics being developed; and (3) developing the pedagogical knowledge needed to integrate an historical perspective in the teaching of school mathematics.

To address the course foci, *Using History* was designed as a 15-week course that included the study of the history of mathematical ideas at the middle and high school level. (Please see Table 19.1 for a sample schedule of topics.) Prior to each class session students completed assigned reading, with readings intended to focus on two aspects to support the acquisition of new knowledge: historical and biographical excerpts to properly situate the topic and primary sources (or their English translations) to illuminate a key mathematical concept.

During class sessions students then examined the mathematical concept from the primary source in detail. Class tasks required students to engage in the mathematics (from a close historical orientation) and to consider the pedagogical implications of their historical investigations.

19.3 The Capstone Project: The First Three Semesters

The culminating task in *Using History* gave students the opportunity to create an instructional unit (Fall 2006) or lesson (Spring 2007 & Fall 2007) based upon new knowledge and skills gained as a result of engaging in the course foci, historical content, and course tasks.

19.3.1 First iteration of the course: The Teaching Unit Assignment

For the first semester of the course I planned for students to draw upon the examples of content, tasks, resources, and readings to create a teaching unit that could be used in a middle or high school classroom. The *Teaching Unit Assignment* was composed of several parts, including (1) a brief history of the topic selected; (2) the student's mathematical interpretation of the topic; (3) a scope and sequence of the unit they designed; (4) lesson plans, accompanying activities, and necessary materials; (5) a rationale for why history was infused in the lessons selected from within the scope and sequence; and (6) a bibliography containing at least 12 resources, several of which were required (*e.g.*, the *Dictionary of Scientific Biography*).

Session	Topic(s)	Assigned Reading (for next class session)
1	Content survey; course introductions	Article: Avital
2	Assignment particulars; What does "*Using History* in the teaching of mathematics" look like?	Articles: Fauvel; Liu
3	Examining history in the classroom, continued;	Article: Wilson & Chauvot
4	*Using History* in the classroom, continued; Trying it out: The Who? How? What? of a sample topic. Overview of resources to be used in the course.	Text: *Math Through the Ages: Expanded Edition* (Berlinghoff and Gouvêa 2004): pp. 5–60, at your leisure; For next class: Sketches 1, 3, & 4
5	Number systems as cultural artifacts	Additional reading: *TBA*
6	Number systems, continued	
7	Working with fractions: A bit more detail	Sketch 5; *Historical Modules* excerpt
8	The unacceptability and unfamiliarity of negative numbers	*The Negative Numbers* (Weizmann Institute of Science materials); Sketches 2 & 8
9	Negative numbers, continued; "text as symbols" or "symbols as text"	Sketch 9; *Historical Modules* excerpt
10	Linear equations	Mazur *Imagining Numbers* (*particularly the square root of minus fifteen*) (Mazur 2003), pp. xiii–76
11	Book Club Discussion #1	Sketches 12 & 14
12	Geometry: The gift from the Greeks (and others!)	*Historical Modules* excerpt; Sketch 7
13	Geometry, continued	Sketch 10
14	Geometry and quadratic equations	
15	Quadratic equations, continued	Sketches 11; *Historical Modules* excerpt
16	Solving cubic equations and the emergence of complex numbers	Sketch 17
17	Complex numbers, continued	Mazur, pp. 77–156
18	Book Club Discussion #2	Sketch 18
19	The development of trigonometry	*Historical Modules* excerpt (trigonometric identities)
20	Trigonometry, continued	
21	The development of pre-calculus ideas; Example: What is the origin of the logarithm?	Mazur, pp. 157–229
22	Book Club Discussion #3	*Historical Modules* excerpt *Quadrature of the parabola*
23	Archimedes and "integration"	Supplemental reading: Pascal's *Arithmetical Triangle* (1654)
24	Pascal and his *Arithmetical Triangle*	
25	Pascal, continued	Supplemental reading: Allen
26	Development of calculus	Supplemental reading: Dunham
27	Development of calculus, continued	

Table 19.1. Sample Course Schedule (Fall 2007)

For several reasons, the Teaching Unit Assignment as originally planned was too ambitious. In one sense, many of the students possessed a negative opinion about having to take the *Using History* course. Ten of the 19 undergraduates enrolled during Fall 2006 had failed or withdrawn due to poor performance at mid-term when taking the course in Spring 2006[2]. Additionally, since the previously unsuccessful students' prior experience with *Using History* was primarily as a mathematics course, it was difficult to fully engage them in two of the three course foci (i.e., cultural and historical aspects of mathematics and the pedagogical knowledge necessary for integrating history of mathematics). Several students' aversion to mathematics—originating from their difficulty with precalculus and calculus concepts and their lack of success in a previous version of *Using History*—was evident in the overall lack of inclusion of mathematical tasks within the teaching units created. Table 19.2 displays the content areas and topic choice descriptors for the teaching units created by students in Fall 2006. Of the topics selected by students, 81% were beginning topics (number, beginning algebra, and some geometry) and only 31% (five of the 16 submitted teaching unit assignments) included significant mathematics content. Two of these contained mathematical errors in either the lessons or accompanying materials (*e.g.*, answer keys).

The hypothesis with which I originally approached the course also guided my reflection of the results of the students' work on the *Teaching Unit Assignment*. If students were not conceptualizing the use of the history of mathematics in teaching as much more than a few historical anecdotes or timeline activities, I believed that the assignment was not providing students with the opportunity to envision the use of the history of mathematics to include actual mathematics. Consequently, I modified the capstone project, and for the Spring and Fall 2007 course iterations a *Model Lesson Assignment* was required as the capstone project in the course.

Content area (number of *Teaching Units* created)	Topic descriptors
Number (4)	multiplication; fractions; square roots; distributive property
Beginning Algebra (3)	slope; linear equations; quadratic equations
Geometry (5)	similar triangles, area and perimeter; parallel lines; Pythagorean Theorem
Advanced Algebra (2)	combinatorics; matrices
Trigonometry (1)	vectors
Other: beginning topic (1)	central tendency

Table 19.2. Teaching Unit Topic Choices: Content Areas and Topic Descriptors (Fall 2006)

19.3.2 Second and third iterations of the course: The Model Lesson Assignment

The adjustment of the *Teaching Unit* into the *Model Lesson Assignment* was designed to enable students to think more deeply on one lesson of a unit, as opposed to trying to conceptualize the use of history of mathematics across an entire unit of instruction in middle or high school mathematics teaching. In many ways, this modification was motivated by the fact that PSMTs at FSU take *Using History* at different times during their two years to complete the program. Consequently, if PSMTs have not taken one of the two mathematics education methods courses, it is difficult to address mathematics history knowledge and instructional planning knowledge across an entire unit—especially if they have not had such experience prior to the *Using History* course. A reasonable compromise entailed requiring students to create a model *lesson* as opposed to a model *teaching unit*. In addition, I anticipated that students' attention to one lesson would engage them in developing mathematics with which they could be successful and that would impact their view that benefits gained from learning mathematics from an historical perspective were worth seeking in their future teaching.

---

[2] Note: Only 16 of the 19 undergraduates are considered in this discussion. Three students did not complete the capstone assignment in the course during Fall 2006.

Minimum resources required for *Teaching Unit* (Fall 2006)	Minimum resources required for *Model Lesson* (Spring & Fall 2007)
7 text resources (one of which must be the *Dictionary of Scientific Biography*)	3 text resources (one of which must be the *Dictionary of Scientific Biography*; not all three can be encyclopedias)
2 website resources (author must be identified)	2 website resources (author must be identified)
2 journal articles (*e.g.*, *Mathematics Magazine*, *Mathematics Teacher*, *ISIS*)	1 journal article (*e.g.*, *Mathematics Magazine*, *Mathematics Teacher*, *ISIS*)
1 "alternative format" resource (*e.g.*, portraits, maps, media files, novels)	1 "alternative resource" (*e.g.*, portraits, maps, media files, novels)

Table 19.3. Required Teaching Unit Resources versus Required Model Lesson Resources

The *Model Lesson Assignment* asked students to spend more time with their topic of choice and use fewer historical resources more deeply in the work of creating a model lesson. Students were tasked with creating a model lesson for which the history of mathematics provides a significantly enhanced perspective in teaching the topic and one which would challenge pre-service teachers' own thinking and understanding. The required elements for the *Model Lesson* included (1) an historical background section, including basic biographical information about mathematicians who contributed to the development of the idea or topic; cultural and societal aspects of the places, people, and events of the major time periods involved; and historico-mathematical information sufficient for "setting the stage" for the topic; (2) the lesson plan and supporting documents, including all of the items needed to complete the lesson, such as maps, copies of original sources, student worksheets, notes to students, PowerPoint presentation slides, and solution guides; and (3) a bibliography containing at least seven resources, several of which were required (*e.g.*, the *Dictionary of Scientific Biography*).

In addition to the concentration on a single model lesson as opposed to an entire unit, the new requirement of seven resources instead of twelve (Table 19.3) was enacted to encourage students to be more selective in the resources that they used in the creation of their model lesson and to spend more time using those resources in its development. This modification emerged from the distinction between learning about resources and learning from resources. In the construction of the teaching units, *Using History* students certainly showed evidence of their ability to access and use a wide variety and a greater number of resources. The intent of the requirement, however, was that students *learn from* the research that they conducted. By reducing the number of resources required for constructing the model lesson I hoped that students would spend more time with the resources that they did access. Consequently, this deeper study would impact their mathematical knowledge for teaching by increasing their historical knowledge in meaningful ways.

The outcomes of the *Model Lesson Assignment* in both Spring 2007 and Fall 2007 were generally more successful than the *Teaching Unit Assignment* in Fall 2006. Neither capstone assignment description in Fall 2006 or Spring 2007 included the requirement that the students emphasize a mathematical component within the unit or model lesson. In Fall 2006 approximately 11% of students chose to include significant mathematics (framed by historical problems) within the content of their teaching unit. In contrast, 46% of Spring 2007 students decided to incorporate significant mathematics informed by historical problems into their model lesson. Beginning in Fall 2007, students were required to include mathematical content within their model lesson. The nature of students' mathematical preparation for teaching and in particular how the study of the history of mathematics influences mathematical knowledge for teaching is beyond the scope of this paper; however, a future publication to examine this very topic is planned.

For the current consideration, however, an example highlights the contrast in quality and content of model lessons submitted in both Spring 2007 and Fall 2007 with lessons submitted within teaching units in Fall 2006. In Fall 2006, no student selected a topic that was related to the concept of infinity. In Spring 2007, however, three students (of the 26 included) focused on topics that included some aspect of the concept (development of π; special constant e; concept of infinity) and in Fall 2007, two of the 18 students included in the study addressed the examination of infinite sequences

19.3. The Capstone Project: The First Three Semesters

Content area (number of *Model Lessons* created)	Topic descriptors
Number (5)	magic squares; fractions; operations with integers
Beginning Algebra (3)	Cartesian plane; linear equations; quadratic equations
Geometry (4)	development of π; area and volume; Pythagorean Theorem
Advanced Algebra (3)	combinatorics; matrices; Fibonacci sequence
Trigonometry (4)	development of sine; development of trigonometry as a field; identities
Calculus (3)	L'Hospital's rule; the derivative
Other: beginning topic (2)	tessellations; building structures
Other: advanced topic (2)	special constants (e); concept of infinity

Table 19.4. Model Lesson Topic Choices: Content Areas and Topic Descriptors (Spring 2007)

and sums throughout history. Mark (Spring 2007) decided to examine the development of the constant e based upon his developing interests in Euler and the concept of infinity while taking *Using History*. His model lesson included historical information to be given to students that focused on "exploring the transcendental number e" (Model Lesson, April 2007), as well as exercises for students to explore the approximation of e and application of the constant in mathematical models. For Mark, it was important to use the history of mathematics to aid in making sense of two concepts that were difficult for him to explore, learn, and accept. Mark now possessed concrete knowledge of the existence of e, as opposed to viewing it as a mysterious constant stored in the calculator's memory. In addition, Mark believed that his knowledge—enhanced by the study of the history of the concept—would in fact impact his future students' learning in similar ways.

Tables 19.4 and 19.5 display the content areas and topic choice descriptors for the model lessons created in Spring 2007 and Fall 2007, respectively. Fifty-four percent of topics chosen by Spring 2007 students and 61% of topics chosen by Fall 2007 students were classified as beginning topics. The decrease in the number of beginning topics chosen in Spring 2007 when compared with Fall 2006 may be a function of the mathematical preparedness of the students enrolled. Although a decrease of beginning topics selected was also evident in Fall 2007 when compared to Fall 2006, it represented a slight increase over the number of beginning topics selected in Spring 2007. This may have been a result of the assignment requirement that lessons must include mathematical content (beginning in Fall 2007).

Although additional investigation of what content PSMTs decided to include in the model lessons (and how that content was presented) is required, it is certain that some PSMTs benefited from the requirement of considering and including mathematics content of their historical research. Lauren observed:

Content area (number of *Model Lessons* created)	Topic descriptors
Number (3)	metric system; fractions; base-number systems (other than base 10); operations with integers
Beginning Algebra (7)	evolution of algebraic notation; linear equations; quadratic equations; geometrical algebra; patterns in Pascal's triangle
Geometry (1)	constructing centers of a triangle
Advanced Algebra (5)	history of the function concept; infinite sequences; logarithms; results from Euler; matrices
Trigonometry (2)	development of sine and cosine; law of sines and law of cosines

Table 19.5. Model Lesson Topic Choices: Content Areas and Topic Descriptors (Fall 2007)

In order to fully understand the historical content of my model lesson I had to do some research on the Babylonian and Greek mathematics, particularly the development of algebra and geometry. Why don't we use more geometry when studying algebra? I think that by doing this, students would be more interested in studying this extremely vital part of mathematics. It would show students why we have and why we need algebra. Algebra used didn't just pop up out of nowhere; it was geometrically inspired. The Greeks' "algebra" was also geometric until late in this time period when Diophantus began using variables/symbols to solve these problems. Therefore, studying the development of geometrical algebra has led me to believe that algebra should not be introduced without the appropriate geometry to explain and show its actual usefulness. (Fall 2007)

Further modifications of the *Using History* course also illuminate interesting potential reasons for PSMTs to engage in challenging but meaningful mathematical ideas.

19.4 Reflections for Further Course Revision

The ability of PSMTs to consider the use of the history of mathematics with their future students is dependent upon their evaluation of the worth of learning mathematics from solving historical problems or investigating alternative algorithms using the historical development of a mathematical concept. Because of the lack of mathematical and pedagogical experiences connecting mathematical topics with their historical development throughout a mathematics teacher preparation program, a course such as *Using History* must provide pre-service teachers with a venue to experience the benefits of historical problems and investigations when learning–or is often the case, re-learning—mathematical concepts found in secondary school mathematics. In many ways, I viewed the PSMTs' work on either a teaching unit or model lesson as a way for them to make sense of mathematical topics while applying a historical perspective. During this sense-making process, I wanted students to develop with respect to their own learning and to consider instances in the secondary school curriculum for which investigating a topic using an historical perspective (*e.g.*, operations with integers) contributes to conceptual understanding. Indeed, many of the prospective teachers benefited from a historical examination of operations with integers because they were confronted with having to explain why algorithms work (*e.g.*, "a negative times a negative is positive"). On many occasions, students revealed that they merely accepted mathematical rules they were told to apply when learning mathematics in grades K–12. Now, however, the history of mathematics provided PSMTs with access to important pedagogical tools to emphasize conceptual understanding of such rules.

To give PSMTs the space to do this in the *Using History* course, it became necessary to modify the requirements of the capstone project. The *Teaching Unit Assignment* was a complicated task for students. The unit potentially covered several ideas related to one topic and required students to navigate a large number of sources in order to identify or create some number of lessons that integrated historical ideas, information, or problems. Many of the Fall 2006 students found the assignment difficult because it required researching historical information, synthesizing and applying mathematical knowledge, and planning instructional tasks. Some PSMTs had not developed the ability to attend to each of these simultaneously to the level required for the assignment. Consequently, the one aspect that was sacrificed was being able to synthesize and apply the new mathematical knowledge that confronted the PSMTs as they investigated the historical development of the topic they chose. This phenomenon was apparent given the small number of teaching units that included a significant mathematical component (11% of Fall 2006 teaching units created).

After the *Teaching Unit Assignment* was modified into the *Model Lesson Assignment*, students taking the *Using History* course could concentrate more on creating one well-considered lesson and if they chose, could highlight a significant mathematics component within their lesson (during the Spring 2007 iteration). Of the 12 individuals (46% of the Spring 2007 students) deciding to emphasize mathematics within their model lesson, two-thirds relied heavily on some form of or the actual course materials when designing their lesson. In many cases, the material students relied upon for their model lesson content came from the *Historical Modules for the Teaching and Learning of Mathematics* [3]. Thus, existing resource materials utilized during the *Using History* course influenced students' conceptions of planning for the use of history in the teaching of mathematics.

Additionally, the *Model Lesson Assignment* continued to evolve in several ways. In previous *Teaching Unit* and *Model Lesson* assignments, the students could draw upon the work (*e.g.*, lesson activities, lesson plans) of others, but the entire unit or lesson could not be the work of others. Students could not merely piece together content from resources. Instead, they were encouraged to construct coherent lessons that incorporated a variety of mathematical, historical, and cultural content. The way in which students combined these elements—selected directly from or built upon the ideas of other resources—was considered as evidence of lesson development. It is worth noting that as a result of the requirement that no entire lesson could be the work of another, an overwhelming majority of students chose the inclusion of historical information or anecdotes as the content of the unit or lesson that they created. In most cases, the self-designed aspect of model lessons did not include mathematical content. Thus, the Fall 2007 iteration of the course included the new requirement that lessons contain significant mathematics content for the *Model Lesson Assignment*. With this new requirement, students were challenged to understand their selected topic and the teaching and learning of mathematics well enough in order to design a coherent lesson incorporating historical and mathematical content.

After the second modification of the course capstone project (requiring students to incorporate a mathematical component), I observed that several PSMTs altered their choice of model lesson topic over the course of the semester, often finally deciding on a topic that was less difficult than their first choice. This observation, coupled with the concern that the course did not allow enough time for PSMTs to teach their model lesson, prompted another version of the course capstone project. For the Fall 2008 course iteration, students were assigned (in pairs) one of the ten course topics. In their teaching pair, PSMTs developed a lesson on one aspect of their assigned topic and implemented their lesson on an assigned date. Initial results from the *Lesson Plan and Delivery Assignment* are varied with respect to the quality and quantity of historical and mathematical ideas incorporated. However, this modification provided a prime benefit in that PSMTs observed almost immediately the outcome of planning and implementing a teaching episode that incorporates history. In the following excerpt, Susan (Fall 2008) reflects on her lesson planning and delivery:

> The day we taught on Sketch 10 was influential in my study of the subject of quadratic equations and completing the square. Doing research on al-Khwārizmī gave me a head start in understanding quadratic equations, but that was a more historical connection between al-Khwārizmī and quadratic equations and I focused more on the biographical parts. When Paul and I were putting together our lesson plan on Sketch 10, I started to look at al-Khwārizmī's explanations of quadratic equations more. I came to understand the concept behind the equations. I have a better appreciation for how I solve equations now with the historical research I did on al-Khwārizmī. Teaching the lesson also helped me understand quadratic equations more by explaining why al-Khwārizmī used "completion" and "balancing" to get the six standard forms of equations.

19.5 Conclusion

The underlying challenge of a *Using History* course is to encourage PSMTs to acquire competence with constructing coherent curriculum and may motivate them to "develop, individually, or, in collaboration, their own material... and to make it available to a wider community" [7, p. 212]. Indeed, I continue to reflect on and revise the capstone project in the course as a way to provide PSMTs with a task in which they can "benefit from both primary and (perhaps more from) secondary materials and [that] they particularly welcome... didactical source material" [7, p. 212].

Motivation for such revision can be found in the PSMTs' expression of their struggles and revelations related to considering the use of the history of mathematics in teaching. At the beginning of this paper I quoted an early entry from James's reflection journal. In his earlier view, James could not quite wrap his head around the idea that "incorporating history into a math class is necessary for a student's education." At the end of the semester, however, James shared a different perspective:

> Regarding the Pascal discussions we've been having lately, I feel that there is just so much depth behind the [arithmetical] triangle that it seems like I could spend an entire semester teaching about its properties. So how would I know what to focus on? My model lesson is basically to teach the students why the triangle is formed the way it is and a few of its properties. I especially want to make sure that the students can make the comparison between the triangle and binomial coefficients, but I also want to teach the students kind of the same thing we learned in class today: probability and combinations. Although it wouldn't be as advanced, you can use the triangle to determine how many different combinations it takes to reach a particular "cell." (James, April 2007)

This excerpt exhibits how James moved from not understanding why he should consider the inclusion of the history of mathematics to struggling to plan for just the right aspects to focus on within his model lesson on Pascal's arithmetical triangle. I considered this shift a successful outcome of the course. I also found James's reflection and those of other *Using History* students as evidence for continuing to craft the best possible capstone task capable of engaging PSMTs in creating model lessons that influence their own learning and that convince them to share their creation with their future students and colleagues.

Bibliography

[1] Conference Board of the Mathematical Sciences, 2001, *The Mathematical Education of Teachers: Vol. 11. Issues in Mathematics Education*, Providence, RI: American Mathematical Society.

[2] Fauvel, J. and J. van Maanen, (eds.), 2000, *History in Mathematics Education: The ICMI Study*, Dordrecht-Boston-London: Kluwer.

[3] Katz, V. J. and K.D. Michalowicz, (eds.), 2005, *Historical Modules for the Teaching and Learning of Mathematics*, Washington DC: The Mathematical Association of America.

[4] National Council for the Accreditation of Teacher Education, 2003, *NCATE/NCTM Program Standards: Programs for Initial Preparation of Mathematics Teachers*, Washington, DC: Author.

[5] Pimm, D., 1983, "Why the History and Philosophy of Mathematics Should Not be Rated X," *For the learning of mathematics* 3(1), 12–15.

[6] Rickey, V. F., 2005, "Teaching a Course in the History of Mathematics," online
`www.math.usma.edu/people/rickey/hm/mini/default.html`

[7] Tzanakis, C. and A. Arcavi, 2000, "Integrating History of Mathematics in the Classroom: An Analytic Survey," in *History in Mathematics Education: The ICMI Study*, J. Fauvel & J. van Maanen (eds.), Dordrecht-Boston-London: Kluwer, pp. 201-240.

About the Author

Kathleen M. Clark is an Assistant Professor of Mathematics Education and a core-faculty member in the FSU-Teach program at The Florida State University. She earned her Ph.D. in Curriculum and Instruction from the University of Maryland, College Park in 2006. From 1987 until 2001, she taught high school mathematics in a variety of contexts in Florida and Mississippi. Kathleen became interested in how to incorporate history of mathematics in the classroom while teaching a history of mathematics course at the Mississippi School for Mathematics and Science and was fortunate to serve as an historical modules field tester (as a participant of the *Institute in the History of Mathematics and Its Use in Teaching* in 1999–2001). Her research interests include: the impact of the study of history of mathematics on teachers' mathematical knowledge for teaching, the ways in which teachers use history in the teaching of mathematics, and the practices and development of mathematics teacher educators.

20

Mapping Our Heritage to the Curriculum: Historical and Pedagogical Strategies for the Professional Development of Teachers

Leo Rogers
Oxford University, United Kingdom

20.1 The English Curriculum: 2008–2009

In 1998/9 the education system in England and Wales began a mathematics curriculum based on 'core skills' that enshrined traditional beliefs about 'levels' of knowledge and encouraged a utilitarian approach with a prescriptive pedagogy. Textbook design followed the syllabus, past test papers became *de facto* part of the curriculum, and the emphasis on examination targets produced little serious engagement with substantial mathematical thinking in most classrooms. The Teacher Training Establishments were inspected for their conformity to the system, and consequently the situation with regard to the history of mathematics has been far from that well-organized plan described by Lingard in Fauvel and van Maanen [11, pp. 117–122]. The latest Inspectors' report on our secondary schools shows that too many pupils are taught formulas that they do not understand, and cannot apply:

> "The fundamental issue for teachers is how better to develop pupils' mathematical understanding. Too often, pupils are expected to remember methods, rules and facts without grasping the underpinning concepts, making connections with earlier learning and other topics, and making sense of the mathematics so that they can use it independently." [21, p. 5]

In complete contrast, the new 'Programmes of Study' for mathematics for the 11–16 age group published in 2007 represent a radical change in approach, encouraging a more thoughtful and problem-based programme, and a "real and relevant context for learning" [23, p. 3]. Within the last year, the government has dropped both the testing at Key Stage 3 (age 14) and the 'Numeracy Framework' that determined much teaching methodology. New examinations for 16 year olds are being piloted and will run in 2010 [24].

The framework of this new curriculum is based on four 'Functional Skills' criteria: *Representing, Analyzing, Interpreting and evaluating, Communicating and reflecting* and one of its "Key Concepts" for all pupils is "*Recognizing the rich historical and cultural roots of mathematics*" [22][1]. These changes open up new possibilities, but we have suffered a target-oriented environment for so long that very few secondary school teachers have had the chance to discover the range and potential of the contributions that history of mathematics could make to pupils' learning. So, what would

---
[1] QCA, the Qualifications and Curriculum Authority, is the Government sponsored body set up to maintain and develop the national curriculum and associated assessments, tests and examinations. Teaching the curriculum is a 'statutory requirement', i.e., it is required by law.

"Recognizing the rich historical and cultural roots of mathematics" mean in practical terms for our teachers? In order to gain acceptance for this new initiative, we need to be able to justify to our teachers the use of history of mathematics as a humanizing and meaningful contribution to pupils' engagement with mathematics. Lack of historical knowledge on the part of the majority of teachers is a significant problem, and the fear of having to learn a 'new' subject when demands on their time are already considerable, does not bode well.

In a recent issue of Educational Studies in Mathematics, colleagues have reviewed the evidence, both theoretical and practical, and renewed their call for the history of mathematics to be taken seriously as an essential part of the mathematics curriculum. In the introduction Radford et al. [25] argue that an important sense of meaning lies within the cultural-epistemic conception of the history of mathematics:

> "The very possibility of learning rests on our capability of immersing ourselves—in idiosyncratic, critical and reflective ways—in the conceptual historical riches deposited in, and continuously modified by, social practices. ...Classroom emergent knowledge is rather something encompassed by the Gadamerian link between past and present. *And it is precisely here, in the unraveling and understanding of this link, which is the topos or place of Meaning, that the history of mathematics has much to offer to mathematics education.*"
> [25, p. 108] (italics mine)

In these terms, history stands in opposition to the utilitarian demands of the old curriculum, but having put history of mathematics into the curriculum, the government has now revealed the problems of resources and training. Changes need to happen not only in the classroom but also, and more importantly, in teacher training. So, *how can we provide material from the history of mathematics that can be integrated in a meaningful and effective way into the everyday activities of the classroom*?

Since there is no time for extra subjects, a useful strategy may be to look at some of the fundamental ideas already there, and produce material that will enrich the mathematical experience of both pupils and teachers. But will this be possible without trivializing the subject we wish to promote?

20.2 A Pedagogical Tradition

There is a tradition of producing materials for teachers and pupils that focuses on an individual's learning process and encourages active engagement in, and discussion of, mathematical problems. This kind of material was introduced by the Association of Teachers of Mathematics (ATM) in the late 1950s, and has been its enduring hallmark. Publications are the result of collaborative research where materials are developed by offering examples of classroom work which require discussion, involve heuristic forms of reasoning, analogy, and inference, encouraging the learner to create and verify their own ideas. Examples like Watson and Mason [33] and Swan [32] now provide practical guidance helping teachers to develop pupils' powers of constructing mathematics for themselves:

> "Our interest is in using mathematical questions as prompts and devices for promoting students in thinking mathematically, and thus becoming better at learning and doing mathematics. ...We hope our work will show how *higher order mathematical thinking* can be provoked and promoted as an integral part of teaching and learning school mathematics, through the teacher's leadership and example." [33, p. 4]

Such publications promote situations that are generic and offer ways for teachers to develop 'Learner Generated Examples' *applicable at all stages* of learning mathematics. The materials promote activities that focus on ambiguity, raise doubts about interpretations, and encourage the learner (and the teacher) to develop a security with mathematical ideas that enables them to engage in intelligent questioning and active discussion of the problems concerned. Teachers engaged in this pedagogy have been raising pupils' learning above mere acquisition of skills by helping pupils to develop their own cognitive tools and achieve a higher order of mathematical activity. 'Rich Tasks' like these are now officially advocated [19]. A practical outcome is that teachers pay more attention to new ideas if they are expressed in the context of a classroom task rather than theoretically. This wealth of experience can provide ways of introducing historical material as activity based learning, using devices already well tested.

20.3 The History of Mathematics and our Mathematical Heritage

A considerable body of literature has been generated to promote and justify the use of the history as an essential aspect of teaching mathematics and a variety of proposals have been put forward to describe how this should happen, but

20.3. The History of Mathematics and our Mathematical Heritage

there has been a long-running dispute between those who wish to maintain the 'purity' of historical research and insist that history of mathematics in school is not proper history, and mathematics educators who want to introduce history of mathematics into their teaching.

Marwick [18] maintains that the job of the historian is to produce *knowledge about the past* based on evidence from interrogating primary sources and interpreting technical terms therein. History, like science, is a cumulative and cooperative activity and accounts should not be accepted as definitive, but always subjected to debate, qualification, and correction. Clearly, these principles apply equally to historians of mathematics.

Research in pupils' learning of mathematics has shown that the interaction of teaching strategies and pupils' achievement relies principally on self-motivation and engagement [17]. Real engagement with mathematical ideas occurs at cognitive (thinking), affective (feeling) and operative (doing) levels, and cultural contexts underlie ideas that are 'embodied' by our personal history and our experiences in the physical world [35, 26]. Much of the evidence from research generated by colleagues in HPM shows that pupils can engage affectively and effectively with historical materials.

The history of mathematics cannot be simply 'applied' to parts of a curriculum, as previously envisaged. Wilder's (1968) *Evolution of Mathematical Concepts* introduced cultural aspects to the study of mathematics, and [36, pp. 355–330] described how the historiography of mathematics has changed. Recent publications give new accounts of the history of mathematics. The collection edited by Katz [16] on the mathematics of ancient cultures, and the new Oxford History of Mathematics (Robson & Stedall [28]) arranged

> "under a series of themes which raise new questions about what mathematics has been, and what it has meant to practise it, [which are neither] descriptive or didactic but investigative, comparing a variety of innovative and imaginative approaches to history" [28, p. 1]

are quite different from earlier writing, appealing to a much wider audience.

Fried [12] considered how teachers are compelled to treat historical approaches given the goals and organizations of school mathematics, suggesting that by the very nature of their activities, history and mathematics education were incompatible. However, there is a confusion between the nature of mathematics *per se* and mathematics education: mathematics education is not mathematics. In teaching mathematics we are initiating pupils into building a mathematical language and becoming able to use its codes and procedures. This process of both building and learning to use the language is quite unlike the mathematical attitudes of Fried's 'working mathematicians' and should not be confused with it [34]. In his 2007 paper Fried suggests that mathematics education forms a 'bridge' between the 'working mathematician's way of knowing mathematics' and the 'historical way of knowing mathematics' and that history can lead us from our own approach to mathematics to appreciating the views of others.

Ivor Grattan-Guiness [15] has made an important distinction between the *History* and the *Heritage* of mathematics. The terms "history" and "heritage" distinguish between different interpretations of a mathematical theory where the corresponding actors are "historians" and "inheritors" (or "heirs") respectively [15, p. 164]. He agrees with Marwick that history of mathematics addresses the question "what happened in the past?" and focuses on the detail of primary sources, the coded language, cultural contexts, negative influences, anomalies, and so on, in order to provide evidence of what happened, and why and how it happened. Such accounts are always open to debate and comparison with other versions from recognized experts.

Heritage, on the other hand, refers to the impact of a theory on later work and addresses the question "how did we get here?" This is where previous ideas are seen in terms of contemporary explanations, and *similarities* with present ideas are sought.

> "The distinction between the history and the heritage of [an idea] clearly involves its relation to its pre-history and its post-history. The historian may well try to spot the historical foresight—or maybe lack of foresight—of his historical figures, ... By contrast, the inheritor may seek historical perspective and hindsight about the ways notions actually seemed to have developed." [15, p. 168] and "...heritage suggests that the foundations of a mathematical theory are laid down as the platform upon which it is built, whereas history shows foundations are dug down, and not necessarily into firm territory." [15, p. 171]

For example, the interpretation of Euclid's work as 'geometrical algebra' by Heath and others has been shown to be quite misguided as history, but as heritage is quite legitimate because it is the form in which Arab mathematicians interpreted the *Elements* as a model for logical justification when they were creating algebra [9, 27]. In a similar manner, talking about 'imaginary numbers' is part of the heritage of Cardano and Bombelli, since that is how complex numbers were thought of by their immediate successors [8, p. 380], even up until the early 19th century. While distinguishing the differences, Grattan-Guinness claims that:

> "Educators can profitably use both history and heritage for their purposes. For example the algebraic version of Euclid, so important in its heritage, is often and well-used in this kind of teaching. But also available is the real Euclid of arithmetic and geometry, including the beautiful theory of ratios ... an excellent route into the notoriously difficult task of teaching ... rational numbers." [15, p. 175]

We have to be careful. Deterministically constructed heritage conveys the impression that the progress of mathematics is cumulative. That may be partly true, mathematics builds on its past achievements, but while we may make stories about the links between the mathematics of the past to the present, the mathematics of the past is not the same as the mathematics of now, even though it may appear so when described in contemporary notation. As Mathematics Educators we have a means of passing on our Heritage by making links between the content we find in the curriculum, and what we know of the history of mathematics. In this way it becomes possible to describe significant landmarks in the history of mathematics in terms that teachers and pupils can understand without making impossible demands on their historical capability or curriculum teaching time.

20.4 Mapping Our Heritage

The idea of a 'heritage map' has evolved from experiences in presenting 'episodes' from the history of mathematics in workshop form, so that interesting problems can arise from the historical contexts. In order to address the question of teachers' lack of knowledge, the focus lies in providing secondary teachers with professional development materials that start from ideas in the curriculum, and open up the possibilities of developing the concepts involved by finding some 'historical antecedents' to support the connections between, and motivations for, these ideas. Askew and Brown [2] showed that one of the key attributes of an effective teacher was the ability to engage pupils by connecting different areas of mathematics.

The material in development has been presented as a series of 'concept maps' that are intended as a metaphor to provide a topographical view of the significant features of a particular mathematical landscape. A concept map is a graphical multi-layered metacognitive tool for organizing and representing knowledge. Novak and Godwin [20] used a hierarchical arrangement of concepts with linking propositions but the 'Mind Map' [5] is more flexible since it allows affective linking through 'affinity diagrams'. For a more fluid way of looking at a map I am indebted to my colleague Jeremy Burke for a discussion of his research in Virtual Learning Environments. Burke & Papadimitriou [4] argue that when faced with non-linear text, a learner would need not only a Map but also a Narrative (the background) and an Orientation that describes the activities provided for students to 'find their bearings' in the map of the topics presented. This idea gives us freedom to consider a map in a virtual environment where the arrangement of concepts, objects, events, propositions and actions may be partially ordered and even multi-layered, crucially breaking up the linear sequence and juxtaposing different ideas. No map is ever 'complete'; what may be chosen to be the principal concept(s) at one stage can be rearranged according to the needs of the learning process, and of the individuals involved. In contrast, most curriculum activities are presented to teachers as a *fixed narrative* of topics to teach, restricted to some imagined 'levels of competence' of the pupils. We can present pupils with a map to be explored and interpreted, some historical narrative, and some guide lines for problems arising from the situation (the orientation), instead of the way they are presented with pieces of a jig-saw without any coherent pattern, and little help to see how the pieces fit together.

Maps clearly have an epistemological function. By organizing concepts and examining the possible links between them in a visual display, maps can be used as scaffolding for learning, leading us to new connections between ideas. Indeed, Barbin [3, p. 18] states that:

> "The history of mathematics shows that mathematical concepts are indeed constructed, modified and extended in order to solve problems. Problems come, as much as into the birth of concepts as into the different meanings attached to concepts as tools for the resolution of problems. This emphasis on the role of problems in the historical construction of knowledge can lead to a new way of conceiving history."

and, I would add, a new way of finding problems amenable to the classroom.

Adapting the map to explore links through the curriculum to historical contexts can act as part of a developable knowledge structure to be offered to a teacher for integrating aspects of our mathematical heritage into a teaching programme. A map can be examined from 'inside-out' and from 'outside-in'; from following particular trails of thought to obtaining a broader overview of a particular set of developments. The history then becomes integral to the exploration of the mathematics, and the 'unraveling and understanding' of the links between the ideas, is the *topos* that Radford and our colleagues are talking about.

A map is there to enable teachers to have the freedom to *make their own narrative*. The map can throw light on certain problems; it can suggest different approaches to teaching and generate didactical questions. It is thus possible to offer ways in which teachers, starting from a particular point in the standard curriculum, could incorporate the teaching of 'key concepts' (QCA 2008) to link with some important developments in the history of mathematics through the use of 'idealized' historical problems and *canonical situations*. By a canonical situation I mean an image, a diagram, a formula, or a way of setting out a problem or process that is developable, has potential to represent more than one idea, and is presented to students to encourage potential links between apparently different areas of mathematics depending on their experience and the opportunities presented.

20.5 Trialling, Qualitative Results, and Development

A large study was undertaken in a secondary school whose objectives for pupils were to appreciate the connection between mathematics and geometric patterns and to be able to build shapes using specially designed templates and explain their mathematical structure [10]. This diagrammatic approach was very successful in raising achievement in the introduction of algebra.

Work with teachers had discussed the Mesopotamian and early Hindu 'cut and paste' geometry that led to the eventual algorithms for 'completing the square'. Templates were not labeled with side lengths and pupils were introduced to arithmetic expressions and equations using area diagrams. Starting with the square, pupils were asked to describe its area. Some gave a side length while others gave a letter, expressing generality.

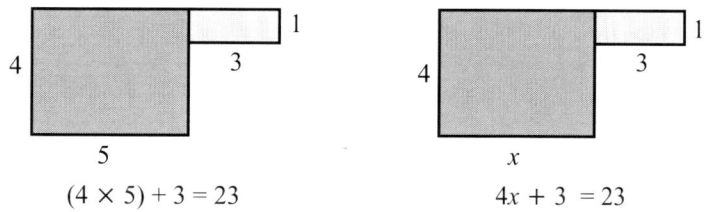

Figure 20.1.

Further studies were carried out in another school from 2003 to 2006 with 'low achievers'. Pupils' final achievement was above expectation and the impact of this approach, sustained over some six weeks, was evident across a whole year group. Pupils were asked to show pieces representing $x^2 + 2x + 1$, to rearrange them into a square and to describe the area in different ways. They were then asked, for example, to make $x^2 + 5x + 6$ into a rectangle and prompted for the dimensions of the rectangle as well as the factorized form.

Students generated new examples and gave the pieces to a neighbor to factorize. In 2008 a group of 14/15 year old students from local schools together with their teachers attended a masterclass where the history of 'completing the

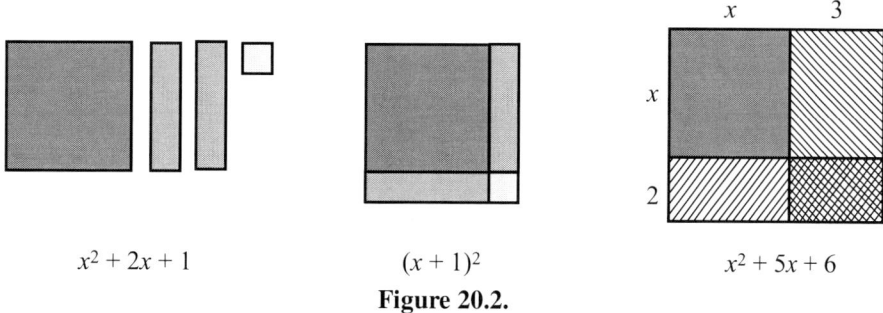

Figure 20.2.

square' was explored through diagrams, tracing the development of the problem from the Babylonians to Renaissance Italy [29].

In this way history of mathematics provided a number of opportunities and material for the development of Key Concepts and Mathematical Processes. Engaging with a problem through its history incorporates the essential aspects of conjecture, justification and generalization, and shows how alternative strategies give rise to new ideas. Recognizing that we have inherited and built on the work of many different cultures showed how various aspects of mathematics are connected, and provided examples of different representations and methods still viable today.

Some comments from pupils:

"Maths is the only lesson that makes you think this much."

"History makes you realize that people in those days were very intelligent."

And from teachers:

"The context of the Maths through history was shown very well, and pupils understanding where their Maths comes from is always interesting and useful for them."

"The historical approach showed a new way of looking at the structure of the quadratic formula. It has given me some new ideas for the classroom"

There are, of course, many more avenues to explore. When asking pupils to make up their own examples using the diagrams, it becomes clear that there are certain problems not easily resolved in this way, motivating the development of new techniques.

Current extension of these ideas now includes experimental work with Concept Maps as a means of supporting teachers in developing ideas from history and ways of expressing them in accessible form. Furthermore, a Working Group has been set up among members of the British Society for Research in Learning Mathematics (BSRLM) to explore these ideas and to produce a series of case studies on their classroom implementation.

20.6 Heritage Maps and Canonical Images

The previous section described the encouraging results obtained from earlier versions of the materials that are undergoing refinement, expansion and trial. Here are descriptions of some possible classroom activities from the two examples of Concept Maps to be see in Fig. 20.3 and 20.4.

Canonical images described earlier can be taken from objects, pictures, diagrams, texts and other materials found in the history of mathematics, or they can be developed from materials found in the contemporary classroom, by finding new ways of using familiar materials, or creating specially designed activities that have the potential to lead the enquirer to more complex ideas, and motivate the development of new techniques. The importance of visualization in these activities is clear: from the representation of objects, to manipulating them physically and learning to do so in the mind to bring out hidden properties. The materials are designed to capitalize on their psychological, pedagogical and epistemological potential [1, 14]. The entry to these exercises needs no special mathematical knowledge, but the success, development, or otherwise, depends entirely on the teacher's pedagogical approach [19]. Since many of the ideas involved can be thought of as 'generic' there is no sense in which an activity is intended to be assigned to any particular 'level' of knowledge in the traditional sense.

Using a square dot lattice is a common activity usually provided for exercises on calculating the areas of shapes made from straight line segments, transforming areas, or exploring ways of building square numbers. Usually the activities stop there, but from a simple beginning we can enter both aspects of our heritage and a range of mathematical problems. Figure 20.3 presents the Concept Map developed to support teachers by showing a range of possible curriculum activities that can be used to connect with various aspects of our mathematical heritage. Narrative and Orientation, such as collections of images, historical notes and references, and links to websites are available for teachers to use [29, 30, 31]. The central image of the square-dot lattice is an example of a canonical situation; it becomes canonical when we see the range of activities and ideas that can be developed from such a simple beginning.

Selecting three possibilities we have:

Pythagorean Arithmetic and its Connections with Geometry. A square of 'dotty paper' has been used here to indicate the potential for pattern forming and recognition that may have gone on in early times, and can certainly

20.6. Heritage Maps and Canonical Images

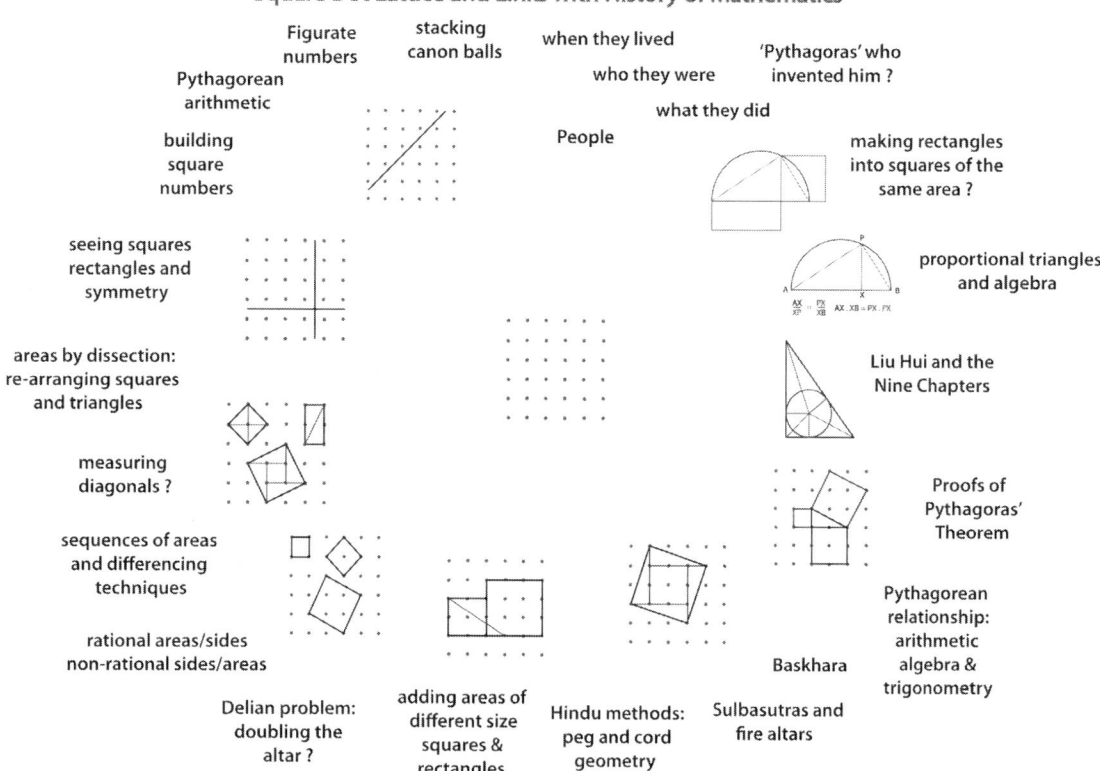

Figure 20.3. Example 1 Map. Curriculum Activity: Square dot Lattice Paper.

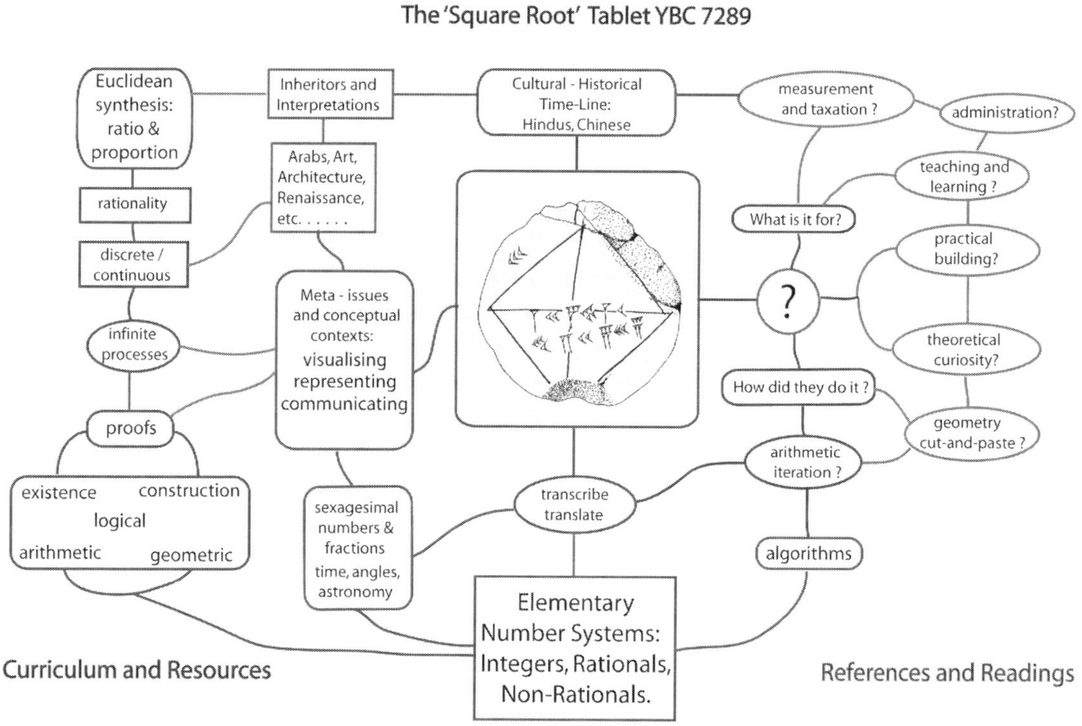

Figure 20.4. Example 2 Map. The 'Square Root' Tablet YBC 7289.

be used in the classroom with good effect. There are many possible explorations: Odd and even squares and rectangles can be divided with straight lines to form squares, rectangles and triangles, and pupils can be asked about the relationships they see. Two triangular numbers make a square number; questions about other figurate numbers and how they can be built up are open for exploration.

Rationality and Irrationality. At some time, pupils meet the diagonal of a square and its non-rational properties. The two examples in this section show how, even with primary pupils, we can enter into discussions about why the square has a 'whole number' area but we cannot find an exact number or fraction for the sides. Building sequences of 'skewed' squares where there is a dot at each corner but no dots on the edges, can lead to other interesting explorations of sequences, differencing techniques, and iteration methods.

The Right Angle in the Semicircle. The potential of this activity is considerable. It offers an alternative introduction to ratio and proportion for early secondary school pupils; it is an illustration of the 'rule of three' still important today; the transformation of ratios into a linear expression is an example of the way in which quadratic formulas were used; and it demonstrates the use of proportional triangles in the work of Viete and Descartes.

The central image in Figure 20.4 is a transcription of the well-known tablet, and links suggest ways of exploiting the idea embodied, both through historical references and through the standard curriculum. Accompanying the Map is a Narrative; 'References and Readings' are a series of notes, suggestions, and weblinks for classroom explorations, including discussion of pedagogical and didactical issues, while 'Curriculum and Resources' indicates the Orientation.

The Ratio of the Side to the Diagonal. From the historical point of view, many questions arise about the context and use of such a calculation; we don't know how people found the value written, but we might speculate on the development of some iterative procedure, and there are plenty of other examples where these people developed algorithms for solving problems. From the point of view of the curriculum there are plenty of opportunities to discuss the development of geometrical and number concepts and the way these were presented in text and diagram form as ratios, proportions, integers, fractions, rationals, and non-rationals. There is obviously a range of meta-issues that can be discussed, even with younger pupils.

Key ideas like the different forms of representation, appropriate notation, and whether a particular procedure is 'allowed' in a given context, can be discussed, and show how finding representations for 'impossible' numbers like $\sqrt{3}$ or π can have a liberating effect, allowing new ideas to flourish. And there is the ever-present idea of 'infinity' to be explored. The material gathered for the Maps comes from many historical documents written by experts and the use of published research to identify some of the significant moments in the evolution of ideas and the ways in which they were understood and handed on. The material is designed so that it can be used in 'episodes' in the normal course of teaching in school and introduced as individual teachers think fit, not necessarily in historical order. Included in the Narrative are notes and references to the historical background, and 'pedagogical notes' aimed to help teachers raise questions and see where the material can be used in their classroom. In this way, selections can also be used as a basis for teachers' professional development both in the historical and mathematical sense. This is where the historical process can be described *in terms communicable to a modern school audience* where the teaching is specifically designed to focus on the pupils' mathematical activity *in the contemplation and discussion of the problems.*

Encouraging teachers and pupils to make their own maps and to compare and combine them is an important learning activity that encourages visualization, linking of apparently different mathematical ideas, and fundamental epistemic activity. It is hoped that by using these Concept Maps, with appropriate Narratives and Orientations, together with further trialling and the development of the pedagogical methodology described above, we may have a chance of truly beginning to realize "the rich historical and cultural roots of mathematics" in our classrooms.

Bibliography

[1] Arcavi, A., 2003, The role of visualisation in the learning and teaching of mathematics. *Educational Studies in Mathematics* 52 (3), 214–241.

[2] Askew, M., M. Brown, et al., 1997, *Effective Teachers of Numeracy: Report of a study carried out for the Teacher Training Agency*. London. King's College London.

20.6. Heritage Maps and Canonical Images

[3] Barbin, E., 1996, "The Role of Problems in the History and Teaching of Mathematics" in Calinger, R. (ed.) *Vita Mathematica*, Washington, MAA, pp. 17–25.

[4] Burke, J. and M. Papadimitriou,, 2002, "Narratives and maps for effective pedagogy in hypermedia learning environments." *Goldsmiths Journal of Education*, 5 (1), 14–25.

[5] Buzan, T. and B. Buzan, 2006, *The Mind Map Book*. Harlow, Essex. BBC Active

[6] Cajori, F., 1896, *The History of Mathematics With Hints on Methods of Teaching*. N. Y.

[7] Dauben, J.W. and C.J. Scriba, 2000, *Writing the History of Mathematics: Its Historical Development*. Basel, Birkhauser Verlag.

[8] Descartes, René, 1637 (1954), *La Geometrie*, New York, Dover.

[9] Djebbar, A., 2005, *L'Algebre Arabe: Genese d'un Art*, Paris, Vuibert.

[10] Fairchild, J., 2001, "Transition from arithmetic to algebra using two-dimensional representations: a school based research study." *Papers on Classroom Research in Mathematics Education*, (Centre for Mathematics Education Research, University of Oxford)

[11] Fauvel, J. and J. van Maanen, 2000, *History in Mathematics Education: the ICMI Study*, Kluwer.

[12] Fried, M. N., 2004, "The Problem of Mathematics Education and History of Mathematics from a Sausserean Point of View." *Proceedings of PME 2004 Bergen Norway*. Paper presented to the PME Semiotics Group, Bergen, 2004.

[13] ——, 2007, "Didactics and History of Mathematics: Knowledge and Self-Knowledge," *Educational Studies in Mathematics* 66 (2), 203–233.

[14] Giaquinto, M., 2007, *Visual Thinking in Mathematics: An Epistemological Study*. Oxford.

[15] Grattan-Guinness, I., 2004, "The mathematics of the past: distinguishing its history from our heritage." *Historia Mathematica* 31 (2), 163-185.

[16] Katz, V., 2008 (ed), *The Mathematics of Egypt, Mesopotamia, China, India and Islam*. Princeton. Princeton University Press.

[17] Martin, A.J., 2007, "Motivation and Engagement," *British Journal for Educational Psychology* 77, 413–440.

[18] Marwick, A., 2001, *The New Nature of History*. Basigstoke, Hampshire. Palgrave.

[19] NCETM, 2009, Rich Tasks www.ncetm.org.uk/mathemapedia/Rich+Tasks

[20] Novak, J.D. and D.B. Gowin, 1984, *Learning How to Learn*, Cambridge University Press.

[21] Ofsted UK, 2008, *Mathematics: Understanding the Score Download*
www.ofsted.gov.uk/Ofsted-home/Publications-and-research/Documents-by-type/Thematic-reports/Mathematics-understanding-the-score

[22] QCDA 2007 (Qualifications and Curriculum Development Authority)
curriculum.qcda.gov.uk/key-stages-3-and-4/
Mathematics KS3: curriculum.qca.org.uk/key-stages-3-and-4/subjects/key-stage-3/mathematics/index.aspx
Mathematics KS4: curriculum.qca.org.uk/key-stages-3-and-4/subjects/key-stage-4/mathematics/index.aspx

[23] QCA 2009a, *Engaging Mathematics for all Learners*. London HMSO

[24] QCA 2009b, QCA/09/4159 April 2009 *Changes to GCSE Mathematics*. London HMSO

[25] Radford, L., F. Furinghetti, and V. Katz, 2007, "The topos of meaning or the encounter between past and present," *Educational Studies in Mathematics* 66, 107–110.

[26] Radford, L., 2005, "Body, Tool, and Symbol: Semiotic Reflections on Cognition" in: E. Simmt and B. Davis (Eds.), *Proc. Canadian Mathematics Education Study Group*, pp. 111–117.

[27] Rashed, R., 2007, *Al-Khwārizmī le Commencement de L'Algebre*. Paris. Blanchard.

[28] Robson, E. and J. Stedall, 2009, *The Oxford Handbook of the History of Mathematics*. Oxford, Oxford University Press.

[29] Rogers, L., 2008, "Completing the Square: Where it comes from and why it is important." Unpublished *PowerPoint presentation and Workshop* shown to teachers, pupils and PGCE students in Oxfordshire during the period 2005–2008.

[30] ——, 2009, NRICH (Enrich) *Mathematics Millennium Project*, Cambridge. (a) The Development of Algebra Part 1—Visualisation http://nrich.maths.org/public/viewer.php?obj_id=6485

[31] ——, 2009, NRICH (Enrich) *Mathematics Millennium Project*, Cambridge. (b) The Development of Algebra Part 2—Representation http://nrich.maths.org/public/viewer.php?obj_id=6546

[32] Swan, M., 2006, *Collaborative Learning in Mathematics: A challenge to our beliefs and practices*. London, NIACE.

[33] Watson, A and J. Mason, 1998, *Questions and Prompts for Mathematical Thinking*. A.T.M.

[34] Watson, Anne, 2008, "School Mathematics as a Special Kind of Mathematics," *For the Learning of Mathematics* 28 (3), 3–19.

[35] Wertsch, J. V., 1998, "Mediated Action," in Becht, W. and Graham, G. A., *Companion to Cognitive Science*, Oxford.

[36] Wilder, R.L., 1968, *Evolution of Mathematical Concepts: An Elementary Study*. N. Y.

About the Author

Leo Rogers' interest in history of science began while at school, and, after reading mathematics and physics, he continued his research in History of Science while working as a secondary school teacher. He became inspired by an original and creative group of mathematics teachers to develop ideas on the relations between the history and the teaching of mathematics, and, as a result, with Philip Jones, he founded HPM in 1972. He has been involved in TV series on mathematics for children, and worked with colleagues throughout Europe on research into teaching, learning, and curriculum development. His published work covers mathematics, mathematics education and history and pedagogy of mathematics. He now works as a researcher in the Education Department at the University of Oxford, UK. His current work is divided between helping to provide practical materials for teachers, and writing about the history of the mathematics curriculum in the UK.

21

Teachers' Conceptions of History of Mathematics

Bjørn Smestad
Oslo University College, Norway

21.1 Introduction

Many researchers are convinced that history of mathematics merits a more important role in mathematics education than it currently occupies. To achieve that goal, more knowledge on teachers' conceptions of history of mathematics is needed.

In this paper, I will describe an interview study on Norwegian teachers' conceptions of the history of mathematics. After giving some background, I will describe the method used before discussing the main findings of the study. These findings provide insights that should be taken into account in further attempts to get more teachers to include history of mathematics in their teaching.

21.2 Background—Norway

Curriculum: In the Norwegian curriculum for primary and lower secondary schools[1] of 1997, the history of mathematics had a prominent place. One of the six main goals of mathematics was "for pupils to develop insight into the history of mathematics and into its role in culture and science" [13]. In addition, the history of mathematics was mentioned in specific goals for several grades.

Textbooks: In Smestad [27] the textbooks for the new curriculum were analyzed. There was not much content involving the history of mathematics in the textbooks. The texts were not relevant to stimulate pupils' inquiring. In addition, a number of errors were found. The texts mostly consisted of dull, biographical information (such as year of birth and country of origin) and they rarely touched upon the development of concepts, mathematicians' motivations for working on mathematics or how mathematics has been used throughout history.

Teacher education: According to the TIMSS 2003 study, many Norwegian mathematics teachers in primary and lower secondary school had few or no credits in mathematics throughout their teacher education [11]. If at all included, the history of mathematics could only have accounted for a small part. The history of mathematics was not included in

---

[1] The age of pupils in primary schools in Norway is 6–13, while the age in lower secondary schools are 13–16.

the in-service courses developed for the new curriculum either [21]. Therefore, most Norwegian mathematics teachers should not be expected to know much more history of mathematics than what they have studied on their own.

Teaching: Little research exists about the actual teaching of the history of mathematics in Norway. An evaluation of the 1997 curriculum [1] only mentions the history of mathematics in passing.

Upper secondary school: The situation in upper secondary school was somewhat different. There was no new curriculum in 1997 and textbooks had for many years included history of mathematics. Most mathematics teachers in upper secondary school had their teacher education from the mathematics department of a university, not in teacher colleges as most teachers in primary and lower secondary school.

21.3 Background—International

There exists no overview of how the history of mathematics is included in mathematics education internationally. However, an analysis of the TIMSS 1999 Video Study [28] showed that only 3% of the 638 mathematics lessons (from seven different countries) included some history of mathematics, and even in these lessons, the time allotted to the history of mathematics was only three minutes on average. The 3% dedicated to the history of mathematics was mostly lectured rather than activity-based. It often consisted of biographical information.

There have been several efforts, both locally and internationally, to help teachers gain insight into the combined fields of the history and the pedagogy of mathematics. For instance, US historical modules [18], an Italian sourcebook [3], Taiwanese teacher training [15] and collaboration with teachers in England [22] are a few recent examples. There have also been empirical studies to measure the effects of including history of mathematics.

The ICMI Study [6] looked at curricula, textbooks and teacher education. It gave many examples, but also pointed to possible problems regarding the teacher and classroom [6, pp. 30–31]. Studies concerning other attempts to change teaching have shown that teacher beliefs have to be taken into account, i.e., [24]. Insight into teachers' conceptions may be useful when discussing how to enhance the use of the history of mathematics in schools.

There is much research to be found on teachers' conceptions (or beliefs) in general [29, 23]. Also, there are studies investigating in which ways working on history of mathematics may influence teachers' conceptions of mathematics [16, 25] or mathematics teaching [9, 2].

However, there is not that much research on teachers' conceptions of history of mathematics. Fraser and Koop [8] collected teachers' opinions on materials connected to the history of mathematics, without the teachers being influenced by enthusiastic teacher-researchers. Most of the teachers found the materials useful, both for motivation and for teaching important concepts, and claimed that they would use the materials in their own classes. However, a significant minority would not, and said that the amount of time taken up would be excessive. Gonulates [10] describes research on prospective teachers wherein a questionnaire was administered before and after instruction based on history of mathematics. The prospective teachers appreciated the role of history of mathematics more for motivational than for conceptual purposes. Even after the instruction, "they see historical materials somehow not very related with the curriculum and they think the use of such materials does not answer the needs of the students directly." In addition, Siu [26] listed reasons teachers give for not including history of mathematics.

21.4 The Research Question

My research question is *"What are mathematics teachers' conceptions of history of mathematics and why?"*

I use the word "conception" in a broad sense, in an attempt to translate the Swedish word "uppfatning" which can be defined like this: *"What a phenomenon basically means to an individual or the fundamental way in which the individual understands the phenomenon, is called the individual's "uppfattning" of the phenomenon."*[2] [30]. I broaden the meaning of "conceptions" to an even greater extent. In my study, I include teachers' experiences with the history of mathematics, their attitudes towards it, and what they think the history of mathematics may be.

---

[2]"Vad detta fenomen i grunden betyder för individen eller det grundläggande sätt på vilket individen förstår fenomenet kallas individens uppfattning *av* fenomenet." The translation into English is mine.

Thus, I am not interested in what history of mathematics "is" or what is "actually" happening in Norwegian classrooms—what Ference Marton would call "first order perspective" ([30]). Instead, I'm interested in teachers' conceptions, including both their conceptions of what history of mathematics is and their conceptions of what they do in their classrooms—what Marton would call "second order perspective." Rather than "a typical view," I'm interested in finding a variety of views.

21.5 Method

I have chosen to do a small case study consisting of qualitative research interviews, within the phenomenological tradition. Due to time restrictions, the number of teachers interviewed is only four. A larger number of teachers would probably have given a broader range of views, but the necessary amount of time was not available.

Earlier studies of textbooks [27] and lessons [28] give us an idea of teachers' (and textbook writers') conceptions of the history of mathematics. This interview study could be seen as part of a bricolage [4], where the new study serves to supplement the findings of my two previous studies.

Fog ([7]) points to the significance of being aware of one's own preconceptions in the field of study. My previous studies [27, 28] form an important part of my preconceptions, along with my strong belief that the history of mathematics may help improve teaching. I have tried to take my preconceptions into account when analyzing.

21.5.1 Generalizability, validity, reliability

Generalizability The simplest form of "generalizability" is the existence proof. For instance, by describing one teacher who claims to have been inspired by a one-day course, I prove that such teachers exist. Such existence proofs may be interesting, but the study makes no attempt to say how many such teachers exist.

Kennedy [19] points out that, in case studies, it must be the information receiver who decides whether analytical generalizations can be made. The researcher "should produce and share the information, but the receivers of the information must determine whether it applies to their own situation." Likewise, it is left to the reader of this study to decide whether the teachers he or she is concerned with, are in a context sufficiently close to the one in which the teachers in this study is situated.

Donmoyer [5] writes that case study research "might be used to expand and enrich the repertoire of social constructions available to practitioners and others; it may help, in other words, in the forming of questions rather than in the finding of answers." My study may be seen in this way, in starting a register of possible conceptions of the history of mathematics which can enrich future discussions.

Reliability One relevant issue concerning the reliability of the findings is the issue of leading questions. By asking the teachers about their own experiences before going on to ask prepared questions, I reduce the risk of the leading questions. However, the fact that I was there in the first place, asking about history of mathematics, may have influenced the teachers. On the more technical side, I believe the transcription reliability is high, as the sound was never a problem, and I had the opportunity to go back over and over to make sure the transcription was okay.

Validity I try to secure the validity in several ways. First of all, I asked follow-up questions during the interview whenever I was uncertain about an interpretation. Secondly, I offered a short summary to the participant at the end of the interview including my interpretation, giving the participant the opportunity to comment. Thirdly, I have analyzed the text in several steps, moving through hermeneutical spirals. By presenting preliminary findings at some meetings, I have also been given feedback which I have brought back to the analysis.

Another form of validity testing concerns the teachers' comments on their own conception of the history of mathematics. When a teacher claims to be an enthusiast, the claim can be supported by having lots of concrete examples on how he uses the history of mathematics in his teaching. If he had claimed to be an enthusiast, but without being able to give any examples, his claim would have been more in doubt.

21.5.2 Choice of participants

I wanted my participants to be different from each other, as I wanted to see a variety of conceptions of history of mathematics. Therefore, I wanted to include teachers both from secondary school (pupils age 13–16), where history of mathematics was introduced with the new curriculum in 1997, and high school (pupils age 16–19), where history of mathematics has been part of the textbooks and curriculum for some years. I also wanted teachers at different ages (because experience may influence their conceptions) and teachers with different education (both from university and from teacher colleges).

I recruited the participants by contacting schools and asking for teachers who would be willing to participate. Although I stressed that an interest in history of mathematics was not necessary, there was at least one teacher who didn't want to be interviewed because she didn't know any history of mathematics. Therefore, it is possible that my four teachers are more interested in history of mathematics than the average teacher. The four interviews were done over a period of several years, which meant that I had the opportunity to analyze previous interviews and improve on my interview guide for the following interviews. For instance, for the first two interviews, I started by asking for reasons to include history of mathematics in the teaching, while for the other two interviews, I started by asking each participant to tell me about the last time he used the history of mathematics in his teaching. I have also confronted participants with opinions given by teachers in earlier interviews to make them discuss further. As this was only done towards the end of the interviews, this has not compromised the reliability of my analyses.

21.6 The Participants

In the following discussion, I will compare the four teachers in key areas, but first I will give an introduction on each participant in the form of short patchworks based on the interviews. The quotes have been translated into English by me.

Teacher 1 (T1) Teacher 1 is in his 40s and has taught in secondary school for about 10 years. On the introduction of the history of mathematics in 1997, he says: *"Well, personally, I think we are struggling with this, we are focused on the mathematics, the four basic operations, on geometry, fractions... We tend to push aside the history of mathematics and not spend enough time on it." "I haven't made much effort to get acquainted with the history, I guess that is something I will have to do eventually."* He also says that it's difficult to work on history of mathematics, because his pupils are not motivated for mathematics. *"It's not easy to motivate students of this age to... (hesitates) I think the students have to be a bit older to understand that [history of mathematics] is important."* Moreover, *"We are working with the exam in mind, and as far as I've seen for the past ten years, history of mathematics is not part of these exams."*

"In teacher education, we had a little history of mathematics, especially connected to numeral systems, but then I worked a lot with small children after becoming a teacher, and it's obvious that you don't work with history of mathematics with them."

On what would be needed to include more history of mathematics, he mentions that as long as the textbooks do not include more of the history of mathematics, it's tempting to skip it.

Teacher 2 (T2) Teacher 2 is also in his 40s, and has been teaching in high school for 20 years.

"When the new curriculum[3] came, a professor came and lectured for a whole day on the history of mathematics. It was an extremely good lecture, and very interesting, and [...] there was little of it that I really had any knowledge of in advance. But I've always been interested, since I was a kid, in Leonardo da Vinci, stories from antiquity, in architecture, science and so on." Apart from this lecture, he has also picked up information from the textbooks.

"I tend to start every new topic with history of mathematics, as an anecdote or as something to spice up the subject. For some, it's just [a short entertainment], for others it is more fundamentally formative." "I may use only five minutes. To make the pupils focus. They come in, lazy. I make a story out of this. If these five minutes make the pupils more motivated or concentrated, it's worth it."

"I would like to know more. Someone could write a concise history of mathematics, popularized. Or I could have had another course."

---
[3] This refers to the new curriculum for high schools in 1994, not the one for primary and secondary schools in 1997 mentioned earlier.

Teacher 3 (T3) Teacher 3 is in his 60s, and has been teaching in high school for 40 years.

> *"I've always been interested in history of mathematics. My first source was an incredibly inspiring mathematics teacher in high school."* *"I got* Mathematics for millions[4] *as a gift from my parents while I was still in high school. And when I'm on holidays, I always visit book stores and buy some book that I enjoy reading. And during my studies, I had many inspiring lecturers. They often used historical bits, especially lives of mathematicians, for instance Abel and Galois."*
>
> *"I try to spice up the teaching with a little history now and then, when it is natural, but I think most mathematics teachers don't do that."* *"When we have finished going through the curriculum and have about a month left before the exam, we start reviewing it. But then they get the chance to try other methods of presenting the subjects, by means of lectures or small projects. Then I encourage them to some historical connection."* *"The main point is to motivate the pupils."*
>
> *"It is enough with one enthusiastic teacher in a school for it to spread to the other teachers, if conditions are favorable."*

Teacher 4 (T4) Teacher 4 is in his 50s, and has been teaching in secondary school for almost ten years.

> *"What I know about history of mathematics is what I have read in books. I saw a book once, was very inspired by it. I thought this sounded very exciting, because I also had an idea that all mathematics came from Greek mathematics. It does not."*
>
> *"I have used history of mathematics to motivate the pupils. My school has many pupils from other parts of the world, and I want to remove the misconception that all mathematics is from the western world."* He asks his pupils a little about mathematics they use in their home countries, and talk about counting methods and calendars. *"We also work on house-building, and discuss the pyramid, African huts that are circular and cylindrical, the Eskimos' igloos, Indian tents, Sami tents. Then we can calculate areas. After a while, my pupils realize that the circular form is better than the western which is quadratic or rectangular. So the African pupils, they get better self-esteem or status by knowing that "we build, in our culture, smarter houses."*

He thinks the pupils like the variation and to discuss their own culture. At the same time, prejudices are dissolved. Both "good" and "less good" students think it is exciting.

He works on the history of mathematics 1–2 times a month, maybe 2–3 lessons each time. He thinks history of mathematics can be used at lower levels of the school system as well.

His impression is that history of mathematics is not used very much. If it is to be used more, courses are needed. In high school, lots of history of mathematics is put into the textbooks, but there is so much mathematics that has to be taught as well.

21.7 Discussion

I will now discuss some of the key issues raised and how the teachers differ.

What is the history of mathematics? Although I have not asked the teachers to define "history of mathematics," different conceptions are apparent in the interviews. In particular, some think that the history of mathematics is about the *development* of mathematics while others think that it is about the *use* of mathematics (throughout history).

Some illustrative examples:

> **T2:** *"Then we talk about how fertile this area [the Nile delta] is, and that it is very important to keep your land, and that they have used trigonometry and surveying to sort everything out after the flooding, when the borders are washed out."*
>
> **T2:** *"We in Europe have been a bit behind when it comes to mathematics and that there are certain things that we think we have discovered, but which have actually been discovered earlier in both Egypt and China."*

---

[4] [14]

T3: *"Geometry is, of course, connected to ancient mathematics. [...We also work on] probability theory, when and how did that come in, and a bit about these erroneous deductions, d'Alembert's error and so on. [...] And a bit later the function concept, which came late when it comes to precision."*

T4: *"I often ask pedagogical questions about mathematics, what is mathematics, where does mathematics come from and [...] what kind of mathematics do they use in their home countries, so for instance we touch upon counting methods, what kinds of techniques are used there?"* T4 also mentions building houses and creating a right angle, but also mentions who developed "our" numeral system.

T4 seems to be more on the use side, particularly in the way he lets pupils look at how mathematics is used in their cultures today. T3 is more on the *development* side, while T2 has both perspectives evenly included. T1 says too little about this to be analyzed. My impression is that the textbooks of upper secondary school (which T2 and T3 have used) have been on the development side, while the new curriculum for primary and lower secondary school (relevant for T1 and T4) was more on the use side. Whether there may be a connection here could be a topic for further study.

How have the teachers acquired their conceptions of history of mathematics? T1 mentions his teacher education as his main source, and Egyptian numeral systems as an example of what he learned. T2 recalls a one-day course by a professor in 1994, which *"was an extremely good lecture, and very interesting, and [...] there was little of it that I really had any knowledge of in advance."* He also mentions that the textbook he uses has historical introductions to the chapters, and that these have been useful. T3 had a very inspiring teacher in high school, and has also read lots of books and had good lecturers while studying at university. T4 has read articles and books, including some by Paulus Gerdes and Ubiratan D'Ambrosio.

This confirms my earlier point: even though the history of mathematics has been included in the curriculum, teachers do not necessarily get any organized help in updating their knowledge. Two of the teachers (T3 and T4) managed fine on their own, while T2 got a one-day course which kindled an interest. T1 never got any input after teacher education, according to him. It would be interesting to speculate on how their conceptions of history of mathematics have been framed by these courses, but that cannot be seen in the interviews. For instance, I cannot determine whether T4's interest in mathematics in different cultures originated in reading Gerdes and D'Ambrosio or whether an interest in culture was already there.

Are the teachers interested in the history of mathematics? Of these four teachers, three claim to be very interested in history of mathematics, while T1 does not. His lack of interest is apparent all through the interview: *"Well, personally I haven't done much with history of mathematics"*; *"I guess that is something that I will have to do eventually."* The other three teachers show their enthusiasm both in what they say and in their way of saying it. T2: *"I think [history of mathematics] is so great..."* T3: *"I've been very interested in [history of mathematics] for many years."* T4: *"I got very excited and thought [history of mathematics] sounded very thrilling."* The interest of T2, T3 and T4 are also apparent in that they can give many examples of history of mathematics relevant for teaching.

How do the teachers include history of mathematics in their teaching? T1 mentions how he talks about Niels Henrik Abel when the Abel competition comes up every year, but does not have many more examples of inclusion of history of mathematics. (He says that the last time he worked on the history of mathematics was "last year.") T2 says that he starts every new chapter with talking a bit about history of mathematics—"maybe just five minutes," he calls it "anecdotes," "stories" and "spice."

T3 also includes history of mathematics in the introduction to a topic, but mostly in projects, topic days and work towards the oral exam—ways of working in which the students are more than passive listeners. He has occasionally cooperated with other subject teachers. T4 includes the history of mathematics once to twice a month in small "projects." Each project lasts 2-3 hours, often applying manipulatives.

The textbooks tend to tell stories from the history of mathematics without giving ideas for further work [27]. This may lead teachers to fall into the pattern of the textbooks, while it may take more interest and enthusiasm to include history of mathematics in a richer variety of ways, such as those suggested by the ICMI Study [6].

What goals do the teachers have when including history of mathematics? One major difference may be 'history as a tool' vs. 'history as a goal' [17]. History can be a tool for understanding the mathematics (i.e by being inspired by

21.7 Discussion

history in working on misconceptions) or as a tool for motivation. It can also be a goal in itself. We could also divide the goals into conceptual, motivational and multicultural [12]. Here are some relevant quotes:

> **T1:** *"It's not easy to motivate students of this age to (hesitates) I think the students have to be a bit older to understand that [history of mathematics] is important."*

> **T2:** *"So for some, it is just a little diversion, and a little refreshment, but for others, it can be more fundamentally formative, a further suggestion for further studies or as a peg to hang it on, to link to other things they have learned."* He also stresses that history of mathematics may explain the importance of mathematics and place *"the subject in our culture and other cultures"*.

> **T3:** *"I have [...] tried to spice up the teaching, if I can call it that, with some historical passages now and then."* *"The main examples I can give have to do with motivating the pupils to work on the topics that may seem dry and boring."* But he also mentions that he has used examples to motivate for deeper understanding, for instance he mentions d'Alembert's misconceptions.

> **T4:** *"I have often used it as a motivation in my teaching, to include the historical mathematics."* He stresses that he both aims to give pupils cultural understanding through the history of mathematics, and to motivate them by referring to their own cultures.

It seems that T1 only sees the "history as a goal" part, while the other three see history both as a goal and as a tool. Only T3 gives concrete examples of how history of mathematics may benefit the learning of mathematics directly. It also seems that T3 sees the conceptual, T2 the motivational and T4 the multicultural aspects as most important, while T1 cannot be put in any of the categories.

The 1997 curriculum was unclear about the point of including history of mathematics. When the new 2006 curriculum was prepared, however, the committee decided that history of mathematics was to be regarded as a tool, not as a goal. And as the new curriculum was supposed to leave the choice of "tools" to the teachers, history of mathematics was not mentioned[5].

What do the teachers think of their pupils' reaction? Unsurprisingly, their opinion on how the pupils react, also varies. While T1 says that *"They try to follow,"* but that *"it's a bit dry"* for them, T2 says that *"They find it fascinating."* T4 says that they find it *"exciting to work on their own cultures and learn about others."*

This could be interpreted as a result of vicious/virtuous cycles: enthusiasm leads to enthusiastic teaching, which then leads to interested pupils, a result which strengthens the teacher's enthusiasm. Whether this is the case in general can certainly not be inferred from this study.

What kind of resources do the teachers feel they need? T1 says that *"This is about the interests of the teachers. If I had more interest in this area, I might have had some books on the shelf and could have used them. But a day has only 24 hours..."* The other three teachers want more courses for teachers, and T2 also mentions that there should have been a "concise history of mathematics, popularized," which could have been sent to teachers.

It should be noted that there existed no such "history of mathematics for teachers" in Norwegian at the time of the interviews.

What do these teachers think of other mathematics teachers' conceptions of history of mathematics? T1 thinks he is representative of teachers at his school at least. T2 says he doesn't know much about what the other teachers do in their classrooms. T3 and T4 feel that other teachers are less interested than themselves. T3: *"I know many teachers who can be characterized as enthusiastic mathematics teachers, with good knowledge of history of mathematics, [...] but I'm afraid they are not among the youngest ones."* T4: *"No, [history of mathematics] is not used much. That is the impression I have."*

So, at least on this point, they mostly seem to agree: the average teacher is not very interested in history of mathematics.

---

[5] Based on personal communication with a member of the committee.

Should history of mathematics be in the curriculum? Bodil Kleve studied mathematics teachers' interpretation of the 1997 curriculum, and showed how three teachers enacted the curriculum in three different ways [20]. With that in mind, it is worth discussing whether curriculum changes matter at all. The four teachers in my study disagree on this. T2 doesn't explicitly address this topic, but refers to the curriculum document when talking about the point of including history of mathematics. T1 notes that history of mathematics has not been included in exams, and uses that as an argument for not stressing that particular part of the curriculum. T3 argues against putting it in the curriculum: "*Then it becomes compulsory, but is that a good idea? Well, possibly it is easier to get it to the pupils in that way, but...then there is a job to do with all the mathematics teachers in Norway, and I'm not sure they have so much knowledge in advance [on history of mathematics].*" T4 argues strongly for having it in the curriculum: "*I say that if it is not there, I'm afraid that it will not be included at all. If it's in the curriculum, it can remind a teacher that there is something. And obviously, we teachers are very different, and if there is something that doesn't appeal to me, I wouldn't teach it either, because then I would teach in a way that's not very positive.*"

Have the curriculum documents influenced these teachers? From these quotes and from what they say elsewhere, my impression is that T1 is not much influenced by the curriculum. T2 is not as much influenced by the curriculum document as by the course and textbooks which may be a result of the curriculum. T3 and T4 would have taught history of mathematics anyway, but T4 feels that other teachers would be influenced.

21.8 Conclusion

Including the history of mathematics into the curriculum in Norway did not work well. One important reason for this is that teachers' conceptions of the history of mathematics were not taken into account. There seems to have been an idea that once it was included in the curriculum, teachers would know what to do. My studies show clearly that this idea is wrong.

The four teachers in this study seem to have different opinions on what history of mathematics is and what the point of including history of mathematics in the curriculum may be. They also include the history of mathematics in different ways and to different degrees. Their levels of enthusiasm are different, they feel differently about what their students think of the history of mathematics, and have different opinions on what is needed when it comes to include more from the history of mathematics in their teaching. As such differences were revealed in a study of only four participants, one may speculate on which breadth of conceptions may be present in the teacher population as a whole.

These different factors should be taken into account when renewed efforts are made to include history of mathematics in mathematics teaching on a national basis.

Bibliography

[1] Alseth, Bjørnar, Gard Brekke, and Trygve Breiteig, 2003, *Endringer og utvikling ved R97 som bakgrunn for videre planlegging og justering: matematikkfaget som kasus, Rapport / Telemarksforsking. Notodden 02/2003.* Notodden: Telemarksforsking.

[2] Charalambous, Charalambos, Areti Panaoura, and George Philippou, 2009, "Using the history of mathematics to induce changes in preservice teachers' beliefs and attitudes: insights from evaluating a teacher education program," *Educational Studies in Mathematics* 71 (2):161–180.

[3] Demattè, Adriano, 2006, *Fare matematica con i documenti storici—una raccolta per la scuola secondaria di primo e secondo grado*: Editore Provincia Autonoma di Trento—IPRASE del Trentino.

[4] Denzin, Norman K., and Yvonna S. Lincoln, 1994, *Handbook of qualitative research*. Thousand Oaks, Calif.: Sage.

[5] Donmoyer, Robert, 1990, "Generalizability and the Single-Case Study," in Elliot W. Eisner and Alan Peshkin (eds.), *Qualitative Inquiry in Education: The Continuing Debate*, New York: Teachers College Press, pp. 175–200.

[6] Fauvel, John, and Jan Van Maanen, 2000, *History in mathematics education, An ICMI study*. Dordrecht: Kluwer Academic Publishers.

21.8. Conclusion

[7] Fog, Jette, 1994, *Med samtalen som udgangspunkt: det kvalitative forskningsinterview*. København: Akademisk Forlag.

[8] Fraser, Barry J., and Anthony J. Koop, 1978, "Teachers' opinions about some teaching material involving history of mathematics," *International journal of mathematical education in science and technology* 9 (2):147–151.

[9] Furinghetti, Fulvia, 2007, "Teacher education through the history of mathematics," *Educational Studies in Mathematics* 66 (2):131–143.

[10] Gonulates, Funda, 2008, "Prospective teachers' views on the integration of history of mathematics in mathematics courses," Paper read at HPM 2008, at Mexico City.

[11] Grønmo, Liv Sissel, 2004, *Hva i all verden har skjedd i realfagene?: norske elevers prestasjoner i matematikk og naturfag i TIMSS 2003*. Oslo: Institutt for lærerutdanning og skoleutvikling, Universitetet i Oslo.

[12] Gulikers, Iris, and Klaske Blom, 2001, "'A Historical Angle', A Survey of Recent Literature on the Use and Value of History in Geometrical Education," *Educational Studies in Mathematics* 47 (2):223–258.

[13] Hagness, Randi, Jorunn Veiteberg, Nasjonalt læremiddelsenter, and Kirke-, utdannings- og forskningsdepartementet, 1999, *The Curriculum for the 10-year compulsory school in Norway*. [Oslo]: National Centre for Educational Resources.

[14] Hogben, Lancelot, 1936, *Mathematics for the million: a popular self educator*. London: Allen & Unwin.

[15] Horng, Wann-Sheng, 2004, "Teacher's professional development in terms of the HPM: A story of Yu," in, *HPM 2004*, Uppsala, Sweden: Uppsala Universitet.

[16] Hsieh, Feng-Jui, 2000, "Teachers' Teaching Beliefs and Their Knowledge about the History of Negative Numbers," in Wann-Sheng Horng and Fou-Lai Lin (eds.), *HPM 2000*, Taipei: National Taiwan Normal University, pp. 88–97.

[17] Jankvist, Uffe Thomas, 2007, "Empirical research in the field of using history in mathematics education," *Nordic Studies in Mathematics Education* 12 (3):83-105.

[18] Katz, Victor J., and Karen Dee Michalowicz, eds., 2004, *Historical Modules for the Teaching and Learning of Mathematics*: Mathematical Association of America.

[19] Kennedy, Mary M., 1979, "Generalizing from single case studies," *Evaluation Quarterly* 3 (4):661–678.

[20] Kleve, Bodil, 2007, *Mathematics teachers' interpretation of the curriculum reform, L97, in Norway*, [Kristiansand]: Agder University College, Faculty of Mathematics and Sciences.

[21] KUF, 2004, *Plan for etterutdanning i matematikk*. Kirke-, utdannings- og forskningsdepartementet, 20.03.1997 1997 [cited 14.06. 2004]. Available from http://odin.dep.no/odinarkiv/norsk/dep/kuf/1997/publ/014005-990176/dok-bu.html.

[22] Lawrence, Snezana, 2008, "History of mathematics making its way through the teacher networks; Professional learning environment and the history of mathematics in mathematics curriculum," in, *International Conference on Mathematics Education*, Monterrey, Mexico.

[23] Leder, Gilah C., Erkki Pehkonen, and Günter Törner, 2003, *Beliefs: a hidden variable in mathematics education?* Boston: Kluwer Academic Publishers.

[24] Manouchehri, Azita, and Terry Goodman, 1998, "Mathematics curriculum reform and teachers: Understanding the connections," *Journal of Educational Research* 92 (1):27.

[25] Philippou, George N., and Constantinos Christou, 1998, "The effects of a preparatory mathematics program in changing prospective teachers' attitudes towards mathematics," *Educational Studies in Mathematics* 35:189–206.

[26] Siu, Man-Keung, 2004, "No, I do not use history of mathematics in my class. Why?," in, *HPM 2004*, Uppsala, Sweden: Uppsala Universitet.

[27] Smestad, Bjørn, 2002, *Matematikkhistorie i grunnskolens lærebøker: en kritisk vurdering, HIF-rapport 2002:1.* [Alta]: Høgskolen i Finnmark Avdeling for nærings–og sosialfag.

[28] ——, 2004, "History of mathematics in the TIMSS 1999 Video Study," in, *HPM2004 & ESU5*, Uppsala, Sweden.

[29] Thompson, Alba G., 1992, "Teachers' beliefs and conceptions: a synthesis of the research," in Douglas A. Grouws (ed.), *Handbook of research on mathematics teaching and learning: a project of the National Council of Teachers of Mathematics*, Reston: National Council of Teachers of Mathematics, pp. 127–146.

[30] Uljens, Michael, 1989, *Fenomenografi: forskning om uppfattningar.* Lund: Studentlitteratur.

About the Author

Bjørn Smestad is a teacher educator at Oslo University College in Norway. He wrote his cand. scient. thesis on British attempts to develop a solid foundation for Newton's theory of fluxions. After becoming a teacher educator in 1998, his main research interests have been history of mathematics in mathematics education as well as technology in mathematics education. In particular, he has been interested in what place history of mathematics currently has in Norwegian schools. He has been a regular participant in and contributor to the ESU and HPM conferences since 2000, and he has been editor of the HPM Newsletter since 2004.

22

The Evolution of a Community of Mathematical Researchers in North America: 1636–1950

Karen Hunger Parshall
University of Virginia, United States of America

22.1 The Seventeenth and Eighteenth Centuries: Mathematics in Colonial Settings[1]

The story of mathematics in colonial North America may be said to begin in 1636 with the founding by the Puritans of the Massachusetts Bay Colony of Harvard College as a Congregationalist institution.[2] It is not by chance that the first colleges in the British colonies south of what would become the border with Canada were Congregationalist, and this includes Harvard and Yale (as well as Dartmouth, Williams, Bowdoin, Middlebury, and Amherst). As heirs of "rational and hierarchical Calvinsim in America" [2, p. 248], Congregationalists valued the intellect and placed considerable emphasis on transplanting from England "the apparatus of civilized life and learning" [17, p. 273]. At Harvard, relative to mathematics, that translated into a curriculum in which mathematics was taught beginning in 1638 in emulation of the English universities like Cambridge on which it was modeled.

Although Harvard trained students for the ministry, it was not a seminary (more than half of its students followed secular pursuits upon graduation). The professional study of theology began only upon completion of the bachelor's degree; what the College was meant to educate were "gentlemen" and to "advance *Learning* and perpetuate it to Posterity" [17, p. 43]. To that end, it was assumed that students knew Latin, the language both of instruction and of most of the textbooks. Undergraduate training then involved a prescribed course in six of the traditional Seven Arts, which included arithmetic, geometry, and astronomy; students also studied philosophy, Hebrew and Greek as well as ancient history. This closely paralleled the curriculum at the Old Cambridge.

Yale had similar educational goals, although it was founded in New Haven in 1701 in part as a reaction to dissatisfaction with what some viewed as the excessively liberal ecclesiastical views of members of the Harvard faculty. According to Yale's charter, its purpose was to constitute a collegiate school "wherein Youth may be instructed in the

---

[1] This chapter has drawn from the unpublished English versions of the following two chapter-length essays: [6] and [7]. Compare also ([21]).
[2] The Collège de Québec, a Catholic institution, was actually founded in Nouvelle-France one year earlier in 1635, but did not begin to teach a complete classical course until 1659. See below. The Congregationalist movement, which dispensed with any organized administrative hierarchy and placed ultimate responsibility for the local church in the hands of its congregation, was greatly influenced by the writings of the English theologian Robert Browne and represented a vocal, dissenting sect from the Church of England. In America, prominent Congregationalists in the colonial period included John Cotton and Jonathan Edwards.

Arts and Sciences who through the blessing of Almighty God may be fitted for Publick employment both in Church and Civil State" [15, p. 7]. Between them, Harvard and Yale exemplified the highest level of mathematical learning and education possible in colonial North America, but that level barely extended beyond arithmetic and geometry in the opening decades of the eighteenth century.

At Harvard, Isaac Greenwood held the first chair for mathematics to be founded in what would become the United States, the so-called Hollis professorship established by the Englishman Thomas Hollis in 1727 [26, pp. 170–182]. In conjunction with his chair, Greenwood wrote in 1729 the first American-produced book of arithmetic to be published in the colonies [31, p. 55] and was the first contributor to a fledgling tradition of science in New England.[3] Two of his papers were published in the *Philosophical Transactions of the Royal Society*, and his teaching left its mark, for among his students was John Winthrop, who later succeeded him in the Hollis professorship.[4]

Winthrop, after Benjamin Franklin, is often regarded as the greatest American contributor to science in the eighteenth century.[5] Born in Boston in 1714, he entered Harvard in 1727 and was officially installed as Hollis Professor in 1739. He immediately undertook a series of astronomical observations that were published in the Royal Society's *Philosophical Transactions* (the first of eleven papers he eventually published there) and that ultimately earned him a fellowship in the Royal Society of London in 1766. Relative to mathematics, Winthrop taught the usual elementary subjects, but in 1751, he also introduced the much more advanced study of fluxions.

The first to leave his mark on mathematics at Yale was Thomas Clap, a Harvard graduate (class of 1722) and Congregationalist minister. Beginning in 1743, freshmen under his guidance studied arithmetic and algebra; sophomores, geometry; and third-year students, algebraic conics and fluxions. In 1758, problems on fluxions began to appear in commencement examinations and became increasingly difficult over subsequent years [27, pp. 217–10]; [29, p. 64]. Unfortunately, Clap was the only member of the staff at Yale who could teach the fluxional calculus, and when he left in 1766, the subject disappeared from the curriculum. It was not until the next century that a chair of mathematics *per se* was successfully inaugurated there.

The fortunes of mathematics at Harvard and Yale were in part a product of their Puritan, Congregationalist foundations.[6] The stricter religious principles of a Quaker variety dominant in Pennsylvania, however, tended to suppress the subject and even opposed the creation of the institutions where it was taught. For the Quakers, the belief in salvation through introspection of one's "inward light" meant that the usual trappings of learning, "Latin, Greek, mathematics, and natural philosophy" were all regarded as useless [12, p. 60]. In Canada, however, the religious divide was even more pronounced.

Although Jesuit missionaries had founded the Collège de Québec in Nouvelle-France in 1635, one year before the founding of Harvard College, the Collège de Québec only began to offer a complete classical course of study–including mathematics–in 1659, and mathematics was taught there primarily as a subject secondary to natural philosophy. Moreover, with the French defeat to the English in 1759, the Collège closed its doors permanently [1, pp. 142–147]. From that point until the union in 1840 of the French and English Canadas, French Canada was an English colony that had its own educational institutions, although these were still colleges or secondary schools in which mathematical education was geared primarily toward commercial mathematics [1, p. 152]. A parallel situation obtained in English Canada. Thus, in Québec, the Jesuits dominated, with their curriculum aimed at training men for the priesthood, while English Canada, following its creation in 1760, was primarily Anglican and followed an English model. In the colonial era, higher education, and hence the possibility of formally acquiring higher-level mathematics, was thus intimately linked to the religious affiliations of the various colonies.[7]

---

[3]The first arithmetics would not appear in French Canada until eighty years later with the publication in 1809 of Jean-Antoine Bouthillier's *Traité d'arithmétique pour l'usage des écoles* [1, pp. 150–151].

[4]Material here concerning the life and works of John Winthrop is drawn from the account given in John Langdon Sibley's *Lives of the Harvard Graduates, College Classes, 1642–1773*, as reprinted in [26, pp. 349–373].

[5]The renowned French scientist Louis-Antoine de Bougainville, known for his work both in mathematics and in natural history, was stationed under Montcalm in Québec in 1756 and remained there until the fall of Nouvelle-France to the English in 1760 [1, p. 145].

[6]The development of mathematics at Princeton was likewise affected by the fact that its founders were Presbyterian.

[7]Here, however, it is critical to realize that "higher" is a relative term. The "higher education" of seventeenth- and eighteenth-century North America was "higher" in the sense that it was beyond basic reading, writing, and arithmetic; the "higher-level mathematics" a student could thus acquire was only exceptionally, as at Yale under Thomas Clap, as "high" as the fluxional calculus. There was in this period no formal differentiation between secondary and what would today be called "higher education." That differentiation would only occur in the nineteenth century (see the next section). On the issue of this structural differentiation in the process of transplanting educational institutions and objectives from European countries to their empires in the Americas, see [25].

22.2 The Nineteenth Century: A Period of General Structure-Building in Higher Education and in Science

With the outbreak of the Revolutionary War in the American–as distinct from the Canadian–colonies in 1776 and the subsequent establishment of a new Republic, there was a strong motivation not only to break from the English mold but also to establish a culture that would ultimately rival those of Europe and especially England. In English Canada, however, the " 'democratic' influences" at play in the United States met with an "ardent opposition" that was "widely shared in loyalist British North America" [1, p. 148]. Early nineteenth-century English Canadian educational activists thus advocated the domestic production of mostly elementary textbooks to serve the needs of an emergent elementary and secondary educational system on an English model. The expectation was then that students who so desired would follow their secondary education in Canada with university education in England [1, p. 149]. This had the effect in English-speaking Canada, at least, of adopting a model and funneling students into a system that was becoming increasingly outmoded in light of educational and scientific developments on the Continent and especially in Prussia. In French-speaking Canada, the educational model remained a Jesuit one oriented toward primary and secondary education, although mathematics began to figure more prominently in that curriculum from the 1830s onward [1, pp. 153–156].

For the United States, however, the change in orientation away from England had major consequences that were reflected in the development of: (1) agencies within the Federal government to handle specific scientific needs such as the U. S. Coast Survey founded in 1807 and the Naval Observatory created in 1842; (2) the American Association for the Advancement of Science spearheaded by a group of scientists in 1848 as a new national organization for the promotion of science; and (3) scientific journals such as the *American Journal of Science and Arts* begun in 1818.[8] Owing to the facts that the relative numbers of scientists were small and that science was not highly specialized, the emphasis in the first half of the nineteenth century was on the general structure-building for science as a whole. These changes were also reflected at many U. S. colleges dating from the colonial period as well as at new colleges that formed in the decades prior to the American Civil War in 1861 [23, pp. 1–51]. The case of Harvard may be taken to exemplify the types of changes that occurred relative to mathematics prior to 1876, although Harvard represents a best-case scenario relative to mathematical training in mid nineteenth-century America.

Harvard began the first full century of American independence from English rule by overhauling its curriculum. Relative to mathematics, knowledge of arithmetic would be required for admission for the first time [5, p. 60]. This served to elevate the level of mathematical ability of Harvard students and to accommodate a new curriculum, this time one that drew its inspiration not from England but from the Continent and particularly from France.

Historians of British mathematics are familiar with the story of how the Cambridge Analytical Society took the first steps to introduce and promote continental achievements, beginning in about 1812. As Charles Babbage, George Peacock, and John Herschel, among others, spearheaded the reform of mathematics at Cambridge, at Harvard the fifth Hollis Professor of Mathematics, John Farrar, translated a number of continental texts–by Silvestre Lacroix, Leonhard Euler, Adrien-Marie Legendre, and others–into English for his students at Harvard beginning in 1818.[9] Of all his publications, however, it was the translation Farrar made (with George Emerson) of the *First Principles of the Differential and Integral Calculus ... Taken Chiefly from the Mathematics of Bézout* (1824) that had the greatest impact. This presentation of the calculus had a distinct advantage over Clap's presentation of fluxions at Yale or over Maclaurin's *Treatise on Fluxions* that had served as a text at the University of Pennsylvania. As Farrar noted in his introduction, he recommended Bézout's approach "on account of the plain and perspicuous manner for which the author is so well known, and also on account of its brevity and adaptation in other respects to the wants of those who have but little time to devote to such studies" [5, p. 130].

Farrar's numerous translations of continental works laid a foundation for teaching and for the development of a curriculum at a level markedly higher than at eighteenth-century Harvard. By 1830, all Harvard freshmen read Legendre's

---

[8] General structure-building also occurred in Canada in the nineteenth century, although, owing first to its colonial status and then to its status after 1867 as a self-governing dominion within the British Empire, the impetus behind such development was perhaps not as strong as in the United States. At least two general scientific societies were founded in Canada in the nineteenth century, the Canadian Institution (later the Royal Canadian Institution) in 1849 and the Royal Society of Canada in 1882. Both supported publications, with the *Canadian Journal* (later the *Proceedings* and then the *Transactions*) of the Canadian Institution being founded in 1852 and the *Transactions of the Royal Society of Canada* following in 1882.

[9] On Farrar and Harvard, see the account given in [5, pp. 127–133].

Elements of Geometry, in addition to what they studied of algebra and solid geometry; sophomores studied trigonometry with applications, topography, and calculus; juniors took natural philosophy, including mechanics, electricity and magnetism; and seniors applied their mathematics in such areas of natural philosophy as optics.

The most immediate beneficiary of this emphasis upon continental mathematics at Harvard was Benjamin Peirce, easily Farrar's most successful student. Owing both to Farrar's instruction and to his work on Nathaniel Bowditch's English translation of Laplace's *Méchanique céleste*, Peirce had a deep appreciation for the scope and power of European mathematics. On assuming the University Professorship of Mathematics and Natural Philosophy in 1833, Peirce "began reformulating Harvard mathematics according to his own vision" [23, p. 17] and that vision entailed producing textbooks for his students which were replete with original results rather than mere summaries or translations of what others had written. In 1838, moreover, Harvard not only adopted an elective system, which meant that all but freshmen might exempt themselves from having to take further mathematics, but also instituted three distinct options for the study of mathematics: a year-long practical course of study; a more theoretical curriculum designed primarily for teachers, and a three-year program aimed at producing professional mathematicians.[10]

The elective system at Harvard not only reduced greatly the number of students Peirce was expected to instruct, but it also liberated him from having to teach a broad range of courses to students across their four years of study. The establishment at Harvard in 1847 of the Lawrence Scientific School for science and engineering at what was basically the graduate level allowed Peirce to elevate his teaching even further and to present more serious mathematics to his better students.[11]

Unfortunately, precious few students benefited from this curriculum; in 1848, for example, there were only two [5, p. 141]. Although he failed to establish a thriving school of mathematical research, Peirce did establish a tradition in applied mathematics at Harvard that was carried on into the twentieth century by his son James Mills Peirce, William Byerly, and Benjamin Osgood Peirce [23, pp. 19–20]. At Harvard, what Peirce accomplished, above all, was the introduction through the Lawrence Scientific School of research-level mathematics. At mid century, only Yale of the American colleges was positioned even marginally to follow suit. In Canada, on the other hand, although McGill College had been officially inaugurated in 1829, the University of Toronto had been established under that name in 1850, and the Université Laval had received a royal charter in 1852, all of these institutions offered mathematics at an exclusively undergraduate level until at least the 1890s [1, pp. 158–159 and 162]. Major changes followed in the United States, however, in the wake of the American Civil War [13, 4].

22.3 A Mathematical Research Community Emerges in the United States: 1876-1900

The American Civil War, which divided the North from the slave-holding South during the bloody four-year period from 1861 to 1865, marked not only a turning point in American political history but also the beginning of a new era in the history of American higher education [30, pp. 1–18]. Before the war, as noted, colleges in the United States had largely been controlled by conservative clergymen of various Protestant religious persuasions and had embraced the traditional classical curriculum–with its focus on Latin, Greek, and Euclidean geometry–as a means for training the mind. Although other subjects–modern foreign languages, the sciences, history–had begun to make inroads into the curriculum [13], the goal of college education remained the production of liberally educated gentlemen, with the duty of the professor being to impart, not to create, knowledge. This was the hallmark of the colonial college.

After the war, a new generation of scholars–some of whom had been trained in Europe, some of whom had actively studied European educational systems, some of whom, like Peirce, were anxious to demonstrate that America was the cultural equal, if not superior, of Europe–sought to transplant their conceptions of the best European educational models to American shores. Their efforts, in some sense, occurred at an auspicious time. The war had created vast fortunes, and some of those newly rich–men like Ezra Cornell and Johns Hopkins–decided that the needs of higher education represented a worthy target of their philanthropy. In 1862, moreover, the Congress had provided Federal support–in the form of the Morrill Act–for the establishment in each of the states of a university that would subsequently be state-supported and that would provide training in practical arts such as agriculture, engineering, and mining for the

---

[10]On Peirce's educational initiatives at Harvard, see [16].
[11]For the complete list of courses, see [5, pp. 137–138] and [23, p. 50, Table 1].

betterment of the nation [8, pp. 149–151]). After the war, this act was extended even to those states that had broken from the Union.

Both of these new models, the privately endowed university and the so-called land-grant universities, provided opportunities for educationally reform-minded scholars–many of whom were scientists of one stripe or another–to implement their fresh ideas about the goals of the American university. For them, the university should not simply provide a liberal, undergraduate education, although this should be one of its goals. It should *also* be a hothouse for research and for the active contribution to the store of knowledge as well as a seedbed for future researchers. The "new education" these reformers ultimately crafted, with its twin ideals of teaching and research, at the new kinds of universities they created, fostered an environment in which many areas–among them, mathematics–blossomed at the research level [4, pp. 326–338].[12] The first university to create and foster a "modern" department of mathematics in the United States was The Johns Hopkins University [18].

Johns Hopkins was an institution born during the so-called "Gilded Age," the period immediately following the Civil War in which many businessmen who had profited from the war–men like John D. Rockefeller and Johns Hopkins– directed significant amounts of their newly amassed fortunes toward philanthropic concerns. In particular, they contributed to higher education and to the creation of the modern university, that is, to institutions devoted to undergraduate studies as well as to specialized training and to the production of original research. For his part, Hopkins bequeathed $7,000,000 for the creation of a university and a medical school, although he specified neither a plan nor a philosophy for the new school. That task fell to Daniel Coit Gilman, the university's first president and one of the new generation of American educational reformers. He was thus in the position of creating a university *de novo*, and he had distinct ideas. In particular, for Gilman, a faculty should consist of "men of acknowledged ability and reputation, distinguished in the special departments of study, capable of advancing these departments & also of inciting young men to study and research" [14, p. 26]. For the department of mathematics, he found a man with these qualities in the sixty-one-year-old English mathematician, James Joseph Sylvester.

Sylvester, who had made a reputation internationally as the developer with Arthur Cayley, of the British approach to invariant theory, animated a program in mathematics at Hopkins that generally centered on his own evolving research in invariant theory, number theory, the theory of partitions, and matrix theory. In his seven-and-a-half year tenure in Baltimore, Sylvester directed the research of some sixteen students, nine of whom ultimately earned the Ph.D.[13] Their work, moreover, tended more often than not to be published in the *American Journal of Mathematics*, a research-level journal founded in 1878, edited by Sylvester, and underwritten financially by the university.

The idea of a university-underwritten research journal was one of Gilman's innovations at Hopkins. As he realized, if the university were going to expect, indeed, require the production of original research from its faculty and students, it would have to provide a vehicle for the publication of such research. Although efforts had been made to sustain specialized mathematics journals in the United States prior to 1878, these had all ended in failure–some after only a volume or two–owing to the lack either of financial support or of sufficient materials or both [23, p. 51]. The *American Journal*, thanks to Sylvester's tireless research efforts as well as to the contributions he actively solicited from Europe, from the United States, and from his Hopkins students and colleagues, was the first specialized journal for research-level mathematics to survive in the United States, and it, in fact, thrived [22, pp. 239–248]. By the late 1870s, the nation was beginning to acquire the critical mass necessary to sustain a research-level mathematical community.

But it was only beginning. When Sylvester left Hopkins in December of 1883 to take up the Savilian Professorship of Geometry at Oxford University back in his native England, his students, although trained at the research level and imbued with the research ethos, found themselves by and large with positions at traditional, teaching-oriented colleges with no resources and precious little time to pursue their own research. For the next decade, American students who might have gone to Hopkins instead went to Europe–and especially to Felix Klein's lecture halls–for their mathematical training [23, pp. 189–259]. On their return to the United States, they, unlike Sylvester's students, found that other

---

[12]Charles Eliot, the president of Harvard from 1869 to 1909, actually coined the phrase "new education" in a pair of articles that appeared in the *Atlantic Monthly* in 1869 [10]. For him, at this moment when he was embarking on what would become his forty-year-long tenure as president, the "new education" meant primarily a curricular emphasis on the sciences, modern languages, and mathematics–as opposed to Latin, Greek, and mathematics–and not a focus on research and the training of future researchers. The latter would come to characterize the "new education" by the turn of the twentieth century.

[13]For a technical discussion and overview of Sylvester's mathematical research, see [18]. For a discussion particularly of his work as well as that of Story and Sylvester's students in the years from 1876 to 1883, see [23, pp. 99–138]. On Sylvester's training of students at the graduate level, see [18] and [22, pp. 235–273].

institutions had begun to follow the Hopkins model. In the 1890s, Yale and Harvard, as well as other universities like Cornell, Clark, and the University of Chicago which had been founded through private munificence, had begun to establish true graduate-level programs. Moreover, the new so-called "land grant" universities also followed that example relative to research and graduate study. Klein's students, unlike Sylvester's, found jobs in each of these kinds of institution in every region of the United States. They continued to do their research; they animated their own graduate-level programs in mathematics; they worked to establish new research-level journals and new societies for the promotion of mathematics [23, pp. 261–275].

Consider, for example, the case of the University of Chicago, a university founded in 1892 thanks to the benefaction of oil magnate, John D. Rockefeller. Its first department of mathematics, consisted of one American and two Germans: the Yale-trained Eliakim Hastings Moore and the German-trained students of Felix Klein, Oskar Bolza and Heinrich Maschke. At the graduate level in particular, these men crafted a curriculum that rivaled those of many of the German universities, offering "courses on the most important branches of modern mathematics such as: Theory of Functions, Elliptic Functions, Theory of Invariants, Modern Analytical Geometry, Higher Plane Curves, Theory of Substitutions, Theory of Numbers, Synthetic Geometry, Quaternions, Theory of the Potential ..., while other courses of a more special character and the Seminars [were] intended to introduce to research work" [23, p. 367].

Moore, Bolza, and Maschke–together with Moore's 1896 Ph.D. student, Leonard Eugene Dicskon, who joined the department in 1900–cooperated to animate the Chicago department of mathematics until Maschke's death in 1908 and Bolza's return to Germany two years later in 1910. Moore, whose early interests were in geometry, switched in the 1890s to group theory, then in the first decade of the twentieth century into axiomatics, and finally from the 1910s until his retirement in 1930 into functional analysis. Dickson took up the algebraic mantel from Moore, focusing on the theory of linear groups during the first of his almost four decades on the Chicago faculty. Bolza brought what he had learned at Klein's feet on the theories of elliptic and hyperelliptic functions and integrals to his Chicago classrooms in the 1890s but then began what would become a dynasty at Chicago in the calculus of variations after 1900. Maschke shifted mathematical gears as well. From the turn of the twentieth century to his untimely death in 1908, he moved from the theory of finite linear groups into the invariant theory of differential forms. Between them, Moore, Bolza, Maschke, and Dickson had guided thirty-one students into mathematical research and to the doctoral degree by 1910 [23, pp. 423–426].

They had also given their students more than an appreciation of mathematical research. They had imbued in them a sense of the importance and desirability of working toward the development of a professional, research-level mathematical community by organizing meetings, editing publications, and fostering societies [23, pp. 321–329]. In particular, in looking for a publisher for the proceedings of the mathematical congress held in Chicago in 1893 in conjunction with the World's Columbian Exposition, Moore and his co-organizers approached the New York Mathematical Society, which had been founded at Columbia University in 1888 and which had begun publishing its *Bulletin* in 1891. The negotiations which ensued not only resulted in the appearance of the Chicago congress volume in 1896 but also in the renaming of the society to the American Mathematical Society in 1894.[14]

Moore and his Midwestern colleagues had set in motion a process that would result in a truly national–as opposed to a Northeastern–society for the promotion of mathematical research by World War I [20]. By 1900, moreover, Moore had also been involved in founding and editing a new, research-level journal, the *Transactions of the American Mathematical Society*, and a year later, he was elected the first Midwestern president of the Society.

That the students in the Chicago program from 1892 to 1910 imbibed from their mentors not only the research ethos but also a sense of the greater community involvement in profession-building so exemplified by Moore can be seen perhaps most stunningly in the cases of four of Moore's early Ph.D.s [20, pp. 330–332]. In addition to Dickson, who came to Chicago during Moore's "algebraic period" and who went on to animate a school of algebraic research at Chicago that thrived well into the twentieth century [19], Moore also guided the research of Oswald Veblen, Robert L. Moore, and George David Birkhoff. Veblen and R. L. Moore came to work with Moore during his "axiomatic period" and wrote dissertations in 1903 and 1905, respectively, which dealt with devising complete and independent sets of axioms for geometry. Veblen moved on to a long and productive career first at Princeton and then at the Institute for Advanced Study, following its founding in Princeton in 1930. His efforts helped to reshape Princeton from a

---

[14]In Canada, specialized mathematical societies were founded later: the Société mathématique de Québec (founded in 1923) and the Canadian Mathematical Society (founded, although not under that name, in 1945).

colonial college into a modern research university on the Chicago model, while his research in differential geometry established Princeton as an internationally recognized center in the field. R. L. Moore had a more peripatetic career, ultimately engendering, through his noted "Moore method" of mathematical instruction, a school of point-set topology from his positions first at the University of Pennsylvania and then at the University of Texas at Austin [32]. Birkhoff came to Chicago from Harvard when Moore was in his "analysis period" and wrote, inspired by both his former Harvard professors and by Moore, a dissertation dealing with boundary value and expansion problems of ordinary linear differential equations. After a short stint with Veblen at Princeton, Birkhoff returned to Harvard, where he embraced and significantly enhanced the research reorientation well under way there. All four of these men also served as editors of major American mathematical research journals and as president of the American Mathematical Society. They followed well the lessons of their mentors at Chicago as they worked to strengthen the American mathematical research community in the decades prior to the outbreak of World War II.

22.4 The Twentieth Century: The Consolidation and Growth of Research-Level Mathematics

As the careers of Moore's Chicago students exemplify, over the period from 1900 to 1950, mathematics at the research level took hold at the new privately endowed universities, at many of the former colonial colleges, at state-supported universities, and at the land-grant institutions. In Canada, this period also witnessed a move, albeit a much slower one, toward mathematical research. One example should suffice to characterize this period of consolidation and growth within the context of mathematics in the United States–that of the land-grant University of California in Berkeley–while the 1924 International Congress of Mathematicians hosted in Toronto will provide a sense of developments in Canada.

When California officially became a state in 1850, its legislature had already made constitutional provisions for establishing what it viewed as the necessary infrastructure for a growing population. Although key among them was the idea of a state-funded university, the University of California was only legislated into existence in 1868 following the decision, made after much negotiation, to use the state's Federal land-grant funds to this end [28, pp. 30–34].

It took a year to put together the first faculty of nine professors and one instructor, but when the university opened its doors to students regardless of gender in the fall of 1869 two among them were mathematicians [28, pp. 50–52]. Research, however, was initially not part of the university's mission. Guided by an acting president and a board of regents with little real expertise in higher education, the university had little sense of mission beyond the prescripts of the Morrill Act to provide an education "somehow useful to the students in their classrooms" [28, p. 61]. That changed briefly in 1872 when the same Daniel Coit Gilman who would accept the presidency of The Johns Hopkins University in 1875 accepted first the presidency of the University of California, yet Gilman was ultimately unable to realize his vision for a research university in California.

Following Gilman's departure, Berkeley found itself in the hands of well-intentioned but lackluster presidents with little vision and little control. They did not, however, prevent the program in mathematics from attempting to capitalize on changes within the emerging American mathematical research community. In 1882, W. Irving Stringham, an 1880 Ph.D. under Sylvester at Hopkins who had pursued postgraduate study under Felix Klein in Leipzig, took up the professorship of mathematics. Eight years later in 1890, Stringham drew another of Klein's students, Mellen Haskell, to the Berkeley faculty as an assistant professor. Although both of these men had done original research–Stringham in elliptic functions and Haskell in Klein's brand of geometry–they found it difficult to maintain their research momentum in an academic atmosphere that failed to encourage it. As Stringham put it in a letter to Klein in 1888, "[t]he plants of intellectual culture grow but slowly, and on new raw ground like that of California they can hardly flourish without very great efforts" [23, p. 266].

Mathematics finally blossomed at Berkeley in the 1930s during the presidency of Robert Gordon Sproul, despite the financial stringencies imposed on the university by the nationwide economic depression. On the advice of a hand-picked and high-powered committee, Sproul plucked Griffith Evans from Rice University in Texas and gave him the explicit task of turning Berkeley's department of mathematics into a major center of mathematical research [24, pp. 287–292]. Evans, a proven and respected researcher who had done seminal work on the theory of integral equations as well as in applied mathematics and mathematical economics, not only had a broad vision of mathematics but also worked tirelessly to make that vision a reality at Berkeley. After his arrival in California, he cooperated with the efforts of the wider American mathematical community to place displaced European scholars, securing for Berkeley the services both of Hans Lewy, a leader in the field of differential equations and formerly at the Mathematics Institute in Göttingen [11, p. 286], and of Alfred Tarski, a major force in mathematical logic. Evans further resolved that Berkeley would build a major program in statistics, and to that end, he hired the Polish statistician, Jerzy Neyman [24, pp. 293–294].

In the fifteen years from 1933 to 1948, seventeen mathematicians, a number of them world-class, joined the Berkeley department. In all, they directed some fifty-five students successfully to the Ph.D., an 83% increase over the previous fifteen-year period. By capitalizing on a supportive administration, on world events, and on his own sense of both mathematical talent and the future directions of the field, Evans "realized the ambition of a major center of mathematical research and teaching at Berkeley" [24, p. 296].

The realization of similar ambitions would also come to Canada, although somewhat later. Writing in 1932, the Hopkins-trained Canadian professor of mathematics at the University of Toronto, John C. Fields, lamented that "progress in mathematics in Canada up to the present has not been all that might have been hoped for, [but] things look more promising for the future. There is a small but increasing group of younger men who are interested in mathematical research, and some of the later appointments have been encouraging" [1, p. 141]. Although only in "the years 1935-1945" would there be "distinct signs of research mathematics beginning to come to Canada" [1, p. 177], such a change may be said to have been foreshadowed by the International Congress of Mathematicians (ICM) organized by Fields and held in Toronto in 1924.

The 1924 ICM had been slated for New York City and would have been, in some sense, the official début of the newly emergent American mathematical research community on the international scene. Instead, when the American Mathematical Society withdrew its support in 1922 for political reasons surrounding the ban by the International Mathematical Union on mathematicians from the former Central Powers, Fields stepped up with an offer to host the event in Canada. By bringing the ICM to Toronto, Fields not only directly exposed his countrymen to some of the best mathematics being done internationally, he also created an opportunity to draw together his far-flung, mathematically-minded countrymen. In all, 107 Canadians and 191 Americans made up the 444 mathematicians present in Toronto, and the work of fifteen Canadians was represented by papers or abstracts in the Congress proceedings. Although much of that work may not have been "at the leading edge of research," it still reflected that mathematical research was being done in Canada in the 1920s [1, p. 173].

22.5 The North American Mathematical Landscape by 1950

The first five decades of the twentieth century witnessed the consolidation and growth of a North American mathematical research community in the context of a reorientation of higher education [23, pp. 427–428]; [1, p. 142]. In the United States, this had been influenced significantly by the endowment of new universities like The Johns Hopkins and the University of Chicago in the closing quarter of the nineteenth century. These universities, led by men influenced by their astute observations of the evolution of the liberal arts colleges in the United States as well as of educational models abroad, set a new standard for American higher education at the same time that they played a critical role in the professionalization of the various academic disciplines. In mathematics, their departments led the way in graduate training and, in the cases of Hopkins and Chicago, in undergraduate training as well. Their leaders, men like Sylvester and E. H. Moore, also recognized the need to establish the accoutrements of a mathematical *profession*–journals, seminars, congresses, societies–and they worked to realize these goals within institutional settings perfectly conducive to and supportive of their efforts.

Long-established institutions like Harvard and Yale reacted to the stimulus these upstarts provided, and this was also true of the newer, nineteenth-century models of state-supported and land-grant schools like Berkeley. In mathematics, this translated into the formation of research-oriented departments from coast to coast actively and successfully involved in training future generations of researchers.

The story, while different chronologically by almost a half-century, was nevertheless similar in Canada. When many of Europe's best mathematicians were forced by the political events of the 1930s and 1940s to flee their homelands, the fact that they were able to establish themselves almost immediately in an environment in North America that fostered their research bore testament to the fact that "[t]he level of mathematical activity in America was comparable to that brought to America by the newcomers" ([3], p. 238) and that Canada was not that far behind [1, p. 177]. The raising of the mathematical bar in North America, particularly in the United States, to a level that had made the country not only competitive but a leader internationally by 1950 owed in large measure to the emergence of the research university and to the efforts of talented individuals dedicated not only to research but also to the establishment of a community of professionals.

Bibliography

[1] Archibald, Thomas and Louis Charbonneau, 2005, "Mathematics in Canada before 1945: A Preliminary Survey." In *Mathematics and the Historians Craft: The Kenneth O. May Lectures*. Ed. Glen Van Brummelen and Michael Kinyon. New York: Springer-Verlag, pp. 141–182.

[2] Baltzell, E. Digby, 1979, *Puritan Boston and Quaker Philadelphia: Two Protestant Ethics and the Spirit of Class Authority and Leadership*. New York: The Free Press.

[3] Bers, Lipman, 1988, "The Migration of European Mathematicians to America." In *A Century of Mathematics in America–Part I*. Ed. Peter L. Duren *et al.* Providence: American Mathematical Society, pp. 231–243.

[4] Bruce, Robert V., 1987, *The Launching of Modern American Science 1846-1876*. New York: Alfred A. Knopf.

[5] Cajori, Florian, 1890, *The Teaching and History of Mathematics in the United States*. Washington, D.C.: Government Printing Office.

[6] Dauben, Joseph W. and Karen Hunger Parshall, 2007, "Dal Liberal Arts College alla Research University: Harvard, Yale e Princeton [From the Liberal Arts College to the Research University: Harvard, Yale, and Princeton]." In *Matematica e Cultura*. Ed. Claudio Bartocci and Piergiorgio Odifreddi. 4 Vols. Turin: Giulio Einaudi Editore S.p.a., 1: 477–504.

[7] ———, 2007, "L'evoluzione della ricerca universitaria: Johns Hopkins, Chicago e Berkeley [Mathematics and the Evolution of the Research University: Johns Hopkins, Chicago, and Berkeley]." In *Matematica e Cultura*. Ed. Claudio Bartocci and Piergiorgio Odifreddi. 4 Vols. Turin: Giulio Einaudi Editore S.p.a., 1: 505–529.

[8] Dupree, A. Hunter, 1986, *Science in the Federal Government: A History of Policies and Activities*. Baltimore: The Johns Hopkins University Press.

[9] Duren, Peter L. *et al.*, Ed. 1988–1989, *A Century of Mathematics in America–Parts I–III*. Providence: American Mathematical Society.

[10] Eliot, Charles, 1896, "The New Education: Its Organization." *Atlantic Monthly* 23, 203–220 and 358–367.

[11] Fermi, Laura, *Illustrious Immigrants: The Intellectual Migration from Europe 1930/41*. Chicago: The University of Chicago Press.

[12] Fisher, Sydney George, 1896, *The Making of Pennsylvania: An Analysis of the Elements of Population and the Formative Influences That Created One of the Greatest of the American States*. Philadelphia: J. B. Lippincott Co.

[13] Guralnick, Stanley M., 1979, "The American Scientist in Higher Education, 1820-1910." In *The Sciences in American Context: New Perspectives*. Ed. Nathan Reingold. Washington, D.C.: Smithsonian Institution Press, pp. 99–141.

[14] Hawkins, Hugh, 1960, *Pioneer: A History of the Johns Hopkins University, 1874-1889*. Ithaca: Cornell University Press.

[15] Kelley, Brooks Mather, 1974, *Yale: A History*. New Haven. Yale University Press.

[16] Kent, Deborah, 2005, "Benjamin Peirce and the Promotion of Research-Level Mathematics in America: 1830-1880." Unpublished Ph.D. Dissertation, University of Virginia.

[17] Morison, Samuel Eliot, 1956, *The Intellectual Life of Colonial New England*. New York: New York University Press.

[18] Parshall, Karen Hunger, 1988, "America's First School of Mathematical Research: James Joseph Sylvester at The Johns Hopkins University 1876-1883." *Archive for History of Exact Sciences* 38(2), 153–196.

[19] ——, 2004, "Defining a Mathematical Research School: The Case of Algebra at the University of Chicago, 1892-1945." *Historia Mathematica* 31, 263–278.

[20] ——, 1984, "Eliakim Hastings Moore and the Founding of a Mathematical Community in America, 1892-1902." *Annals of Science* 41, 313–333; Reprinted in *A Century of Mathematics in America–Part II*. Ed. Peter L. Duren *et al.* Providence: American Mathematical Society, 1989, pp. 155-175.

[21] ——, 2003, "Historical Contours of the American Mathematical Research Community." In *A History of School Mathematics*. 2 Vols. Ed. George M. A. Stanic and Jeremy Kilpatrick. Reston, VA: National Council of Teachers of Mathematics, 1:113–157.

[22] ——, 2006, *James Joseph Sylvester: Jewish Mathematician in a Victorian World*. Baltimore: The Johns Hopkins University Press.

[23] Parshall, Karen Hunger and David E. Rowe, 1994, *The Emergence of the American Mathematical Research Community 1876-1900: J. J. Sylvester, Felix Klein, and E. H. Moore*. Providence: American Mathematical Society and London: London Mathematical Society.

[24] Rider, Robin, 1989, "An Opportune Time: Griffith C. Evans and Mathematics at Berkeley." In *A Century of Mathematics in America–Part II*. Ed. Peter L. Duren *et al.* Providence: American Mathematical Society, pp. 282–302.

[25] Schubring, Gert, 2002, "A Framework for Comparing Transmission Processes of Mathematics to the Americas." *Revista Brasileira de História da Matemática* 2, 45–63.

[26] Shipton, Clifford K., 1963, *New England Life in the 18th Century: Representative Biographies from Sibley's Harvard Graduates*. Cambridge, MA: Harvard University Press.

[27] Simons, Lao Genevra, 1936, "Short Stories in Colonial Geometry." *Osiris*, 1, 584–605.

[28] Stadtman, Verne A., 1970. *The University of California 1868-1968*. New York: McGraw-Hill Book Company.

[29] Tucker, Louis Leonard, 1962, *Puritan Protagonist: President Thomas Clap of Yale College*. Chapel Hill: University of North Carolina Press.

[30] Veysey, Laurence R., 1965, *The Emergence of the American University*. Chicago: The University of Chicago Press.

[31] Wagner, Charles A., 1950, *Harvard: Four Centuries and Freedoms*. New York: E.P. Dutton.

[32] Zitarelli, David E., 2004, "The Origin and Impact of the Moore Method." *American Mathematical Monthly* 111, 465–486.

About the Author

Karen Parshall is Professor of History and Mathematics and Associate Dean for the Social Sciences at the University of Virginia. She has served as the Book Review Editor, Managing Editor, and Editor-in Chief of *Historia Mathematica*, as the chair of the International Commission for the History of Mathematics, and on the governing boards of both the History of Science Society and the American Mathematical Society. In 1996–1997, she was awarded a John Simon Guggenheim fellowship for a two-pronged biographical project on the nineteenth-century English mathematician, James Joseph Sylvester. The author of six books and some fifty scholarly articles, her special field is the history of nineteenth- and twentieth-century mathematics.

23

The Transmission and Acquisition of Mathematics in Latin America, From Independence to the First Half of the Twentieth Century[1]

Ubiratan D'Ambrosio
UNICAMP and UNIBAN, Brazil

23.1 Introduction

Recent scholarship traces the occupation of the Americas, justifiably called the New World, to about 40,000 years ago. The search for explanations (religions, arts and sciences), systems of values and behavior styles (communal and societal life), the psycho-emotional and the imaginary, and the models of production and of property were developed, in these cultures, in a way which is completely different of what was developed in the so-called Old World.

The major encounter identified as the Conquest of the Americas, initially by Spanish and Portuguese navigators, later followed by British, French and Dutch, brought European knowledge systems to the New World. In the period known as the Colonial Era, which extended from the early 16th century through the transition from the 18th to the 19th century, European knowledge laid firm institutional grounds.

The process of submitting of the entire planet to European powers reflects the ethos of these powers and thus characterizes the different objectives of the conquest and the way the colonies were ruled. This must be taken into account when trying to understand the specificities of mathematics development in the various countries of the region [3].

To consolidate the conquest, Spain and Portugal organized the administration of their colonies into Viceroyalties. They divided the conquered lands into five major areas: *Virreinado de Nueva España* (Viceroyalty of New Spain, roughly what is today Mexico and upper Central America), *Virreinado de Nueva Granada* (Viceroyalty of New Granada, southern Central America, approximately Costa Rica, Colombia, Venezuela, Ecuador), *Virreinado de Peru* (Viceroyalty of Peru, roughly what is today Peru, Bolivia and Northern Chile), *Virreinado de La Plata* (Viceroyalty of La Plata, roughly what is now Chile, Paraguay, Argentina and Uruguay) and the Viceroyalty of Brazil, which was a Portuguese conquest. In the process of independence, in the early 19th century, a new division was established, which is approximately, 200 years later, the current political map of Latin America.

---
[1] Based on the talk given in the HPM Satellite Meeting of ICME 11, Mexico City, 2008.

In this paper I will focus on the developments of the countries in the region called Latin America after they acquired their independence from Spain and Portugal, in the early 19th century.[2]

23.2 Independence

The Independence of the Thirteen Colonies from England greatly influenced and supported independence movements in the colonies of Spain and Portugal. The independence of the Viceroyalties of New Spain (Nueva España), New Granada (Nueva Granada), Peru, La Plata and Brazil was achieved in the first quarter of the 19th century. The political division in countries, following the independence, is practically the same as today. Spain retained sovereignty only in Puerto Rico and Cuba until the Spanish-American War (1898).

The pattern of cultural dynamics since colonial times did not change much, except for the fact that the hegemony of Spain and Portugal was challenged by the influence of other European countries.

A different History of Mathematics in Latin America begins. During the colonial era, the influence was, essentially, from Portugal and Spain. After independence, France became very influential, and French textbooks were amply translated. The newly independent countries attracted immigrants from other European countries. New waves of immigrants arrived, mainly from Italy and Germany. This changed considerably the academic atmosphere and, of course, the presence of mathematics in it was intensified.

It was soon recognized in the newly independent countries that [Western] Science and Mathematics are essential in the modern world. Public opinion was ready to support investment in scientific and mathematical research, in spite of being absolutely unable to guess what kind of research was supported.

Mexico, as a new independent country, continued to be a very important scholarly center, reflecting the importance of New Spain in the colonial times. But the independence of Guatemala, in 1821, lessened the influence of Mexico in Central and South America. The establishment of new universities and the renewal of the old ones, immediately preceding and after independence, generated open attitudes with respect to sources of knowledge on which to build up the newly established countries of Latin America. Formerly restricted almost exclusively to influences coming from Spain and Portugal, the new countries attracted considerable attention from the rest of Europe, and a new wave of scientific expeditions came to South America. The scientific exploratory missions had a great influence in creating a new intellectual elite throughout the region. This new source of intellectual interest is seen very strongly in the building up of large and diversified libraries, both public and private, and the acquisition of modern literature.

In Costa Rica the colonial authorities established the *Casa de Enseñanza de Santo Tomas* in 1814, in which the most influential teacher was Rafael Francisco Osejo, born in 1780. He wrote in 1830 *Lecciones de aritmética*, written in the form of questions and answers, a common feature in that period. In 1843 the *Casa de Enseñanza* [House of Teaching] was transformed into the *Universidad de Santo Tomas*, where careers in Engineering were established, but no career in Pure Sciences and Mathematics.

Colombia soon attracted foreign mathematicians. The Frenchman Bergeron introduced Descriptive Geometry in the country, and the Italian Agustín Codazzi (1793–1859) was influential in creating the *Colegio Militar*. Lino Pombo (1797–1862), who was particularly influential in founding the *Academia de Matematicas de Venezuela*, wrote a complete course of Mathematics.

In Brazil, the decision of the royal family of Portugal to transfer the capital of the Kingdom of Portugal from Lisbon to Rio de Janeiro, in 1808, to escape Napoleon's invasion, was decisive in changing the cultural life in the colony. The Portuguese court settled in Rio de Janeiro, where it was necessary to create an infrastructure to run, from a colonial town, the vast Kingdom of Portugal. They founded a major Library and a *Escola Militar* [Military School], among the first institutions of higher learning in the colony. Both were influential in the development of Mathematics in Brazil. In the school, a doctorate in Mathematics was established and a number of theses were submitted and defended.[3] Translations of the textbooks of Euler, of Lacroix, of Legendre and others were quite important in generating what we might call a mathematical style in Brazil.

Particularly interesting is the case of Joaquim Gomes de Souza (1829–1863), known as "Souzinha," the first Brazilian mathematician with a European visibility. He presented his results in the *Académie des Sciences de Paris* and in

---

[2] For a brief study of the pre-Columbian, the Conquest and the Colonial period, see [4]
[3] Clovis Pereira da Silva analyses these theses in [12]

the *Royal Society*. Only short notice of the papers were given,[4] and they were posthumously published as *Mélanges du Calcul Intégral*, as an independent printing by Brockhaus, of Leipzig, in 1889. This work, dealing mainly with partial differential equations, is permeated by very interesting historical and philosophical remarks, revealing Souzinha's access, before moving to Rio, to the most important literature then available. How was this possible in Maranhão, his home state in the North of Brazil? Most probably this was possible thanks to the existence of important private collections in Maranhão. A major task of historians of Brazilian Mathematics is to identify these libraries and, possibly, to recover the major work of Souzinha, a general treatise on Cosmos, much in the style of Humboldt. The book, described by Souzinha and others, is lost.[5]

Argentina, independent since 1816, experienced a remarkable intellectual development. In 1822, the ephemeral *Sociedad de Ciencias Físicas y Matemáticas* was founded in Buenos Aires. (See [10] for details.) There was an emergence of private libraries in Buenos Aires. Particularly important is the private library of Bernardino Speluzzi (1835–1898), which listed the main works of Newton, D'Alembert, Euler, Laplace, Carnot and several other modern classics. Another intellectual, Valentin Balbin (1851–1901), proposed, while Rector of the National College of Buenos Aires a new study plan, in 1896, which included History of Mathematics as a distinct discipline. This is probably the first formal interest in the History of Mathematics in South America, which eventually led to an important school of History of Science in Argentina. The French mathematician Jorge Duclout (1853–1929), who emigrated to Argentina in the decade of 1880, participated in the ICMs of Heidelberg (1904) and of Cambridge (1912), as an Argentinian delegate. He was responsible for inviting Albert Einstein to Argentina, in 1925, and initiated a Mathematics Seminar in the *Sociedad Científica Argentina*, in Buenos Aires, founded in 1872 and still active.

In Peru, it is to be mentioned an interest in Statistics, beginning with the book *Ensayo de estadística completa de los ramos económico-políticos de la provincia de Azángaro...* by José Domingos Choquechuanca (1789–1858), published in 1833.

In Chile, the Universidad de Chile was created in 1842, with a Faculty of Physical and Mathematical Sciences. A most distinguished member of the Faculty was Ramón Picarte, a lawyer, who had his paper *La división reducida a una adición*, accepted and published by the Academy of Sciences of Paris in 1859.[6] Much emphasis was given to teacher training. An agreement with the government of Germany, in 1882, provided the pedagogical support to reforming the education in the country. In 1889, fifteen mathematicians emigrated to Chile, to create a National Pedagogical Institute and to teach in Higher Secondary Schools. Surely, this is a factor in the excellence of Secondary Education in Chile, until nowadays. The German *Professor Doctors* were T. Kausel, R. Pöenish, R. von Lilienthal, A. Tafelmacher, F. Krüger, H. Stringe, G. Weidmann, H. Brockmann, O. Dörr, A.Heisler, A. Kredler, V.Misckssch, A. Ruckold, E. Thanheiser, R. Lenscher. A study of their mathematical background and their activities before and after immigration to Chile is due.

The influence of Auguste Comte (1798–1857) towards the end of the century was very important in all Latin America, particularly in Mexico and in Brazil. Although the main reason was the demand of the emerging political elites to build up the ideological framework of the new countries, Comtian ideas influenced a considerable development of Mathematics and the Sciences in general.[7]

23.3 The 20th Century up to the End of World War II

In the beginning of the 20th century, there was remarkable progress in Science and Technology in Latin America. In 1901, the Brazilian Alberto Santos Dumont (1873–1932), living in Paris, succeeded in manning a balloon around the Eiffel Tower, winning the Deutsch prize, and flying a heavier-than-air device (1904). (See [8] for details.)

The Health Sciences went through a significant development. An important contribution of Juan Carlos Finlay (1833–1915), from Cuba, was the discovery of the mosquito vector of yellow fever, in 1884. In 1902, the Brazilian physician Oswaldo Cruz (1872–1917), considered the founder of experimental medicine in Brazil, implemented programs of eradication of these vectors and of mass vaccination.

---

[4] See [7]. It is quite interesting to read the referee's reports and the reaction of Gomes de Souza to the fact that Liouville did not give an appraisal of the paper, according to Gomes de Souza, because of "la petite jalousie." A thorough study of the scientific works of J. Gomes de Souza is still due.

[5] For a more detailed study of the development of mathematics in Brazil, see [5]

[6] I did not have personal access to these papers and to the records of Picarte's presence in Paris.

[7] See the important doctoral dissertation [17].

The visit of Albert Einstein to Argentina and Brazil, in 1925, was a big support to the emergent scientific establishment in both countries. Particularly, in Brazil, the visit was a sort of final coup to the supporters of Positivism.

As the century advanced, a major recognition of Latin America science achievements was the granting of the Nobel Prize in Physiology or Medicine, in 1947, to Bernardo Houssay (1887–1971), from Argentina, for the discovery of the role played by the hormone of the pituitary lobe in the metabolism of sugar. Houssay, who graduated in Argentina, remained most of his life as Professor of the University of Buenos Aires and was most influential in forming a number of Latin American scientists.

Another international recognition was due to the Brazilian nuclear physicist Cesare G.M.Lattes (1925–), who in 1947, while in the University of California in Berkeley, played a major role in the discovery of the meson pi.

In mathematics, there were also remarkable advances. In the turn of the 19th century to the 20th century, we register important events which directly influenced mathematics in the region. We notice particularly an important effort of Germany to establish areas of influence in the Southern part of South America. What Lewis Pyenson called the "German Cultural Imperialism" is clearly seen in the development of the so-called Exact Sciences in Argentina, the same as in Chile. A major step to consolidate this influence was the efforts for the development of the Astronomical Observatory of La Plata. Richard Gans (1890–1954), a physicist who emigrated to Argentina in 1912, was very influential in the development of Science. (See [13] for more details.)

In 1917, the Spanish mathematician Julio Rey Pastor (1888–1962) visited Argentina and decided to stay there. Curricular innovation was going on in several institutions in Argentina. The Italian Hugo Broggi (1880–1965) emigrated to Argentina in 1912 to teach at the *Universidad de La Plata*. He emphasized numerical calculations and was responsible for the introduction of modern mathematical economics in Argentina. (See [9].) Particularly interesting are his 1919 lectures on Mathematical Analysis. In the preface, he thanks his colleague and friend, J. Rey Pastor, of the Universidad de Madrid, for "a work [of reviewing the book] which the ancients forgot to list among those of Hercules". ([1, p.3])

Rey Pastor remained in Argentina for most of his life, although he frequently visited Spain. These visits influenced greatly the development of Spanish mathematics. In addition to making important contributions to mathematics, mainly to Projective Geometry, Rey Pastor is essentially noteworthy for his contributions to the history of mathematics, specially of Iberian mathematics in the 16th century. Rey Pastor also marked new directions in historiography, drawing attention to the mathematical achievements that made possible the great age of navigation. A representative of his contribution is the 1942 book *La Ciencia y la Técnica en el Descubrimiento de América*, (Espasa-Calpe Argentina S.A., Buenos Aires).

José Babini (1897–1983), who was a disciple of Rey Pastor in Argentina, became one of the most distinguished historians of science and mathematics in Latin America. His career as a driving force of Mathematics in Argentina is significant. He was a founder of the *Unión Matemática Argentina*, and, in 1920, he became Professor at the *Universidad Nacional del Litoral*. Babini published important papers in Mathematics and contributed considerably to scholarship on the Jewish medieval contributions to Mathematics. Besides, he was concerned with the popularization of science and mathematics and wrote many books and published articles in non-specialized periodicals. His major work was doubtless the book he co-authored with Julio Rey Pastor, on the History of Mathematics [14]. I consider this a well written book, in a good Spanish literary style. The authors cover the important facts and personalities in the conventional histories of mathematics plus some aspects of the Muslim influence in Iberic Mathematics and of the mathematics supporting the Iberic navigation, as well as interesting references to Indian Mathematics. It is important to observe that the book was finished and first published in 1951, when these topics were not commonly seen in similar books. Also interesting is the annotated Bibliography, which hints on the scholarship of the History of Mathematics in that period. The study of this Bibliography and the comments would be, in itself, quite interesting. Later, Babini was a major collaborator in a multi volume History of Science, written by Aldo Mieli, a distinguished Italian scholar who had immigrated to Argentina.

In the 1930s, some European mathematicians immigrated to Argentina. Among them was the distinguished Italian mathematician Beppo Levi (1875–1961), who established an important research center in Rosario and founded an influential journal, *Mathematica Notae*. Well known for his seminal theorem on the theory of integral, Beppo Levi devoted much of his research to the history of mathematics. Particularly to be noticed is his 1947 book *Leyendo a Euclides*, (Editorial Rosario S.A., Rosario), a critical analysis of the general organization of the *Elements*.

One of the important and influential mathematicians in Latin America was a disciple of Rey Pastor, Luis Alberto Santaló. Born in 1911, Santaló had studied with W. Blaschke in Germany and already acquired an international visibility when he emigrated to Argentina during the Spanish Civil war. Santaló reached world renown as a founder of modern Integral Geometry and for his contributions to the History of Geometric Probabilities and has published relevant studies on Buffon. He became a most influential scholar in Mathematics, Mathematics Education and the History of Mathematics, in all of Latin America.

In neighboring Uruguay, an important tradition of mathematical research was established early in the 20th century. A representative of this movement, particularly devoted to the history of mathematics, was Eduardo García de Zuñiga (1867–1951), who succeeded in creating a most complete library in the history of mathematics at the *Facultad de Ingeniería de la Universidad de la Republica*, in Montevideo. His research was mainly in Greek Mathematics.[8]

In Brazil, the proclamation of the Republic in 1889 reinforced the influence of positivism. In the beginning of the 20th century, a number of young mathematicians were absorbing the most recent progress of Europe. Among them were Otto de Alencar, Manuel Amoroso Costa, Teodoro Augusto Ramos and Lelio I. Gama. In 1916 the *Academia Brasileira de Ciências* was founded. With the inauguration of the *Universidade de São Paulo*, in 1934, the first fully operational university in Brazil, we see a new direction in Science and Mathematics. We might say this is the beginning of academic research in Brazil. The University of São Paulo recruited promising researchers in Europe. Among several young faculty, from Italy came Luigi Fantapiè, Giàcomo Albanese and Gleb Wataghin, in Functional Analysis, Algebraic Geometry and Physics, respectively. (See [18] for more information.)

23.4 After the End of the Second World War

After World War II, a number of European scientists emigrated to Latin America, among them many mathematicians. The distinguished Portuguese mathematician, Antonio Aniceto Monteiro (1907–1980), joined the *Universidade do Brasil*, in Rio de Janeiro in 1945 and in 1947 moved to Argentina, where he was responsible for establishing a very important research center in Bahia Blanca. (See [15, 16]). Of fundamental importance was the presence of André Weil and Jean Dieudonné as Professors at the University of São Paulo, Brazil, from 1946 to 1948, and other French mathematicians of the Bourbaki group, for shorter periods. (See [2].) Although not as distinguished, German mathematicians went to South America. Chile gave continuity to the previous collaboration with German mathematicians and the *Centro de Investigaciones Matematicas de la Universidad de Chile* hired, in 1952, K. Legrady, of the University of Hamburg, a specialist in Integral Geometry.

Before WW II, Europe was the source of visitors and the place for Latin Americans to go abroad for studies. After World War II, we see a large number of European scientists and mathematicians in Latin America, some looking for employment and some as part of the efforts of former colonial powers, specifically France and England, to preserve their cultural presence in what became known as the Third World. Instrumental in these efforts were organizations such as the British Council, ORSTOM and the *Coopération française*. UNESCO played an important role in supporting the efforts of former colonial empires.

With respect to American influence, we see that the cultural and economic interest of the United States in Latin America were, at best, moderate before World War II. This changed radically after the war. Particularly, this resulted in an increasing influence of the United States in the development of Science and Mathematics in Latin America. Technological development has different characteristics, since it is subordinated to more immediate economic interests.

The growth of American influence is of a different nature, clearly focusing on continental hegemony. The Organization of American States was instrumental in favoring United States influence and exchanges. The United States became the main destination of a generation of young students pursuing their doctorates abroad. The creation of the National Science Foundation set up the model to be soon followed by practically every Latin American country through the national research councils in most countries, usually called the CONICYTs, the CONACYTs and the like.[9] An effort of the American Association for the Advancement of Sciences to cooperate with homologous organizations in Latin America is also to be noted.

In the fifties, a view of the state of Science and Mathematics in Latin American as a whole was due. The initiative focused on Mathematics and an important meeting was convened by the *Oficina Regional de Ciencia y Tecnologia para America Latina y el Caribe* [Regional Office of Science and Technology for Latin America and the Caribbean] of UNESCO, in Montevideo, Uruguay, in 1951, to report on Mathematics research going on in the region. This was

---
[8] His collected works have been published as [6].
[9] These are the initials of *National Council of Scientific and Technological Developments*, in Spanish.

the *Symposium sobre Algunos problemas matemáticos que se están estudiando en Latino America* [Symposium about some problems in Mathematics being studied in Latin America].

The Proceedings of the Symposium gives an idea of some of the areas of mathematical research deserving interest in Latin America. We learn there of the work of Leopoldo Nachbin, of the *Universidade do Brasil*, who was then doing very advanced research on the Theorem of Stone-Weierstrass and launching the basis of a major school on Holomorphy and Approximation theory in Brasil; of the advances on Integral Geometry by Luis Santaló, one the most distinguished researchers in this area, in *Facultad de Ciencias de la Plata*, Argentina; of the presence of Francis D. Murnangham in Brazil with the mission of building up a research group on modern applied mathematics and matrix theory in the *Instituto Tecnológico de Aeronáutica*, a model institution of advanced technology sponsored by the Brazilian Armed Forces and academically modeled upon M.I.T., in São José dos Campos, Brasil; of Mischa Cotlar, who in the *Facultad de Ciencias de Buenos Aires* was doing important work on Ergodic Theory in cooperation with R. Ricabarra; of Mario O. González, of the *Universidad de la Habana*, working on Differential Equations; of Alberto González Dominguez, of the *Facultad de Ciencias de Buenos Aires*, working on distributions and analytic functions; of Carlos Graeff Fernández, of the *Universidad de México*, working on Birkhoff's gravitational theory; of Godofredo Garcia, of the *Facultad de Matemáticas de Lima*, on General Relativity; of Rafael Laguardia, of the *Instituto de Matematica y Estadistica de la Facultad de Ingeneria de Montevideo*, on Laplace transforms.

We also learn about the presence of Wilhelm Damköhler, a German specialist in the Calculus of Variations, who emigrated to the *Universidad Nacional de Tucuman*, Argentina, and later went to the *Universidad de Potosí*, Bolívia; of Peter Thullen, of the O.I.T. [International Organization of Labor] office in Paraguay, working on Several Complex Variables; of Kurt Fraenz, of the *Facultad de Ciencias de Buenos Aires*, on the mathematical theory of electric circuits. Invited discussants were Augustin Durañona y Vedia, of the *Facultad de Ciencias de La Plata*; Roberto Frucht, of the *Facultad de Matematicas y Fisica de Santa Maria*, Chile; Pedro Pi Calleja, of the *Facultad de Ciencias de La Plata*; Cesario Villegas Mañe, of the *Facultad de Ingeneria de Montevideo*. Some visitors who had been lecturing in South America also participated in the Symposium, among them Paul Halmos, from the USA.

An important, although very incomplete, account of research in Mathematics going on in Latin America was given by Julio Rey Pastor in the paper "*La matemática moderna en Latino América* [Modern Mathematics in Latin America]," which appeared in the *Segundo Symposium sobre Algunos problemas matemáticos que se están estudiando en Latino América, Villavicenzio-Mendoza, 21–25 Julio 1954*, UNESCO, Montevideo; p. 9–20.

This "dropping of names" should not be regarded as an account of what was going on in South America in 1950. Many more individuals were active in Mathematics. Although the UNESCO Symposium was an attempt to have a picture of what was going on in mathematics research in Latin America, a number of mathematicians quite active in several countries of the region were not invited to the meeting. But every one of the mathematicians attending the Symposium deserves a study of their life and work and of their influence in the respective countries. This should be a priority theme for research in the History of Mathematics in Latin America. Looking into the *Mathematical Reviews* we might be able to find how representative was this group of invitees and to identify several other active mathematicians who were not invited.

In mid-century, Rafael Laguardia and José Luiz Massera were responsible for the creation of a distinguished research group in the stability theory of differential equations in the *Instituto de Matemática y Estadistica de la Facultad de Ingeneria de la Universidad de la Republica*, in Montevideo. This research group attracted young mathematicians from all of Latin America and abroad and achieved world recognition. The military dictatorship established in Uruguay in 1971 saw, in the declared political position of José Luiz Massera and Rafael Laguardia, a reason to simply close the excellent mathematical library of the university, thus interrupting all mathematics research in the country. Uruguayan mathematicians immigrated to several countries where they became very influential. Massera spent all the period of the military regime in jail and on his release he did not resume mathematics research and decided to pursue a political career. About the Library, Massera commented, in 1986, that

> "Without taking into account certain merits in other sciences, the library (referring to one in the Facultad de Ingeniería, Montevideo) was not long ago one of the best libraries in Latin America, if not the best, and it had a level comparable to that of the great libraries in Europe and in the USA."

Rafael Laguardia died in Montevideo during the political repression. Maybe more than in any other country under military dictatorship in South America in the sixties, Uruguay is an example of how a flourishing research group can be immobilized by a governmental decision.

23.5 Concluding Remarks on Contemporary Developments

A good perception of the developments of mathematics in Latin America after the fifties can be obtained by the study of the *Colóquios Brasileiros de Matemática* [Brazilian Mathematics Colloquia], held every two years, beginning in 1957, organized by the distinguished IMPA /*Instituto de Matemática Pura e Aplicada* [Institute of Pure and Applied Mathematics], of Rio de Janeiro. Also the ELAM /*Escuela Latinoamericana de Matemáticas* [Latin American School of Mathematics], held in different countries, is an indicator of the evolution of mathematics research in the region. It is important to analyze Latin American participation in the International Congresses of Mathematicians and other international meetings.

To analyze Contemporary History is a difficult task, since we have to refer to processes still going on and we risk stumbling into personal and political sensibilities. Several academians were active in the period when military regimes took control of the governments of countries that were showing the strongest vitality in research in Sciences and Mathematics. The military coups, which occurred sequentially in the four countries that were active in research: Brazil 1964, Argentina 1966, Uruguay 1971, Chile 1973, caused an important migratory flux of scientists in all areas, initially among these countries, and soon directed to the few Latin American countries which were able to keep democratic regimes, particularly Mexico and Venezuela. After the re-democratization of Argentina (1983), Brazil (1984), Uruguay (1984) and Chile (1989), some scientists returned and reclaimed their positions. Others were able to maintain their positions during the military regimes, and kept these positions after the end of military dictatorships.

The dividing line between opponents and sympathizers or collaborators of the military regimes is very difficult to draw. Obviously, personal conflicts are still latent. An International Symposium on "*La Migración de Científicos en los Países del Cono Sur: determinaciones económicas y políticas* [The Migration of Scientists in the Countries of the Southern Cone: Economical and Political Determinants]," was convened by the FEPAI: *Fundación para el Estudio del Pensamiento Argentino* [Foundation for the Study of Argentinian Thought], in July 1986. The interventions and debates revealed open wounds which remain from the period of military dictatorship. Although unpleasant and somewhat painful, it is important to look into this period and its consequences while some of the protagonists are still alive. I esteem this as an important and needed research project.

Although we recognize some interest in the History of Science and Mathematics since colonial times, in the last decades it became a growing area of academic interest throughout Latin America. The founding of the *Sociedad Latinoamericana de Historia de las Ciencias y la Tecnologia* [Latin American Society of History of Science and Technology], in 1983, in Mexico, stimulated the organization of national societies devoted to the History of Science. Some national societies emerged. Young scientists have recently obtained doctorates in history of science in both Europe and in North America, which is a hopeful sign of maturity and gives a professional status for research throughout Latin America. Among the research areas more frequently seen are the traditional areas studied in Europe and USA universities, but there a growing interest in studying the History of local Science, Technology and Mathematics.

Although the scenario of mathematics research in Latin America is not homogeneous, we recognize a great effort, even in the less favored countries, to develop mathematics research. Many mathematicians from Latin America, some with graduate degrees obtained in their countries, immigrated to the USA, where they are permanent faculty. The role of Marshall Stone in the welcoming of Latin American mathematicians to the international mathematical community was of great importance. (See [11] for details.) Membership in the International Mathematical Union/IMU is also a significant indicator. Members from Latin America increased. They are now Argentina, Brazil, Chile, Colombia, Cuba, Mexico, Peru, Uruguay and Venezuela. [10] A relevant indicator of the quantitative and qualitative development of Mathematics in Latin America is the participation in the ICM/International Congresses of Mathematicians and other major events, where mathematicians from Latin America have been invited as major speakers.

It is impossible to give in a short paper an account of the important recent developments in Science, Technology and Mathematics in Latin America. By electing only a few events and names, it is inevitable that some very important individuals, institutions and events are not mentioned, which does not imply lessening their academic standing.

---

[10] For details, see www.mathunion.org/

Bibliography

[1] Broggi, Hugo, 1919. *Análisis Matemático*, 2 vols., Facultad de Ciencias Físicas, Matemáticas y Astronómicas de La Plata.

[2] Da Cunha Pires, Rute, 2006. *A Presença de Nicolas Bourbaki na Universidade de São Paulo*, Doctoral Thesis, Pontifícia Universidade Católica de São Paulo.

[3] D'Ambrosio, Ubiratan, 1988. The Ethos of Spanish and Portuguese Colonial Enterprises, in *Revolutions in Science. Their Meaning and Relevance*, William R. Shea editor, Watson Publishing International/Science History Publications, Canton, MA, pp. 284–291.

[4] D'Ambrosio, Ubiratan, 2006. A Concise View of the History of Mathematics in Latin America, *Ganita Bharati* (Bulletin of the Indian Society for the History of Mathematics), vol. 28 (2006), 1–2, June, December, pp.111–128.

[5] D'Ambrosio, Ubiratan, 2008. *Uma História Concisa da Matemáticano Brasil*, Editora Vozes, Petrópolis.

[6] García de Zuñiga, E., 1992. *Lecciones de Historia de las Matemáticas* (ed. Mario H. Otero), Facultad de Humanidades y Ciencias de la Educación, Montevideo.

[7] Gomes de Souza, J., 1856. *Comptes-Rendus de l'Académie des Sciences de Paris*, tomes XL, 1310, and XLI, 100 and *Proceedings of the Royal Society*, 146–149.

[8] Hoffman, Paul, 2003. *Wings of Madness. Alberto Santos-Dumont and the Invention of Flight*, Hyperion, New York.

[9] López, M Fernández, 2003. Ugo Broggi: a precursor in mathematical economics, *European Journal of the History of Mathematical Thought*, vol 10, (2), 303–328.

[10] Nicolau, Juan Carlos, 1996. La Sociedad de Ciencias Fisicas y Matematicas de Buenos Aires (1822–1824), *Saber y Tiempo*, 2, 149–160.

[11] Parshall, Karen H., 2009. Marshall Stone and the Internationalization of the American Mathematical Research Community, *Bulletin of the AMS* (New Series), 46, 459–482.

[12] Pereira da Silva, Clovis, 2003. *A Matemática no Brasil. Uma história de seu desenvolvimento*, 3ª ed., Editora Edgard Blücher, São Paulo.

[13] Pyenson, Lewis, 1985. *Cultural Imperialism and Exact Sciences. German Expansion Overseas 1900–1930*, Peter Lang, New York.

[14] Rey Pastor, Julio and José Babini, 1951. *Historia de la Matemática*, Espasa-Calpe Argentina S.A., Buenos Aires.

[15] Rezende, Jorge; Monteiro, Luiz; and Amaral, Elsa (co-eds.), 2007. *António Aniceto Monteiro: Uma fotobiografia a várias vozes*, Sociedade Portuguesa de Matemática.

[16] Saraiva (ed.), Luís, 2008. *António Aniceto Monteiro (1907-1980)*, Número especial do Boletim da Sociedade Portuguesa da Matemática.

[17] Silva da Silva, Circe Mary, 1991. *Positivismus und Mathematikunterrichte: Portugiesche und franzsische Einflüsse in Brasilien im 19. Jahrundert*, IDM, Bielefeld.

[18] Zornoff Taboas, Plínio, 2005. *Luigi Fantappiè: Influência na Matemática Brasileira. Um Estudo de História como contribuição para a Educação Matemática*, Doctoral Thesis, IGCE/UNESP/Rio Claro, Brazil.

About the Author

Ubiratan D'Ambrosio is Emeritus Professor of Mathematics, State University of Campinas/UNICAMP, São Paulo, Brazil (retired in 1994). He received the degree of Doctor (Mathematics) from University of São Paulo/USP, Brazil, in 1963, with a thesis on Geometric Measure Theory and the Calculus of Variations. He was elected Fellow of the American Association for the Advancement of Science with the citation "For imaginative and effective leadership in Latin American Mathematics Education and in efforts towards international cooperation (1983)." He received the Kenneth O. May Medal in the History of Mathematics, granted by the International Commission of History of Mathematics, an affiliate of the IUHPS and IMU (2001). He was also the recipient of the Felix Klein Medal of Mathematics Education, granted by the International Commission of Mathematics Instruction/ICMI, an affiliate of the IMU (2005). He is a member of the International Academy of the History of Science and is currently Professor of Mathematics Education and History of Mathematics at UNIBAN/Universidade Bandeirantes de São Paulo, Brazil.

24

In Search of Vanishing Subjects: The Astronomical Origins of Trigonometry

Glen Van Brummelen
Quest University, Canada

24.1 Introduction

As any high school math teacher will tell you, the word "trigonometry" means "triangle measurement." The more perspicacious teacher might even know that the word was first coined by Bartholomew Pitiscus with his *Trigonometriae* [16], a study of the so-called "science of triangles" (Figure 24.1). This sounds familiar, even comfortable to modern teachers and researchers; we feel that we know what trigonometry is about, what it's for, and where it came from. For most of us, we couldn't be more wrong.

By 1600, much of the trigonometry that we saw in school had been known for well over a millennium. It had traveled through several major mathematical cultures, taking on different forms as it went. Trigonometry was not even properly

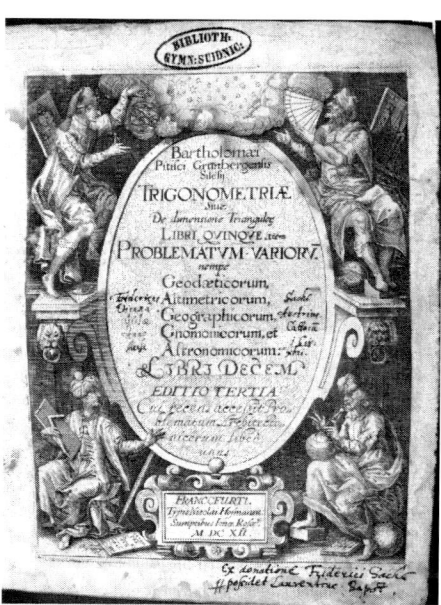

Figure 24.1. The title page of Bartholomew Pitiscus's *Trigonometriae*.

a mathematical subject for most of this time, existing and taking its purpose mostly as a helpmate to astronomy. And this implies that what was really important to most trigonometric practitioners during its age of discovery—working with problems on the surface of the celestial sphere—is so marginal to us that (with a couple of minor exceptions) it has not been taught for several decades. How many of us today can state a single theorem in spherical trigonometry?

It is always a valuable experience to relive the mathematics of our past; this is especially true for trigonometry. Exploring the shadowy edges of a subject that recedes and eventually vanishes entirely into other disciplines and cultures as we move back in time reminds us of a couple of points that we are wont to forget. Firstly, mathematics has always gained much inspiration from outside intellectual endeavors. Secondly, and just as significantly, mathematics develops much as other great ideas do: in fits and starts, with sudden advances but also with changes in direction and byways abandoned. The upward march of mathematical progress is far from monotonic.

24.2 False Beginnings

As one might expect for a subject vanishing gradually into the historical mist, different authors have different ideas about when trigonometry began. Claims start as early as the Egyptian Rhind papyrus (1850 BC; see [18]), which contains five problems that correspond roughly to finding the slope of the side of a pyramid, given various facts about its dimensions. Then there is the infamous and nearly contemporaneous Babylonian tablet Plimpton 322 see [19], which contains a collection of Pythagorean triples. The way in which these triples are ordered has tempted some to speculate that the unknown author was dealing systematically with the relative dimensions of sides of right-angled triangles.

But is this trigonometry? Such a simple question deserves, but is not going to get, a simple answer. The Plimpton 322 trigonometric theory has been scorned by recent historians: there are explanations of the tablet (in terms of school problems involving reciprocal pairs) that are a better fit with other tablets that have been excavated. Beyond this, neither Old Babylon nor ancient Egypt had any concept of an angle or arc. This is deeply significant because, as we shall see shortly, what distinguished trigonometry as a fundamental and cosmos-altering idea when it really did get started was the relation between measurements of two qualitatively different sorts of objects: arcs on circles on the one hand, and line segments on the other. So, we may exclude these very ancient episodes as outside of what identifies trigonometry as a unique art: the capacity to convert arbitrary angles/arcs on a circle to lengths of line segments.

This reasoning is immediately satisfying, but as tempting as it is, it's suspect. Unless you believe that there is just one way to codify our knowledge—namely, how we moderns do it—then, using a definition thought up after more than two millennia of progress doesn't do justice to the past. It is the basic historical error sometimes called presentism, or Whig history: using our modern understanding to shape the past to our liking. If we allow ourselves to do that, then in terms coined by Ivor Grattan-Guinness [9], our project becomes the *heritage* of *our* mathematics, rather than the *history* of *theirs*.

But if we mathematicians cannot begin with definitions, then what can we do? We must go back to the axioms. What is the ultimate purpose of history in general, and of the history of mathematics in particular? As a small corner of the history of ideas, we are here part of what seems to be an eternal debate, framed nicely by two 20th century historians. Arthur Lovejoy pursued his historical quarry armed with the concept of a *unit-idea* (most famously in *The Great Chain of Being*, 1936/1964), revisiting a timeless concept as it is itself revisited by different communities and cultures. Quentin Skinner [21], Lovejoy's main critic, rejects altogether the notion of a "perennial idea" independent of culture, in favor of "contextualism": recovering the original author's intent through an appreciation of how language, schools of thought, and other influences shaped a unique fabric of intellectual life, incommensurable with apparently similar ideas elsewhere that are separated in place or time. The difference between the two men comes down to opposing answers to the "why" question: do we study history primarily to understand timeless concepts, or rather to understand human behavior?

At first glance mathematics seems to lend itself to Lovejoy's unit-idea, as timeless as theorems appear to be. But the defining trend in the recent practice of the history of mathematics is toward contextualism. Studying Plimpton 322 to understand and derive new results about Pythagorean triples is a silly thing to do. Studying it to wrestle with how mathematics affected the Babylonians, and how the Babylonians affected the growth of the subject, is meaningful. So, if history is relevant to us as a study of the changing human condition, then the Rhind papyrus (as interesting as it is) has nothing to do with the history of trigonometry—not because it doesn't fit a disciplinary definition, but because it

24.3 Heavenly Foreshadowing

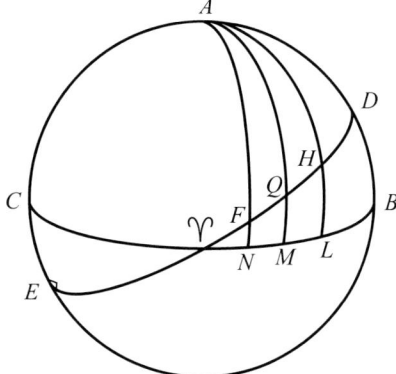

Figure 24.2. The diagram for III.6 of Theodosius's *Spherics* (simplified slightly)

did not affect any later developments in the subject. Missing a linked narrative to join the Rhind papyrus to the rest of our account, we are relegated to cross-cultural comparison. We might learn something interesting from such an exercise, but we would no longer be doing *history*. The essence of the discipline is the thread of *transmission*, and because of this we can rule the Rhind papyrus and Plimpton 322 out of the story.

24.3 Heavenly Foreshadowing

To find the earliest frayed, yet legitimate ends of the trigonometric thread, we fast forward a millennium to the late Babylonian period, where astrologers/astronomers spent prodigious efforts predicting celestial events. But they did so using arithmetical patterns that they recognized in their prognostications, extrapolations of their observations rather than relying on an underlying physics. Their field of play was the celestial sphere, the dome of the heavens that contains the stars on its surface, and us at the center. Whether or not the Babylonians thought of this dome in geometric terms as we do is a matter for debate. They did follow the sun's journey through the heavens over the course of the year, and this path, the *ecliptic*, was eventually divided into units roughly equal to one day's travel yet convenient for their base 60 number system. And here we have the first appearance of a 360° circle.

Meanwhile, in Greece the astronomers were thinking geometrically, but not directly in a computational manner. Scientists such as Autolycus (late 4th century BC; see [3]), Theodosius (2nd century B.C.; see [22]), and even Euclid (3rd century BC; see [4]) wrote a number of treatises in the geometrical tradition known as "spherics." As the name suggests, these works contained mathematical theorems about constructions on the surface of a sphere—but these theorems were not what one might expect from a book of geometry. Consider this statement from Theodosius's *Spherics*, simplified to avoid verbiage: in Figure 24.2, draw two great circles BC (with pole A) and DE. Cut equal contiguous arcs HQ and QF on DE, both above BC, and draw arcs from A downward to the equator as illustrated. Then $LM > MN$.

No doubt this theorem is true, but what use could it possibly have in a geometric context? On the other hand, suppose we consider Figure 24.2 to be the celestial sphere, BC the celestial equator, and DE the ecliptic. Suddenly the arcs LM and MN are significant: they represent projections of arcs of the ecliptic onto the equator; in other words, they are parts of a coordinate system. Measured from ♈ (the so-called vernal equinox, because that is where the sun is on that day), $\alpha = L\text{♈}$ is the *right ascension* of H, and $\delta = LH$ is its *declination*. And now, Theodosius's result can be seen to be equivalent to the statement that right ascensions not only increase as the sun moves along the ecliptic, but at an ever-faster rate.

So spherics, far from being geometry, was a thinly disguised form of astronomy: mathematical, but not quantitative. We are getting closer to trigonometry, but without numbers we are still in a completely different intellectual space.

We come perilously close in the third century BC, with the appearance of what might be described as a trigonometric lemma in Aristarchus's *On the Sizes and Distances of the Sun and Moon* (see [11]) and Archimedes's *Sand Reckoner* (3rd century BC; see [6]). In the right-angled triangles of Figure 24.3, where $AC = DF$ and $\alpha > \beta$,

$$\frac{EF}{BC} < \frac{\alpha}{\beta} < \frac{DE}{AB}.$$

Aristarchus uses this theorem to determine bounds on the ratio of the sun's distance from the earth to the moon's

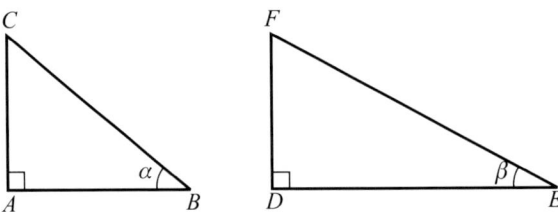

Figure 24.3. An ancient trigonometric lemma

distance; Archimedes uses it to get a convenient lower bound for the sun's diameter, on his way to bounding the number of grains of sand it would take to fill the universe. The trigonometric content in the lemma can be seen more clearly by us moderns with this equivalent statement:

$$\frac{\sin \alpha}{\sin \beta} < \frac{\alpha}{\beta} < \frac{\tan \alpha}{\tan \beta}.$$

When the angles are small, this formula produces reasonably tight bounds. But this lemma is not very close to a systematic tool converting arbitrary arcs to precise lengths.

24.4 A Union of Opposites

What happens next is, sadly, lost to us. Somewhere in the second century BC and later, a genuinely different activity began. Rather than merely capturing trigonometric quantities between bounds, scientists began to assert them to have particular values. We may only hypothesize, but surely it cannot be a coincidence that this is the century that brought Babylonian base 60 numeration, including the 360° circle, to Greek geometry. Our earliest evidence for this union of opposites is in reports of the works of the shadowy, yet foundational figure of Hipparchus of Rhodes. Although all but one of his books is lost, we do have an ancient report that Hipparchus wrote a treatise determining the lengths of chords in a circle. The first table of chord lengths that we actually have in our possession is in Claudius Ptolemy's *Almagest* (AD 140; see [17]) almost three centuries later, excerpted in Figure 24.4. It must be admitted that we have no idea whether Hipparchus used base 60 numeration, but we nevertheless assert that Hipparchus holds the laurel of originator of trigonometry.

How can we say this? Following Skinner and the contextual model, a beginning would imply a clean break with what went before: the birth of a new astronomical conversation. It is clear that Hipparchus's work fits the bill. He posed geometric models of the motions of the sun and moon, which is par for the Greek course to this point; but he took these models in a new quantitative direction. For instance, Ptolemy quotes Hipparchus as having derived accurate numerical estimates for lunar parameters, for instance, that the ratio of the radius of the moon's epicycle to the radius of its deferent circle is $247\frac{1}{2}/3122\frac{1}{2}$. These precise specifications extend the power of the geometric model and give it a startlingly modern appearance: for the first time, a geometric model conforming to an assumed physics (namely, Aristotle's) could be used to predict the future behavior of an astronomical object.

Precisely how Hipparchus did this is not known, but we have Ptolemy's methods, which cannot be much different. We begin with Figure 24.5, Ptolemy's determination of the eccentricity of the sun's orbit around the earth—which has some claim to being the oldest trigonometric problem. From our vantage point on the earth, the sun travels around us with speeds that vary slightly throughout the year. For instance, in Hipparchus's time it took $94\frac{1}{2}$ days to travel from the vernal equinox to the summer solstice, and $92\frac{1}{2}$ days to proceed onward to the autumnal equinox. To allow the sun actually to change its speed on the circle would force a descent into the mathematical unknown, so instead Ptolemy (following Hipparchus) moves the earth away from the center of the circle Z, to a new position E. But how far should the earth move; i.e., how long is ZE?

Knowing the season lengths and a year length of $365\frac{1}{4}$ days, we can find θK and KL in degrees, from which it is easy to find $\theta NY = 4; 20° = 4 + \frac{20}{60}°$. The key step is to convert this arc length into the length of the chord θK, which Ptolemy does easily enough with his table of chords: $4; 32 = 4 + \frac{32}{60}$ units, where the circle has radius 60. From this, we know that $XE = 2; 16$. We can apply a similar procedure to go from QPK to QFK, and onward to $ZX = 1; 2$. Pythagoras takes care of the rest; Ptolemy finds $ZE = 2; 29\frac{1}{2}$. Clearly, this procedure places Ptolemy and Hipparchus in a world very different from the lower and upper bounds of Aristarchus and Archimedes, not too long before.

24.4. A Union of Opposites

Arcs	Chords
½°	0;31,25
1°	1;2,50
1½°	1;34,15
⋮	⋮
36°	37;4,55
⋮	⋮
60°	60;0,0
⋮	⋮
72°	70;32,3
⋮	⋮
90°	84;51,10
⋮	⋮
120°	103;55,23
⋮	⋮
179½°	119;59,56
180°	120;0,0

Figure 24.4. An extract from Ptolemy's table of chords (interpolation column omitted). The entries are in sexagesimal (base 60) notation, so for instance, $\mathbf{1; 2, 50} = \mathbf{1} + \frac{2}{60} + \frac{50}{60^2}$.

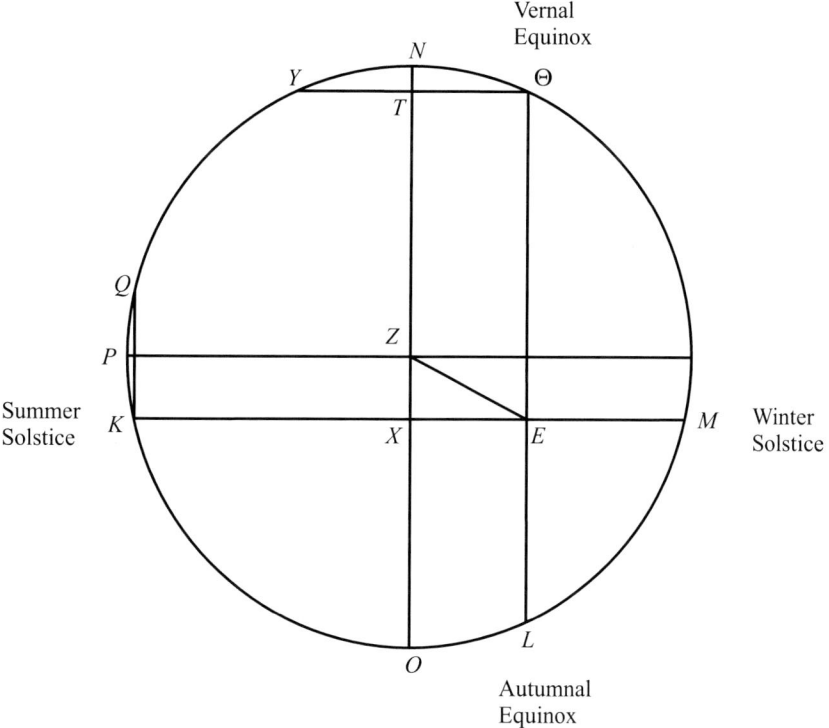

Figure 24.5. Hipparchus's solar model, and perhaps the first trigonometric problem. The goal is to find the orbit's eccentricity ZE given the lengths of the seasons and the radius of the circle.

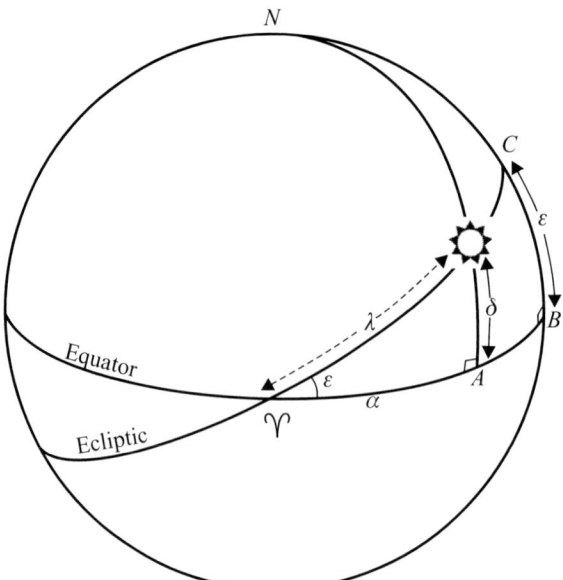

Figure 24.6. Using Menelaus or the Rule of Four Quantities to determine the declination

But we have blurred over the blessed moment. Just how did Ptolemy transform arcs into chord lengths; i.e., how did he come up with his chord table in the first place? Some entries are easy: for instance, in a circle of radius 60, the chord of 60° (i.e., Crd 60°) is simply 60 units long, because if one joins the endpoints of the chord to the circle's center, an equilateral triangle results. Chords of other arcs such as 36°, 72°, 90°, and 120°—those corresponding to the sines that some high school students memorize on the unit circle—are almost as easy. But to go further one must turn to more general geometrical tools, in the Euclidean style. Using Ptolemy's Theorem (in a quadrilateral inscribed in a circle, the product of the diagonals is equal to the sum of the products of the opposite sides), Ptolemy proves equivalents of the sine difference and cosine summation formulas. With the help of the Pythagorean theorem and a half-angle identity, Ptolemy is able to calculate chord lengths of all multiples of $1\frac{1}{2}°$.

Clearly, with the tools already at Ptolemy's disposal, all he needs now to fill in the gaps and end up with a table of chords for $\frac{1}{2}°$- arc increments is a good value for Crd 1°- but at this pivotal moment, geometry fails him. To accomplish this would be equivalent to trisecting the angle. So Ptolemy resorts to an equivalent of the same centuries-old lemma used by Aristarchus and Archimedes: if $\alpha > \beta$, then

$$\frac{\operatorname{Crd}\alpha}{\operatorname{Crd}\beta} < \frac{\alpha}{\beta}.$$

Substituting $\alpha = 1°, \beta = \frac{3}{4}°$ and $\alpha = 1\frac{1}{2}°, \beta = 1°$, Ptolemy magically gets $1; 2, 50 < \operatorname{Crd} 1° < 1; 2, 50($ $= 1 + \frac{2}{60} + \frac{50}{60})$; so with a guilty glance in Euclid's direction, he asserts that Crd 1° actually equals $1; 2, 50$, more or less.

But what of the sphere, which because of its astronomical application is supposed to dominate trigonometry? It is close behind. Some time between Theodosius and Menelaus (an astronomer about half a century before Ptolemy), the science of spherics was similarly transformed by the infusion of quantitative results. We are not sure when this happened; our first record of the new science is Menelaus's *Spherics* (see [13]). The first of its three books already overturns the discipline. Instead of dealing with astronomical concepts at arm's length, Menelaus approaches spherical triangles genuinely as geometrical objects, modeling his presentation on the sections of *Elements* Book I dealing with plane triangles. Book II (and part of the middle of Book III) brings us back to familiar Theodosian ground, with demonstrations of various inequalities of arcs.

However, Book III begins with a transformative bang. Its first theorem, the most frequently used in all of spherical trigonometry for the next 900 years, works with the configuration in Figure 24.6, called a *spherical quadrilateral*. It asserts two relations between chords of great circles in the diagram: namely,

$$\frac{\operatorname{Crd} 2AZ}{\operatorname{Crd} 2BZ} = \frac{\operatorname{Crd} 2AG}{\operatorname{Crd} 2GD} \cdot \frac{\operatorname{Crd} 2DE}{\operatorname{Crd} 2EB} \text{ and } \frac{\operatorname{Crd} 2AB}{\operatorname{Crd} 2AZ} = \frac{\operatorname{Crd} 2BD}{\operatorname{Crd} 2DE} \cdot \frac{\operatorname{Crd} 2GE}{\operatorname{Crd} 2GZ}.$$

24.4. A Union of Opposites

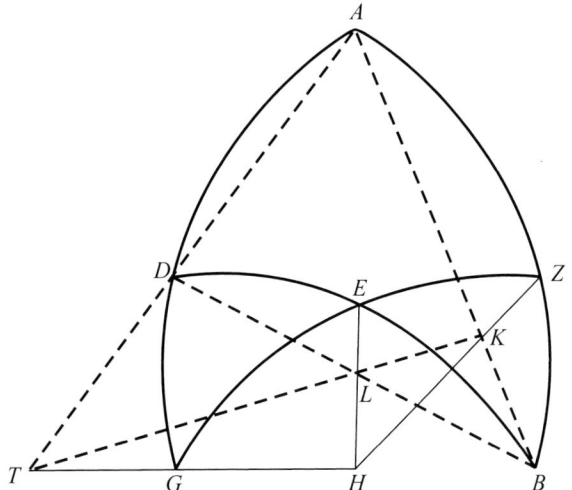

Figure 24.7. Menelaus's Theorem, plane and spherical

(To read the theorem in modern terms, replace all the occurrences of "Crd 2" with "sin.") With these two seemingly obscure statements it is possible to solve almost all problems in spherical astronomy. For instance consider Figure 24.6, in which the sun's position on the ecliptic is given by λ, and the task is to determine its equatorial coordinates—namely, the right ascension α and the declination β. If we apply the first theorem above to the configuration ♈ABC☼ and note that several of the arcs are equal to 90°, we arrive at the standard formula $\sin \delta = \sin \lambda \sin \varepsilon$. (The derivation of α is left to the interested reader, with the hint to use the second of the Menelaus theorems above.)

In these unenlightened times, the term "Menelaus's Theorem" is most often associated with one of its planar equivalents, in Figure 24.7

$$\frac{AK}{KB} = \frac{AT}{TD} \cdot \frac{DL}{LB}.$$

Menelaus proves the first of the spherical formulas above from this theorem; the reader is challenged to see how it can be done from the diagram. (The second Menelaus Theorem can be proved similarly, from another planar statement similar to the above.)

It is no exaggeration to assert that the spherical Menelaus's Theorem was the most important trigonometric result in the ancient and medieval world. Almost all spherical problems were reduced to it until turn-of-the-millennium Muslim scientists found less cumbersome theorems on which to base their work. Most common of these replacements was the Rule of Four Quantities, which asserts (Figure 24.8; $\angle A = \angle D$; $\angle G = $ either $\angle Z$ or $180° - \angle Z$)

$$\frac{\sin AB}{\sin BG} = \frac{\sin DE}{\sin EZ}.$$

To see how much easier this is to use than Menelaus, apply it to configuration ♈ABC☼ in Figure 24.7. The result $\sin \delta = \sin \lambda \sin \varepsilon$ pops out almost immediately.

It is startling to notice that the Rule of Four Quantities, and several other important spherical trigonometric theorems besides, actually appear immediately after Menelaus's Theorem in Book III of his *Spherics*. But, for whatever reason, it took 900 years for astronomers to adopt these more elegant results to their work. So, Menelaus's *Spherics* was not

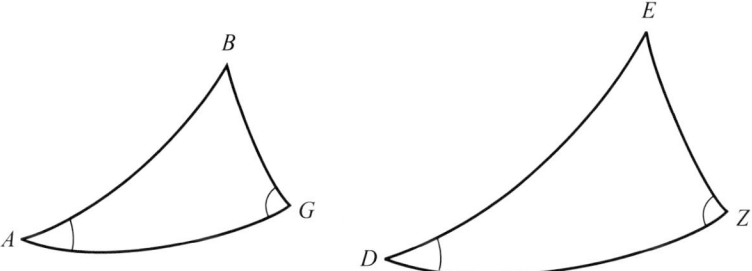

Figure 24.8. The Rule of Four Quantities

only a genre-breaker for spherics and a foundational work for spherical astronomy; it was also centuries ahead of its time.

24.5 Cultural Divergence: Trigonometry in India

If we are to take Skinner's contextualism at face value, it would seem to be a fool's errand to write a history that crosses cultural gaps as vast as that separating ancient Greece from medieval India. What common ground could they possibly have? And yet the origins of trigonometry in India, murky and foreign though they appear, echo distinctly of Greek ancestry. Among other tell-tale traces, we find the division of the circle into 360° and epicyclic models for the motions of the planets. But these models are not employed in the same way. The Greeks did something like what we would do: begin with the physics and work out from it a geometry that "saves the phenomena," then work from the geometry to the predictions. Indian astronomy concerns itself more with the predictions, and is much less bothered about a direct connection to physical reality. Thus one finds for each planet a pair of epicycles representing two different phenomena, but the planet cannot even be located on the diagram as a single identifiable point.

What we know of early Indian astronomy and its accompanying trigonometry seems to have come from a phase of Greek astronomy in the lost period before Ptolemy. In fact, the theory has recently been revived that some of the earliest Indian trigonometric tables take some inspiration from Hipparchus himself. The most peculiar feature of some of these tables (see Figure 24.9) is the use of a base circle with radius $R = 3438$. This seemingly bizarre choice comes from dividing the circle into $60 \cdot 360 = 21,600$ minutes of arc, and choosing one minute of the circle's arc as a unit of *length*. Thus $R = 21,600/2\pi \approx 3438$. Now, a recent reconstruction by Dennis Duke [7] of calculations of numerical parameters for the lunar model due to Hipparchus and quoted by Ptolemy in the *Almagest* (including the $247\frac{1}{2}/3122\frac{1}{2}$ ratio we saw earlier) provides compelling evidence that $R = 3438$ is precisely what Hipparchus used. This innovation has the highly desirable property that sines of small arcs (measured in minutes) are almost exactly equal to the arcs themselves—a feature that we Westerners exploited when we adopted radian measure of angles, and which we use implicitly every time we take a derivative of a trigonometric function. This property gave Indian trigonometers a leg up over the Greeks in one important respect: the problem that Ptolemy (as well as Muslim and Western astronomers, all the way through Copernicus's collaborator and trigonometric table-making prodigy Rheticus) faced in determining geometrically inaccessible chords or sines of certain small angles was neatly bypassed.

θ	Sin θ
3¾°	225
7½°	449
11¼°	671
15	890
⋮	⋮
30	1719
⋮	⋮
45	2431
⋮	⋮
60	2978
⋮	⋮
75	3321
⋮	⋮
86;15	3431
90	3438

Figure 24.9. Extracts from Āryabhata's sine table

24.5. Cultural Divergence: Trigonometry in India

i	Sin i	$\Delta(i)$	$\Delta^2(i)$	Sin $i/\Delta^2(i)$
1	59.9969	59.9969	-0.018276	-0.00030461
2	119.9756	59.9787	-0.036546	-0.00030461
3	179.9178	59.9421	-0.054805	-0.00030461
4	239.8051	59.8873	-0.073047	-0.00030461
5	299.6194	59.8143	-0.091267	-0.00030461
6	359.3424	59.7230	-0.109459	-0.00030461
7	418.9559	59.6136	-0.127618	-0.00030461
8	478.4419	59.4859	-0.145738	-0.00030461
9	537.7821	59.3402	-0.163814	
10	596.9585	59.1764		

Figure 24.10. Sines, their differences, and a relationship that leads to a new method of calculating them.

Calculating the rest of the sine table was done in India in a manner alien to the West. Consider for instance Āryabhata's method in his Āryabhatīya (AD 499; see [10]). The values of the first few first differences of the entries, $\Delta(i)$ in Figure 24.10, decreases in a manner reminiscent of a similar trend in a cosine table. Likewise, the second differences $\Delta^2(i)$ increase in a manner reminiscent of the sine function (if we ignore the negative sign). Any introductory calculus student can tell you why. But if we divide the second differences by the original sines, we find not only that they are close to each other, but that they are in fact precisely the same number.

How does this lead to a completed sine table? Simply by assuming that the last column remains always constant, one can work one's way backward through the columns from right to left, adding an extra entry to each column as one goes. Eventually, row by row, the entire grid can be filled in, and the sine table emerges as the leftmost column.

But if the title of this article refers to *origins*, why talk of India at all? Clearly India was not the first to think trigonometrically. As our earlier rejection of the Egyptian and Babylonian episodes indicates, being first off the mark is not *sufficient* to be identified with the origins of the subject, particularly if what one does has no interaction with later developments. In this case there is little doubt of a continued conversation, both backward (as we have seen) and forward. Indian astronomy was the first inspiration for the beginnings of Muslim science in the late 8th century AD, and many Indian innovations (including the sine) found their way to Islam around this time. Only in the next century, in the era of translations of Greek scientific texts, did Islam inherit the Greek geometric tradition. And even then, Indian achievements (especially in computation) continued to resonate through Islam and eventually into the West.

So India is beyond all doubt part of the trigonometric story. But is it *necessary* to be first in line when considering the origins of an idea? Consider the obvious differences in style and values between Ptolemy's and Āryabhata's trigonometric table construction methods. Ptolemy demonstrates theorems. As much as he can, he lays out in full Euclidean glory the logical reasons to accept the statements that he sets out. The original chords are verified, and the identities are proved relying on propositions that either can be found in Euclid, or are proved based on Euclid. Only when he encounters the problem with Crd 1° does he find himself forced outside of geometric reasoning, and even then he strays as little as he can. This minor deviation from the geometric straight and narrow, repeated by a number of Muslim mathematicians, met with explicit critiques and repudiations, for instance from the 12th century Islamic scientist al-Samaw'al (see [24, pp. 145–146]) and the Renaissance philosopher Giordano Bruno (see [8]).

What al-Samaw'al and Bruno would have thought of Āryabhata's work, I shudder to think. The contrast between Āryabhata and Ptolemy begins in what is committed to paper. Āryabhata prescribes the entire calculation in a short paragraph. There is no geometric proof, no attempt to justify the calculation even in a heuristic way. In early Indian astronomy, the Euclidean dream of basing knowledge on an axiomatic (or at least a logical) foundation is simply not valued. On the other hand, we find a keen awareness of the values of successive differences between entries in trigonometric tables. These patterns, likely first discovered empirically, led to Āryabhata's and many other insights— eventually, to the amazing 14th century derivations of formulas resembling Taylor series for the sine and cosine by Mādhava's Keralese school (see [24], pp. 113–120). This is close to an understanding and manipulation of what we might call first- and higher-order derivatives (although paradoxically these are derivatives without calculus), and it took until the Renaissance and later for these potent ideas to be rediscovered elsewhere. These mathematical mannerisms, traits, and commitments are foreign to what went before, almost entirely incommensurable with Ptolemy and his tradition. Nevertheless, computational reasoning in trigonometry went on to add richness when it transmitted to Muslim astronomy, and onward to the West.

So Āryabhata and his colleagues, while not at the front of the chronological queue, originated a way of thinking trigonometrically that was new, foundational, to have long consequences, and extremely effective. This effectiveness must be evaluated on its own terms, for values of geometric precision and logical clarity simply don't apply here. By the standard of what Indian astronomy wanted from its trigonometry, namely, accuracy and predictive power, the bar is easily cleared. It was especially in Muslim science that Indian and Greek trigonometry came together: by synthesizing them, Islamic scientists fused the advantages of both and produced a discipline that, other than the lack of symbolic notation, looks very familiar to our modern eyes.

24.6 Conclusion

The question of origins in the history of mathematics can be the tip of a rather large iceberg, full of interpretive problems and easily missed sensitivities. It is deceptively easy to allow our own point of view to organize and place values on the historical literature; we must be constantly on our guard against this. But we can be assertive about where trigonometry came from intellectually: it was motivated almost entirely by the needs of astronomy, and for its first 1500 years was motivated mostly by its link to that science.

This raises interesting and difficult issues for the teacher of trigonometry who wants to import some of its history. How is it possible, in the limited classroom time available, to incorporate the admittedly substantial amount of astronomy needed to truly teach the subject with a historical bent? Even more troubling, if history is to be used as a template, what do we do with the fact that what was important historically (applications on the sphere) is utterly forgotten today? Clearly a trigonometry unit based on history would be a complete rebuild from what is taught now, and such a challenge seems almost insurmountable for an individual teacher within an existing curriculum. Some individual historical episodes have great classroom potential: for instance, the use of basic angle sum/difference and half angle identities to build early sine tables. But a significant part of the story comes with too much overhead to be incorporated in the classroom as more than a sidelight. Does this mean that the big picture is unimportant for mathematics teachers interested in history? Hardly. While the spherical law of cosines is likely to remain in a dark closet, a teacher who is aware of the astronomical origins of trigonometry can provide surprising motivations for specific topics, even if what s/he is able to say cannot be very specific. When we teach our children to use trigonometry to find the heights of trees and distances across lakes, it is best to remember that it is, and has always been, a much more powerful enterprise.

Acknowledgment This article was inspired in part by the Kenneth May lecture given at the annual meeting of the Canadian Society for History and Philosophy of Mathematics in June 2008 at the University of British Columbia in Vancouver, and by a plenary lecture given at the History and Pedagogy of Mathematics satellite meeting of the International Congress of Mathematics Education in July 2008 in Mexico City. Those lectures were inspired by the author's book, *The Mathematics of the Heavens and the Earth: The Early History of Trigonometry*, Princeton 2009.

Bibliography

[1] Aaboe, Asger, 1964, *Episodes from the Early History of Mathematics*, New York: Random House / The L. W. Singer Company.

[2] Āryabhata, 1976, *Āryabhatīya of Āryabhata*, eds. K. S. Shukla and K. V. Sarma, New Delhi: Indian National Science Academy.

[3] Autolycus, 1971, *The Books of Autolykos. On a Moving Sphere and On Risings and Settings*, Frans Bruin and Alexander Vondjidis (tr.), American University of Beirut.

[4] Berggren, J. Lennart, and Robert Thomas, 1996, *Euclid's "Phaenomena": A Translation and Study of a Hellenistic Treatise in Spherical Astronomy*, New York/London: Garland. Reprinted Providence, RI: American Mathematical Society, 2006.

[5] Björnbo, Axel Anthon, 1902, "Studien über Menelaos' *Sphärik*," *Abhandlungen zur Geschichte der mathematischen Wissenschaften* 14, 1–154.

[6] Dijksterhuis, E. J., 1956, *Archimedes*, Copenhagen: Munksgaard. Reprinted Princeton: Princeton University Press, 1987.

[7] Duke, Dennis, 2005, "Hipparchus' eclipse trios and early trigonometry," *Centaurus* 47, 163–177.

[8] Gatti, Hilary, 1999, *Giordano Bruno and Renaissance Science*, Ithaca/London: Cornell University Press.

[9] Grattan-Guinness, Ivor, 2005, "History or heritage? An important distinction in mathematics and for mathematics education," in *Mathematics and the Historian's Craft: The Kenneth O. May Lectures*, Glen Van Brummelen and Michael Kinyon (eds.), New York: Springer, pp. 7–21. Reprinted from *American Mathematical Monthly* 111 (2004), 1–12.

[10] Hayashi, Takao, 1997, "Āryabhaṭa's rule and table for sine-differences," *Historia Mathematica* 24, 396–406.

[11] Heath, Thomas Little, 1913, *Aristarchus of Samos. The Ancient Copernicus*, Oxford: Clarendon Press. Reprinted New York: Dover, 1981.

[12] Knorr, Wilbur, 1985, "Ancient versions of two trigonometric lemmas," *Classical Quarterly* 35, 362–391.

[13] Krause, Max, 1936, "Die Sphärik von Menelaus aus Alexandrien in der Verbesserung von Abū Naṣr Manṣūr b. ᶜAli ibn ᶜIraq," *Abhandlungen der Gesellschaft der Wissenschaften zu Göttingen*, Philologisch-Historische Klasse (3) 17, 1–254, 1–110 (Arabic numbering).

[14] Lovejoy, Arthur O., 1936/1964, *The Great Chain of Being: A Study in the History of an Idea*, Cambridge, MA: Harvard University Press.

[15] Pingree, David, 1978, "History of mathematical astronomy in India," in *Dictionary of Scientific Biography*, vol. 15, Charles Coulston Gillispie (ed.), New York: Charles Scribner's Sons, pp. 533–633.

[16] Pitiscus, Bartholomew, 1600, *Trigonometriae*, Augsburg: D. Custodis.

[17] Ptolemy, Claudius, 1984, *Ptolemy's Almagest*, Gerald J. Toomer (tr.), London: Duckworth / New York: Springer Verlag. Reprinted Princeton: Princeton University Press, 1998.

[18] Robins, Gay, and Charles Shute, 1987, *The Rhind Mathematical Papyrus: An Ancient Egyptian Text*, London: British Museum Publications.

[19] Robson, Eleanor, 2001, "Neither Sherlock Holmes nor Babylon: a reassessment of Plimpton 322," *Historia Mathematica* 28, 167–206.

[20] Sidoli, Nathan, 2006, "The sector theorem attributed to Menelaus," *SCIAMVS* 7, 43–79.

[21] Skinner, Quentin, 2002, "Meaning and understanding in the history of ideas," Chapter 4 of *Visions of Politics*, vol. 1: *Regarding Method*, Cambridge, UK: Cambridge University Press, pp. 57–89.

[22] Theodosius, 1927, *Les Sphériques de Théodose de Tripoli*, Paul ver Eecke (tr.), Bruges: Desclée de Brouwer.

[23] Toomer, Gerald J., 1973, "The chord table of Hipparchus and the early history of Greek trigonometry," *Centaurus* 18, 6–28.

[24] Van Brummelen, Glen, 2009, *The Mathematics of the Heavens and the Earth: The Early History of Trigonometry*, Princeton/Oxford: Princeton University Press.

About the Author

Glen Van Brummelen is a historian of mathematics and astronomy, particularly in ancient Greece and medieval Islam. He is author of *The Mathematics of the Heavens and the Earth: The Early History of Trigonometry* (Princeton, 2009) and co-editor of *Mathematics and the Historian's Craft* (Springer, 2005). Glen has served as president of the Canadian Society for History and Philosophy of Mathematics, and as abstracts editor of *Historia Mathematica*. He is currently on a quixotic quest to write a book on spherical trigonometry that he hopes will return the subject to the popular imagination.

About the Editors

Victor J. Katz received his Ph.D. in mathematics from Brandeis University in 1968 and was for many years Professor of Mathematics at the University of the District of Columbia. He has long been interested in the history of mathematics and, in particular, in its use in teaching. The third edition of his well-regarded textbook, *A History of Mathematics: An Introduction*, appeared in 2008. A brief version of this text was published in 2003. Katz is also the editor of *The Mathematics of Egypt, Mesopotamia, China, India and Islam: A Sourcebook*, published in 2007. Professor Katz has written many articles on the history of mathematics and its use in teaching and has spoken widely on the subject. He presented an invited lecture, "Stages in the History of Algebra with Implications for Teaching," at ICME-10 in Copenhagen in 2004. He has edited or co-edited two recent books dealing with this subject, *Learn from the Masters* (1994) and *Using History to Teach Mathematics* (2000). He also co-edited two collections of historical articles taken from journals of the Mathematical Association of America in the past 90 years, *Sherlock Holmes in Babylon and other Tales of Mathematical History* (2004) and *Who Gave You the Epsilon and Other Tales of Mathematical History* (2009). He has directed two NSF-sponsored projects that helped college teachers learn the history of mathematics and how to use it in teaching and also involved secondary school teachers in writing materials using history in the teaching of various topics in the high school curriculum. These materials, *Historical Modules for the Teaching and Learning of Mathematics*, were published on a CD by the MAA in 2005. Professor Katz was the founding editor of *Loci: Convergence*, the MAA's online magazine in the history of mathematics and its use in teaching, serving from 2004 until 2009.

Constantinos Tzanakis graduated from the mathematics department of the University of Athens, Greece, holds a master's degree in Astronomy from the Department of Astronomy, Sussex University, UK and a Ph.D. in Theoretical Physics, from the Université Libre de Bruxelles, Belgium. His research interests and publications are in Statistical Physics, Relativity Theory and related areas, as well as in mathematics and science education, especially the integration of history and epistemology in teaching, the relation between mathematics and physics and its didactical implications, and the didactics of statistics. He has been the chair of the ICMI affiliated International Study Group on the Relations between the History & Pedagogy of Mathematics (the HPM Group) from 2004. to 2008 and co-organizer of several meetings of this group, including the European Summer University on the History and Epistemology in Mathematics Education (ESU). He is currently professor of mathematics and physics at the Department of Education of the University of Crete, Greece. Together with M. Kourkoulos, they have been organizing the International Colloquium on Didactics of Mathematics at the University of Crete, Greece, since 1998.